# Forecast Verification

# Forecast Verification

## A Practitioner's Guide in Atmospheric Science

**SECOND EDITION**

*Edited by*
**Ian T. Jolliffe**
**David B. Stephenson**
*University of Exeter, UK*

**⊛WILEY-BLACKWELL**

A John Wiley & Sons, Ltd., Publication

Wiley-Blackwell is an imprint of John Wiley & Sons, formed by the merger of Wiley's global Scientific, Technical and Medical business with Blackwell Publishing.

*Registered office*
John Wiley & Sons, Ltd, The Atrium, Southern Gate, Chichester, West Sussex, PO19 8SQ, UK

*Editorial offices*
9600 Garsington Road, Oxford, OX4 2DQ, UK
The Atrium, Southern Gate, Chichester, West Sussex, PO19 8SQ, UK
111 River Street, Hoboken, NJ 07030-5774, USA

For details of our global editorial offices, for customer services and for information about how to apply for permission to reuse the copyright material in this book please see our website at www.wiley.com/wiley-blackwell.

*Library of Congress Cataloging-in-Publication Data*

Forecast verification : a practitioner's guide in atmospheric science / edited by Ian T. Jolliffe and David B. Stephenson. – 2nd ed.
        p. cm.
    Includes index.
    ISBN 978-0-470-66071-3 (cloth)
  1. Weather forecasting–Statistical methods–Evaluation.    I. Jolliffe, I. T.    II. Stephenson, David B.
    QC996.5.F67 2011
    551.63–dc23

                                                                                                    2011035808

A catalogue record for this book is available from the British Library.

This book is published in the following electronic formats: ePDF 9781119960010; Wiley Online Library 9781119960003; ePub 9781119961079; Mobi 9781119961086

Set in 10/12 pt Times by Aptara Inc., New Delhi, India


First Impression    2012

# Contents

**8 Ensemble forecasts**     **141**
*Andreas P. Weigel*

**9 Economic value and skill**     **167**
*David S. Richardson*

**10 Deterministic forecasts of extreme events and warnings**     **185**
*Christopher A.T. Ferro and David B.*
*Stephenson*

# List of contributors

**Dr Jochen Broecker**
Max-Planck-Institute for the Physics of Complex Systems, Noethnitzer Str. 38, 01187 Dresden, Germany
broecker@pks.mpg.de

**Dr Barbara G. Brown**
Research Applications Laboratory, National Center for Atmospheric Research, P.O. Box 3000, Boulder CO 80307-3000, USA
bgb@ucar.edu

**Michel Déqué**
Météo-France CNRM,CNRS/GAME, 42 Avenue Coriolis, 31057 Toulouse Cedex 01, France
deque@meteo.fr

**Dr Elizabeth E. Ebert**
Centre for Australian Weather and Climate Research (CAWCR), Bureau of Meteorology, GPO Box 1289, Melbourne, Victoria 3001, Australia
e.ebert@bom.gov.au

**Dr Christopher A.T. Ferro**
Mathematics Research Institute, College of Engineering, Mathematics and Physical Sciences, University of Exeter, Harrison Building, North Park Road, Exeter EX4 4QF, UK
c.a.t.ferro@exeter.ac.uk

**Dr Eric Gilleland**
Research Applications Laboratory, National Center for Atmospheric Research, P.O. Box 3000, Boulder CO 80307-3000, USA
ericg@ucar.edu

**Professor Robin Hogan**
Department of Meteorology, University of Reading, P.O. Box 243, Reading RG6 6BB, UK
r.j.hogan@reading.ac.uk

**Professor Ian Jolliffe**
30 Woodvale Road, Gurnard, Cowes, Isle of Wight, PO31 8EG, UK
ian@sandloch.fsnet.co.uk

**Dr Robert E. Livezey**
5112 Lawton Drive, Bethesda, MD 20816, USA
bobbilbo@msn.com

**Dr Ian B. Mason**
32 Hensman St., Latham, ACT, Australia, 2615
ibmason@grapevine.com.au

**Dr Simon J. Mason**
International Research Institute for Climate and Society (IRI), Columbia University, 61 Route 9W, P.O. Box 1000, Palisades, NY 10964-8000, USA
simon@iri.columbia.edu

**Dr Matt Pocernich**
Research Applications Laboratory, National Center for Atmospheric Research, P.O. Box 3000, Boulder CO 80307-3000, USA
e-mail: matt_pocernich@hotmail.com

**Dr Jacqueline M. Potts**
Biomathematics and Statistics Scotland, Craigiebuckler, Aberdeen AB15 8QH, UK
jackie@bioss.ac.uk

**David S. Richardson**
European Centre for Medium-Range Weather Forecasts (ECMWF), Shinfield Park, Reading, RG2 9AX, UK
david.richardson@ecmwf.int

**Professor David B. Stephenson**
Mathematics Research Institute, College of Engineering, Mathematics and Physical Sciences, University of Exeter, Harrison Building, North Park Road, Exeter EX4 4QF, UK
d.b.stephenson@exeter.ac.uk

**Dr Andreas P. Weigel**
Federal Office of Meteorology and Climatology MeteoSwiss, Kraehbuehlstr. 58, P.O. Box 514, CH-8044 Zurich, Switzerland
andreas.weigel@alumni.ethz.ch

# Preface

In the eight years since the first edition was published, there has been considerable expansion of the literature on forecast verification, and the time is ripe for a new edition. This second edition has three more chapters than the first, as well as a new Appendix and substantially more references. Developments in forecast verification have not been confined to the atmospheric science literature but, as with the first edition, we concentrate mainly on this area.

As far as we are aware, there is still no other book that gives a comparable coverage of forecast verification, although at least two related books have appeared outside the atmospheric science area. Pepe (2003) is concerned with evaluation of medical diagnostic tests, which, although essentially concerned with 'forecast verification', has a very different emphasis, whilst Krzanowski and Hand (2009) is more narrowly focused on ROC curves.

We have retained many of the authors from the first edition, as well as bringing in a number of other experts, mainly for the new chapters. All are well-regarded researchers and practitioners in their fields. Shortly after the first edition was published, an extended and constructive review appeared (Glahn, 2004; Jolliffe and Stephenson, 2005). In this new edition we and our authors have attempted to address some of the issues raised by Glahn.

Compared with the first edition, the introductory and scene-setting Chapters 1 and 2 have only minor changes. Chapter 3 on 'Deterministic forecasts of binary events' has gained an additional author and has been rewritten. Much material from the first edition has been retained but has been restructured, and a non-trivial amount of new material, reflecting recent developments, has been added. Chapters 4 and 5 on, respectively, 'Deterministic forecasts of multi-category events' and 'Deterministic forecasts of continuous variables' have only minor improvements.

One of the biggest areas of development in forecast verification in recent years has been for spatial forecasts. This reflected by a much-expanded Chapter 6 on the topic, with three new authors, all of whom are leaders in the field.

In the first edition, probability forecasts and ensemble forecasts shared a chapter. This is another area of active development and, as suggested by Glahn (2004) and others, the two topics have been separated into Chapters 7 and 8 respectively, with two new authors. Chapter 9 on 'Economic value and skill' has only minor changes compared to the first edition.

Chapters 10 and 11 are both new, covering areas that have seen much recent research and are likely to continue to do so. Chapter 10 covers the related topics of verification of forecasts for rare and extreme events, and verification of weather warnings. By their nature the latter are often extreme, though many types of warnings are issued for events that are not especially rare. Impact rather than rarity is what warrants a warning. One context in which extremes are of particular interest is that of climate change. Because of the lack of verifying observations, the topic of verification of climate projections is still in its infancy, though likely to develop. There has been more activity on verification of seasonal and decadal forecasts, and these together with verification of climate projections, are the subject of Chapter 11.

The concluding Chapter 12 reviews some key concepts, summarizes some of the verification/evaluation activity in disciplines other than atmospheric sciences, and discusses some of the main developments since the first edition. As with the first edition, a Glossary is provided, and in addition there is an Appendix on available software. Although such an Appendix inevitably becomes

out of date more quickly than other parts of the text, it is arguably the most useful part of the book to practitioners for the first few years after publication. To supplement the Appendix, software and data sets used in the book will be provided via our book website: http://emps.exeter.ac.uk/fvb. We also intend to use this website to record errata and suggestions for future additions.

We hope you enjoy this second edition and find it useful. If you have any comments or suggestions for future editions, we would be happy to hear from you.

Ian T. Jolliffe
David B. Stephenson

# Preface to the first edition

Forecasts are made in many disciplines, the best known of which are economic forecasts and weather forecasts. Other situations include medical diagnostic tests, prediction of the size of an oil field, and any sporting occasion where bets are placed on the outcome. It is very often useful to have some measure of the skill or value of a forecast or forecasting procedure. Definitions of 'skill' and 'value' will be deferred until later in the book, but in some circumstances financial considerations are important (economic forecasting, betting, oil field size), whilst in others a correct or incorrect forecast (medical diagnosis, extreme weather events) can mean the difference between life and death.

Often the 'skill' or 'value' of a forecast is judged in relative terms. Is forecast provider A doing better than B? Is a newly developed forecasting procedure an improvement on current practice? Sometimes, however, there is a desire to measure absolute, rather than relative, skill. Forecast verification, the subject of this book, is concerned with judging how good is a forecasting system or single forecast.

Although the phrase 'forecast *verification*' is generally used in atmospheric science, and hence adopted here, it is rarely used outside the discipline. For example, a survey of keywords from articles in the *International Journal of Forecasting* between 1996 and 2002 has no instances of 'verification'. This journal attracts authors from a variety of disciplines, though economic forecasting is prominent. The most frequent alternative terminology in the journal's keywords is 'forecast *evaluation*', although *validation* and *accuracy* also occur.

Evaluation and validation also occur in other subject areas, but the latter is often used to denote a wider range of activities than simply judging skill or value – see, for example, Altman and Royston (2000).

Many disciplines make use of forecast verification, but it is probably fair to say that a large proportion of the ideas and methodology have been developed in the context of weather and climate forecasting, and this book is firmly rooted in that area. It will therefore be of greatest interest to forecasters, researchers and students in atmospheric science. It is written at a level that is accessible to students and to operational forecasters, but it also contains coverage of recent developments in the area. The authors of each chapter are experts in their fields and are well aware of the needs and constraints of operational forecasting, as well as being involved in research into new and improved methods of verification. The audience for the book is not restricted to atmospheric scientists – there is discussion in several chapters of similar ideas in other disciplines. For example ROC curves (Chapter 3) are widely used in medical applications, and the ideas of Chapter 8 are particularly relevant to finance and economics.

To our knowledge there is currently no other book that gives a comprehensive and up-to-date coverage of forecast verification. For many years, The WMO publication by Stanski *et al.* (1989) and its earlier versions was the standard reference for atmospheric scientists, though largely unknown in other disciplines. Its drawback is that it is somewhat limited in scope and is now rather out-of-date. Wilks (2006b [formerly 1995], Chapter 7) and von Storch and Zweirs (1999, Chapter 18) are more recent but, inevitably as each comprises only one chapter in a book, are far from comprehensive. The current book provides a broad coverage, although it does not attempt to be encyclopedic, leaving the reader to look in the references for more technical material.

Chapters 1 and 2 of the book are both introductory. Chapter 1 gives a brief review of the history and current practice in forecast verification, gives some definitions of basic concepts such as skill and value, and discusses the benefits and practical considerations associated with forecast verification. Chapter 2 describes a number of informal descriptive ways, both graphical and numerical, of comparing forecasts and corresponding observed data. It then establishes some theoretical groundwork that is used in later chapters, by defining and discussing the joint probability distribution of the forecasts and observed data. Consideration of this joint distribution and its decomposition into conditional and marginal distributions leads to a number of fundamental properties of forecasts. These are defined, as are the ideas of accuracy, association and skill.

Both Chapters 1 and 2 discuss the different types of data that may be forecast, and each of the next five chapters then concentrates on just one type. The subject of Chapter 3 is binary data in which the variable to be forecast has only two values, for example {Rain, No Rain}, {Frost, No Frost}. Although this is apparently the simplest type of forecast, there have been many suggestions of how to assess them, in particular many different verification measures have been proposed. These are fully discussed, along with their properties. One particularly promising approach is based on signal detection theory and the ROC curve.

For binary data one of two categories is forecast. Chapter 4 deals with the case in which the data are again categorical, but where there are more than two categories. A number of skill scores for such data are described, their properties are discussed, and recommendations are made.

Chapter 5 is concerned with forecasts of continuous variables such as temperature. Mean squared error and correlation are the best-known verification measures for such variables, but other measures are also discussed including some based on comparing probability distributions.

Atmospheric data often consist of spatial fields of some meteorological variable observed across some geographical region. Chapter 6 deals with verification for such spatial data. Many of the verification measures described in Chapter 5 are also used in the spatial context, but the correlation due to spatial proximity causes complications. Some of these complications, together with some verification measures that have been developed with spatial correlation in mind, are discussed in Chapter 6.

Probability plays a key role in Chapter 7, which covers two topics. The first is forecasts that are actually probabilities. For example, instead of a deterministic forecast of 'Rain' or 'No Rain', the event 'Rain' may be forecast to occur with probability 0.2. One way in which such probabilities can be produced is to generate an ensemble of forecasts, rather than a single forecast. The continuing increase of computing power has made larger ensembles of forecasts feasible, and ensembles of weather and climate forecasts are now routinely produced. Both ensemble and probability forecasts have their own peculiarities that necessitate different, but linked, approaches to verification. Chapter 7 describes these approaches.

The discussion of verification for different types of data in Chapters 3–7 is largely in terms of mathematical and statistical properties, albeit properties that are defined with important practical considerations in mind. There is little mention of cost or value – this is the topic of Chapter 8. Much of the chapter is concerned with the simple cost-loss model, which is relevant for binary forecasts. However, these forecasts may be either deterministic as in Chapter 3, or probabilistic as in Chapter 7. Chapter 8 explains some of the interesting relationships between economic value and skill scores.

The final chapter (9) reviews some of the key concepts that arise elsewhere in the book. It also summarises the aspects of forecast verification that have received most attention in other disciplines, including Statistics, Finance and Economics, Medicine, and areas of Environmental and Earth Science other than Meteorology and Climatology. Finally, the chapter discusses some of the most important topics in the field that are the subject of current research or that would benefit from future research.

This book has benefited from discussions and help from many people. In particular we would like to thank the following colleagues for their particularly helpful comments and contributions: Barbara Casati, Martin Goeber, Mike Harrison, Rick Katz, Simon Mason, Buruhani Nyenzi and Dan Wilks. Some of the earlier work on this book was carried out while one us (I.T.J.) was on research leave at the Bureau of Meteorology Research Centre (BMRC) in Melbourne. He is grateful to BMRC and its staff, especially Neville Nicholls, for the supportive environment and useful discussions; to the Leverhulme Trust for funding the visit under a Study Abroad Fellowship; and to the University of Aberdeen for granting the leave.

Looking to the future, we would be delighted to receive any feedback comments from you, the reader, concerning material in this book, in order that improvements can be made in future editions (see www.met.rdg.ac.uk/cag/forecasting).

# 1

# Introduction

**Ian T. Jolliffe and David B. Stephenson**

*Mathematics Research Institute, University of Exeter*

Forecasts are almost always made and used in the belief that having a forecast available is preferable to remaining in complete ignorance about the future event of interest. It is important to test this belief *a posteriori* by assessing how skilful or valuable was the forecast. This is the topic of *forecast verification* covered in this book, although, as will be seen, words such as 'skill' and 'value' have fairly precise meanings and should not be used interchangeably. This introductory chapter begins, in Section 1.1, with a brief history of forecast verification, followed by an indication of current practice. It then discusses the reasons for, and benefits of, verification (Section 1.2). The third section provides a brief review of types of forecasts, and the related question of the target audience for a verification procedure. This leads on to the question of skill or value (Section 1.4), and the chapter concludes, in Section 1.5, with some discussion of practical issues such as data quality.

## 1.1 A brief history and current practice

Forecasts are made in a wide range of diverse disciplines. Weather and climate forecasting, economic and financial forecasting, sporting events and med-

ical epidemics are some of the most obvious examples. Although much of the book is relevant across disciplines, many of the techniques for verification have been developed in the context of weather, and latterly climate, forecasting. For this reason the current section is restricted to those areas.

### 1.1.1 History

The paper that is most commonly cited as the starting point for weather forecast verification is Finley (1884). Murphy (1996a) notes that although operational weather forecasting started in the USA and Western Europe in the 1850s, and that questions were soon asked about the quality of the forecasts, no formal attempts at verification seem to have been made before the 1880s. He also notes that a paper by Köppen (1884), in the same year as Finley's paper, addresses the same binary forecast set-up as Finley (see Table 1.1), though in a different context.

Finley's paper deals with a fairly simple example, but it nevertheless has a number of subtleties and will be used in this and later chapters to illustrate a number of facets of forecast verification. The data set consists of forecasts of whether or not a tornado will occur. The forecasts were made from

*Forecast Verification: A Practitioner's Guide in Atmospheric Science*, Second Edition. Edited by Ian T. Jolliffe and David B. Stephenson.
© 2012 John Wiley & Sons, Ltd. Published 2012 by John Wiley & Sons, Ltd.

**Table 1.1** Finley's tornado forecasts

|  | Observed | | |
| --- | --- | --- | --- |
| Forecast | Tornado | No Tornado | Total |
| Tornado | 28 | 72 | 100 |
| No tornado | 23 | 2680 | 2703 |
| Total | 51 | 2752 | 2803 |

10 March until the end of May 1884, twice daily, for 18 districts of the USA east of the Rockies. Table 1.1 summarizes the results in a table, known as a $(2 \times 2)$ contingency table (see Chapter 3). Table 1.1 shows that a total of 2803 forecasts were made, of which 100 forecast 'Tornado'. On 51 occasions tornados were observed, and on 28 of these 'Tornado' was also forecast. Finley's paper initiated a flurry of interest in verification, especially for binary (0–1) forecasts, and resulted in a number of published papers during the following 10 years. This work is reviewed by Murphy (1996a).

Forecast verification was not a very active branch of research in the first half of the twentieth century. A three-part review of verification for short-range weather forecasts by Muller (1944) identified only 55 articles 'of sufficient importance to warrant summarization', and only 66 were found in total. Twenty-seven of the 55 appeared before 1913. Due to the advent of numerical weather forecasting, a large expansion of weather forecast products occurred from the 1950s onwards, and this was accompanied by a corresponding research effort into how to evaluate the wider range of forecasts being made.

For the $(2 \times 2)$ table of Finley's results, there is a surprisingly large number of ways in which the numbers in the four cells of the table can be combined to give measures of the quality of the forecasts. What they all have in common is that they use the joint probability distribution of the forecast event and observed event. In a landmark paper, Murphy and Winkler (1987) established a general framework for forecast verification based on such joint distributions. Their framework goes well beyond the $(2 \times 2)$ table, and encompasses data with more than two categories, discrete and continuous data, and multivariate data. The forecasts can take any of these forms, but can also be in the form of probabilities.

The late Allan Murphy had a major impact on the theory and practice of forecast verification. As well as Murphy and Winkler (1987) and numerous technical contributions, two further general papers of his are worthy of mention here. Murphy (1991a) discusses the complexity and dimensionality of forecast verification, and Murphy (1993) is an essay on what constitutes a 'good' forecast.

Weather and climate forecasting is necessarily an international activity. The World Meteorological Organization (WMO) published a 114-page technical report (Stanski *et al.*, 1989) that gave a comprehensive survey of forecast verification methods in use in the late 1980s. Other WMO documentation is noted in the next subsection.

### 1.1.2   Current practice

The WMO provides a Standard Verification System for Long-Range Forecasts. At the time of writing versions of this are available at a number of websites. The most up-to-date version is likely to be found through the link to the User's Guide on the website of the Lead Centre for the Long Range Forecast Verification System (http://www.bom.gov.au/wmo/lrfvs/users.shtml). The document is very thorough and careful in its definitions of long-range forecasts, verification areas (geographical) and verification data sets. It describes recommended verification strategies and verification scores, and is intended to facilitate the exchange of comparable verification scores between different centres. An earlier version is also available as attachments II-8 and II-9 in the WMO *Manual on the Global Data-Processing System* (http://www.wmo.int/pages/prog/www/DPS/Manual/WMO485.pdf). Attachment II-7 in the same document discusses methods used in standardized verification of NWP (Numerical Weather Prediction) products. Two further WMO documents can be found at http://www.wmo.int/pages/prog/amp/pwsp/pdf/TD-1023.pdf and http://www.wmo.int/pages/prog/amp/pwsp/pdf/TD-1103.pdf. These are respectively Guidelines (and Supplementary Guidelines) on Performance Assessment of Public Weather Services. The latter is discursive in nature, whilst the guidelines in the former are more technical in nature.

European member states report annually on verification of ECMWF (European Centre for Medium Range Weather Forecasts) forecasts in their national weather services, and guidance on such verification is given in ECMWF Technical Memorandum 430 by Pertti Nurmi (http://www.ecmwf.int/publications/library/ecpublications/_pdf/tm/401-500/tm430.pdf).

At a national level, verification practices vary between different National Services, and most use a range of different verification strategies for different purposes. For example, verification scores used at the time of writing by the National Climate Centre at the Bureau of Meteorology in Australia range through many of the chapters that follow, for example proportion correct (Chapter 3), LEPS scores (Chapter 4), root mean square error (Chapter 5), anomaly correlation (Chapter 6), Brier skill score (Chapter 7) and so on (Robert Fawcett, personal communication).

There is a constant need to adapt practices, as forecasts, data and users all change. An increasing number of variables can be, and are, forecast, and the nature of forecasts is also changing. At one end of the range there is increasing complexity. Ensembles of forecasts, which were largely infeasible 30 years ago, are now commonplace (Chapter 8), and the verification of spatial forecasts has advanced significantly (Chapter 6). At the other extreme, a wider range of users requires targeted, but often simple (at least to express), forecasts. The nature of the data available with which to verify the forecasts is also evolving with increasing use of remote sensing by satellite and radar, for example.

An important part of any operational verification system is to have software to implement the system. As well as the widely available software described in Appendix, national weather services often have their own systems. For example, the Finnish Meteorological Institute has a comprehensive operational verification package, which is regularly updated (Pertti Nurmi, personal communication).

A very useful resource is the webpage of the Joint Working Group on Forecast Verification Research (http://www.cawcr.gov.au/projects/verification/). It gives a good up-to-date overview of verification methods and issues associated with them, together with information on workshops and other events related to verification.

## 1.2 Reasons for forecast verification and its benefits

There are three main reasons for verification, whose description dates back to Brier and Allen (1951), and which can be described by the headings *administrative, scientific* and *economic*. Naturally no classification is perfect and there is overlap between the three categories. A common important theme for all three is that any verification scheme should be *informative*. It should be chosen to answer the questions of interest and not simply for reasons of convenience.

From an administrative point of view, there is a need to have some numerical measure of how well forecasts are performing. Otherwise, there is no objective way to judge how changes in training, equipment or forecasting models, for example, affect the quality of forecasts. For this purpose, a small number of overall measures of forecast performance are usually desired. As well as measuring improvements over time of the forecasts, the scores produced by the verification system can be used to justify funding for improved training and equipment and for research into better forecasting models. More generally they can guide strategy for future investment of resources in forecasting.

Measures of forecast quality may even be used by administrators to reward forecasters financially. For example, the UK Meteorological Office currently operates a corporate bonus scheme, several elements of which are based on the quality of forecasts. The formula for calculating the bonus payable is complex, and involves meeting or exceeding targets for a wide variety of meteorological variables around the UK and globally. Variables contributing to the scheme range from mean sea level pressure, through precipitation, temperature and several others, to gale warnings.

The scientific viewpoint is concerned more with *understanding,* and hence improving the forecast system. A detailed assessment of the strengths and weaknesses of a set of forecasts usually requires more than one or two summary scores. A larger investment in more complex verification schemes will be rewarded with a greater appreciation of exactly where the deficiencies in the forecast lie, and with it the possibility of improved understanding of the physical processes that are being forecast. Sometimes there are unsuspected biases in either the forecasting models, or in the forecasters' interpretations, or both, which only become apparent when more sophisticated verification schemes are used. Identification of such biases can lead to research being targeted to improve knowledge of why they occur. This, in turn, can lead to improved scientific understanding of the underlying processes, to improved models, and eventually to improved forecasts.

The administrative use of forecast verification certainly involves financial considerations, but the third, 'economic', use is usually taken to mean something closer to the users of the forecasts. Whilst verification schemes in this case should be kept as simple as possible in terms of communicating their results to users, complexity arises because different users have different interests. Hence there is the need for different verification schemes tailored to each user. For example, seasonal forecasts of summer rainfall may be of interest to both a farmer, and to an insurance company covering risks of event cancellations due to wet weather. However, different aspects of the forecast are relevant to each. The farmer will be interested in total rainfall, and its distribution across the season, whereas the insurance company's concern is mainly restricted to information on the likely number of wet weekends.

As another example, consider a daily forecast of temperature in winter. The actual temperature is relevant to an electricity company, as demand for electricity varies with temperature in a fairly smooth manner. In contrast, a local roads authority is concerned with the value of the temperature relative to some *threshold*, below which it should treat the roads to prevent ice formation. In both examples, a forecast that is seen as reasonably good by one user may be deemed 'poor' by the other. The economic view of forecast verification needs to take into account the economic factors underlying the users' needs for forecasts when devising a verification scheme. This is sometimes known as 'customer-based' or 'user-oriented' verification, as it provides information in terms more likely to be understood by the 'customer' or 'user'

than a purely 'scientific' approach. Forecast verification using economic value is discussed in detail in Chapter 9. Another aspect of forecasting for specific users is the extent to which users prefer a simple, less informative forecast to one that is more informative (e.g. a probability forecast) but less easy to interpret. Some users may be uncomfortable with probability forecasts, but there is evidence (Harold Brooks, personal communication) that *probabilities* of severe weather events such as hail or tornados are preferred to crude *categorizations* such as {Low Risk, Medium Risk, High Risk}. User-oriented verification should attempt to ascertain such preferences for the user or 'customer' at hand.

A benefit common to all three classes of verification, if it is informative, is that it gives the administrator, scientist or user concrete information on the quality of forecasts that can be used to make rational decisions.

This section has been written from the viewpoint of verification of forecasts issued by National Meteorological Services. Virtually all the points made are highly relevant for forecasts issued by private companies, and in other subject domains, but it appears that they may not always be appreciated. Although most National Weather Services verify their forecasts, the position for commercially provided forecasts is more patchy. Mailier *et al.* (2008) reported the findings of a survey of providers and users of commercial weather forecasts in the UK. The survey and related consultations revealed that there were 'significant deficiencies in the methodologies and in the communication of forecast quality assessments' and that 'some users may be indifferent to forecast quality'.

## 1.3 Types of forecast and verification data

The wide range of forecasts has already been noted in the Preface when introducing the individual chapters. At one extreme, forecasts may be binary (0–1), as in Finley's tornado forecasts; at the other extreme, ensembles of forecasts will include predictions of several different weather variables at different times, different spatial locations, different vertical levels of the atmosphere, and not just one forecast but a whole ensemble. Such forecasts are extremely difficult to verify in a comprehensive manner but, as will be seen in Chapter 3, even the verification of binary forecasts can be a far-from-trivial problem.

Some other types of forecast are difficult to verify, not because of their sophistication, but because of their vagueness. Wordy or descriptive forecasts are of this type. Verification of forecasts such as 'turning milder later' or 'sunny with scattered showers in the south at first' is bound to be subjective (see Jolliffe and Jolliffe, 1997), whereas in most circumstances it is highly desirable for a verification scheme to be objective. In order for this to happen it must be clear what is being forecast, and the verification process should ideally reflect the forecast precisely. As a simple example, consider Finley's tornado forecasts. The forecasts are said to be of occurrence or non-occurrence of tornados in 18 districts, or subdivisions of these districts, of the USA. However, the verification is done on the basis of whether a funnel cloud is seen at a reporting station within the district (or subdivision) of interest. There were 800 observing stations, but given the vast size of the 18 districts, this is a fairly sparse network. It is quite possible for a tornado to appear in a district sufficiently distant from the reporting stations for it to be missed. To match up forecast and verification, it is necessary to interpret the forecast not as 'a tornado will occur in a given district', but as 'a funnel cloud will occur within sight of a reporting station in the district'.

As well as an increase in the types of forecasts available, there have also been changes in the amount and nature of data available for verifying forecasts. The changes in data include changes of observing stations, changes of location and type of recording instruments at a station, and an increasing range of remotely sensed data from satellites, radar or automatic recording devices. It is tempting, and often sensible, to use the most up-to-date types of data available for verification, but in a sequence of similar forecasts it is important to be certain that any apparent changes in forecast quality are not simply due to changes in the nature of the data used for verification. For example, suppose that a forecast of rainfall for a region is to be verified, and that there is an unavoidable change in the set of stations used for verification. If the mean or variability of rainfall is different for the new set of stations, compared to the old, such differences can affect many of the scores used for verification.

Another example occurs in the seasonal forecasting of numbers of tropical cyclones. There is evidence that access to a wider range of satellite imagery has led to redefinitions of cyclones over the years (Nicholls, 1992). Hence, apparent trends in cyclone frequency may be due to changes of definition, rather than to genuine climatic trends. This, in turn, makes it difficult to know whether changes in forecasting methods have resulted in improvements to the quality of forecasts. Apparent gains can be confounded by the fact that the 'target' that is being forecast has moved; changes in definition alone may lead to changed verification scores.

As noted in the previous section, the idea of matching verification data to forecasts is relevant when considering the needs of a particular user. A user who is interested only in the position of a continuous variable relative to a threshold requires verification data and procedures geared to binary data (above/below threshold), rather than verification of the actual forecast value of the variable.

The chapters of this book cover all the main types of forecasts that require verification, but less common types are not covered in detail. For example, forecasts of wind direction lie on a circle rather than being linearly ordered and hence need different treatment. Bao *et al.* (2010) discuss verification of directional forecasts when the variable being forecast is continuous, and there are also measures that modify those of Chapter 4 when forecasts fall in a small number of categories (Charles Kluepfel, personal communication)

## 1.4 Scores, skill and value

For a given type of data it is easy enough to construct a numerical score that measures the relative quality of different forecasts. Indeed, there is usually a whole range of possible scores. Any set of forecasts can then be ranked as best, second best, . . . , worst, according to a chosen score, though the ranking need not be the same for different choices of score. Two questions then arise:

• How to choose which scores to use?
• How to assess the absolute, rather than relative, quality of a forecast?

In addressing the first of these questions, attempts have been made to define desirable properties of potential scores. Many of these will be discussed in later chapters, in particular Chapter 2. The general framework of Murphy and Winkler (1987) allows different 'attributes' of forecasts, such as *reliability, resolution, discrimination* and *sharpness* to be examined. Which of these attributes is most important to the scientist, administrator or end-user will determine which scores are preferred. Most scores have some strengths, but all have weaknesses, and in most circumstances more than one score is needed to obtain an informed picture of the relative merits of the forecasts.

'Goodness' of forecasts has many facets: Murphy (1993) identifies three types of goodness:

- Consistency (the correspondence between forecasters' judgements and their forecasts).
- Quality (the correspondence between the forecasts and matching observations).
- Value (the incremental economic and/or other benefits realized by decision-makers through the use of the forecasts).

It seems desirable that the forecaster's best judgement and the forecast actually issued coincide. Murphy (1993) describes this as 'consistency', though confusingly the same word has a narrower definition in Murphy and Daan (1985) – see Chapter 2. The choice of verification scheme can influence whether or not this happens. Some schemes have scores for which a forecaster knows that he or she will score better on average if the forecast made differs (perhaps is closer to the long-term average or climatology of the quantity being forecast) from his or her best judgement of what will occur. In that case, the forecaster will be tempted to *hedge,* that is, to forecast something other than his or her best judgement (Murphy, 1978), especially if the forecaster's pay depends on the score. Thus administrators should avoid measuring or rewarding forecasters' performance on the basis of such scoring schemes, as this is likely to lead to biases in the forecasts.

The emphasis in this book is on quality – the correspondence between forecast and observations. Value is concerned with economic worth to the user. Chapter 9 discusses value and its relationship to quality.

### 1.4.1 Skill scores

Turning to the matter of how to quantify the quality of a forecast, it is usually necessary to define a baseline against which a forecast can be judged. Much of the published discussion following Finley's (1884) paper was driven by the fact that although the forecasts were correct on 2708/2803 = 96.6% of occasions, it is possible to do even better by always forecasting 'No Tornado', if forecast performance is measured by the percentage of correct forecasts. This alternative unskilful forecast has a success rate of 2752/2803 = 98.2%. It is therefore usual to measure the performance of forecasts relative to some 'unskilful' or reference forecast. Such relative measures are known as *skill scores*, and are discussed further in several of the later chapters (see, e.g., Sections 2.7, 3.4, 4.3 and 11.3.1).

There are several baseline or reference forecasts that can be chosen. One is the average, or expected, score obtained by issuing forecasts according to a random mechanism. What this means is that a probability distribution is assigned to the possible values of the variable(s) to be forecast, and a sequence of forecasts is produced by taking a sequence of independent values from that distribution. A limiting case of this, when all but one of the probabilities is zero, is the (deterministic) choice of the same forecast on every occasion, as when 'No Tornado' is forecast all the time.

'Climatology' is a second common baseline. This refers to always forecasting the 'average' of the quantity of interest. 'Average' in this context usually refers to the mean value over some recent reference period, typically of 30 years length.

A third baseline that may be appropriate is 'persistence'. This is a forecast in which whatever is observed at the present time is forecast to persist into the forecast period. For short-range forecasts this strategy is often successful, and to demonstrate real forecasting skill, a less naive forecasting system must do better.

### 1.4.2 Artificial skill

Often when a particular data set is used in developing a forecasting system, the quality of the system is then assessed on the same data set. This will invariably lead to an optimistic bias in skill scores. This inflation of skill is sometimes known as 'artificial skill', and is a particular problem if the score itself has been used directly or indirectly in calibrating the forecasting system. To avoid such biases, an ideal solution is to assess the system using only forecasts of events that have not yet occurred. This may be feasible for short-range forecasts, where data accumulate rapidly, but for long-range forecasts it may be a long time before there are sufficient data for reliable verification. In the meantime, while data are accumulating, any potential improvements to the forecasting procedure should ideally be implemented in parallel to, and not as a replacement for, the old procedure.

The next best solution for reducing artificial skill is to divide the data into two non-overlapping, exhaustive subsets, the *training set* and the *test set*. The training set is used to formulate the forecasting procedure, while the procedure is verified on the test set. Some would argue that, even though the training and test sets are non-overlapping, and the observed data in the test set are not used directly in formulating the forecasting rules, the fact that the observed data for both sets already exist when the rules are formulated has the potential to bias any verification results. A more practical disadvantage of the test/training set approach is that only part of the data set is used to construct the forecasting system. The remainder is, in a sense, wasted because, in general, increasing the amount of data or information used to construct a forecast will provide a better forecast. To partially overcome this problem, the idea of *cross-validation* can be used.

Cross-validation has a number of variations on the same basic theme. It has been in use for many years (see, e.g., Stone, 1974) but has become practicable for larger problems as computer power has increased. Suppose that the complete data set consists of $n$ forecasts, and corresponding observations. In cross-validation the data are divided into $m$ subsets, and for each subset a forecasting rule is constructed based on data from the other $(m-1)$ subsets. The rule is then verified on the subset omitted from the construction procedure, and this is repeated for each of the $m$ subsets in turn. The verification scores for each subset are then combined to give an overall measure of quality. The case $m = 2$ corresponds to repeating the test/training set approach with the roles of test and training sets reversed, and then combining the results from the two analyses. At the opposite extreme, a commonly used special case is where $m = n$, so that each individual forecast is based on a rule constructed from all the other $(n-1)$ observations.

The word 'hindcast' is in fairly common use, but can have different meanings to different authors. The cross-validation scheme just mentioned bases its 'forecasts' on $(n-1)$ observations, some of which are 'in the future' relative to the observation being predicted. Sometimes the word 'hindcast' is restricted to mean predictions like this in which 'future', as well as past, observations are used to construct forecasting procedures. A wider definition includes any prediction made that is not a genuine forecast of a *future* event. With this usage, a prediction for the year 2010 must be a hindcast, even if it is only based on data up to 2009, because year 2010 is now over. The term *retroactive forecasting* is used by Mason and Mimmack (2002) to denote the form of hindcasting in which forecasts are made for past years (e.g. 2006–2010) using data prior to those years (perhaps 1970–2005).

The terminology *ex ante* and *ex post* is used in business forecasting. *Ex ante* means a prediction into the future before the events occur (a genuine *fore*cast), whereas *ex post* means predictions for historical periods for which verification data are already available at the time of forecast. The latter is therefore a form of hindcasting.

### 1.4.3  Statistical significance

There is one further aspect of measuring the absolute quality of a forecast. Having decided on a suitable baseline from which to measure skill, checked that the skill score chosen has no blatantly undesirable properties, and removed the likelihood of artificial skill, is it possible to judge whether an observed improvement over the baseline is statistically significant? Could the improvement have arisen by chance? Ideas from statistical inference, namely hypothesis testing and confidence intervals, are needed to address this question. Confidence intervals for a number of measures or scores are described in Section 3.5.2, and several other chapters discuss tests of hypotheses in various contexts. A difficulty that arises is that many standard procedures for confidence intervals and tests of hypothesis assume independence of observations. The temporal and spatial correlation that is often present in environmental data means that adaptations to the usual procedures are necessary – see, for example, Section 4.4.

### 1.4.4  Value added

For the user, a measure of value is often more important than a measure of skill. Again, the value should be measured relative to a baseline. It is the *value added*, compared to an unskilful forecast, which is of real interest. The definition of 'unskilful' can refer to one of the reference or baseline forecasts described earlier for scores. Alternatively, for a situation with a finite number of choices for a decision (e.g., protect or don't protect a crop from frost), the baseline can be the best from the list of decision choices ignoring any forecast (e.g., always protect or never protect regardless of the forecast). The avoidance of artificially inflated value and assessing whether the 'value added' is statistically significant are relevant to value, as much as to skill.

## 1.5  Data quality and other practical considerations

Changes in the data available for verification have already been mentioned in Section 1.3, but it was implicitly assumed there that the data are of high quality. This is not always the case. National Meteorological Services will, in general, have quality control procedures in place that detect many errors, but larger volumes of data make it more likely that some erroneous data will slip through the net. A greater reliance on data that are indirectly derived via some calibration step, for example rainfall intensities deduced from radar data, also increases the scope for biases in the inferred data. Sometimes the 'verification observations' are not observations at all, but are based on analyses from very-short-range forecast models. This may be necessary if genuine observations are sparse and not conveniently spaced geographically in relation to the forecasts. A common problem is that forecasts may be spatially continuous or on a grid, but observations are available only for an irregular set of discrete spatial points. This is discussed further in Section 6.2.

When verification data are incorrect, the forecast is verified against something other than the truth, with unpredictable consequences for the verification scores. Work on discriminant analysis in the presence of misclassification (see McLachlan, 1992, Section 2.5; Huberty, 1994, Section XX-4) is relevant in the case of binary forecasts. There has been some work, too, on the effect of observation errors on verification scores in a meteorological context. For example, Bowler (2008) shows that the apparent skill of a forecasting system can be reduced by the equivalent of one day in forecast lead time.

In large data sets, missing data have always been commonplace, for a variety of reasons. Even Finley (1884) suffered from this, stating that '... from many localities [no reports] will be received except, perhaps, at a very late day.' Missing data can be dealt with either by ignoring them, and not attempting to verify the corresponding forecast, or by estimating them from related data and then verifying using the estimated data. The latter is preferable if good estimates are available, because it avoids throwing away information, but if the estimates are poor, the resulting verification scores can be misleading.

Data may be missing at random, or in some non-random manner, in which particular values of the variable(s) being forecast are more prone to be absent than others. For randomly missing data the mean verification score is likely to be relatively unaffected by the existence of the missing data, though the variability of the score will usually increase. For data that are missing in a more systematic way, the verification scores can be biased, as well as again having increased variability.

One special, but common, type of missing data occurs when measurements of the variables of interest have not been collected for long enough to establish a reliable climatology for them. This is a particular problem when extremes are forecast. By their very nature, extremes occur rarely and long data records are needed to deduce their nature and frequency. Forecasts of extremes are of increasing interest, partly because of the disproportionate financial and social impacts caused by extreme weather, but also in connection with the large amount of research effort devoted to climate change.

It is desirable for a data set to include some extreme values so that full coverage of the range of possible observations is achieved. However, a small number of extreme values can have undue influence on the values of some types of skill measure, and mask the quality of forecasts for non-extreme values. To avoid this, measures need to be robust or resistant to the presence of extreme observations or forecasts. Alternatively, measures may be devised specifically for verification of forecasts or warnings of extreme events – see Chapter 10.

A final practical consideration is that there can be confusion over terminology. This is partly due to the development of verification in several different disciplines, but even within atmospheric science different terms can be used for the same thing, or the same term (or very similar terms) used for different things. For example, *false alarm rate* and *false alarm ratio* are different measures for binary deterministic forecasts (see Chapter 3), but are easily confused. Barnes *et al.* (2009) found that of 26 peer-reviewed articles published in American Meteorological Society journals between 2001 and 2007 that used one or both of the measures, 10 (38%) defined them inconsistently with the currently accepted definitions. The glossary in this book will help readers to avoid some of the pitfalls of terminology, but care is still needed in reading the verification literature.

Even the word 'verification' itself is almost unknown outside of atmospheric science. In other disciplines 'evaluation' and 'assessment' are more common. It seems likely that Finley's use of the phrase 'verification of predictions' in 1884 is the historical accident that led to its adoption in atmospheric science, but not elsewhere.

## 1.6  Summary

As described in Section 1.2, verification has three main uses:

- Administrative: to monitor performance over time and compare the forecast quality of different prediction systems.
- Scientific: to diagnose the drivers of performance and inform improvements in prediction systems.
- Economic: to build credibility and customer confidence in forecast products by demonstrating that predictions have economic value to users.

Verification is therefore an indispensible part of the development cycle of prediction systems. With increasing complexity and sophistication of forecasts, verification is an active area of scientific research – see, e.g., the review by Casati *et al.* (2008), which is part of a special issue of *Meteorological Applications* on forecast verification. Subsequent chapters of the book give an introduction to some of the exciting developments in the subject, as well as giving a clear grounding in the more established methodology.

# 2

# Basic concepts

**Jacqueline M. Potts**

*Biomathematics and Statistics Scotland, Aberdeen, UK*

## 2.1 Introduction

Forecast verification involves exploring and summarizing the relationship between sets of forecast and observed data and making comparisons between the performance of forecasting systems and that of reference forecasts. Verification is therefore a statistical problem. This chapter introduces some of the basic statistical concepts and definitions that will be used in later chapters. Further details about the use of statistical methods in the atmospheric sciences can be found in Wilks (2006b) and von Storch and Zwiers (1999).

## 2.2 Types of predictand

The variable for which the forecasts are formulated is known as the *predictand*. A *continuous* predictand is one for which, within the limits over which the variable ranges, any value is possible. This means that between any two different values there are an infinite number of possible values. For discrete variables, however, we can list all possible values. Variables such as pressure, temperature or rainfall are theoretically continuous. In reality, however, such variables are actually discrete because measuring devices have limited reading accuracy and variables are usually recorded to a fixed number of decimal places. Verification of continuous predictands is considered in Chapter 5. *Categorical* predictands are discrete variables that can only take one of a finite set of predefined values. If the categories provide a ranking of the data, the variable is *ordinal*; for example, cloud cover is often measured in oktas. On the other hand, cloud type is a *nominal* variable since there is no natural ordering of the categories. The simplest kind of categorical variable is a *binary* variable, which has only two possible values, indicating, for example, the presence or absence of some condition such as rain, fog or thunder. Verification of binary forecasts is discussed in Chapter 3, and forecasts in more than two categories are considered in Chapter 4.

Forecasts may be *deterministic* (e.g. *rain tomorrow*) or *probabilistic* (e.g. *70% chance of rain tomorrow*). There is more than one way in which a probability forecast may be interpreted. The frequentist interpretation of *70% chance of rain tomorrow* is that rain occurs on 70% of the occasions when this forecast is issued. However, such forecasts are usually interpreted in a subjective way as expressing the forecaster's degree of belief that the event will occur (Epstein, 1966). Probability forecasts are

*Forecast Verification: A Practitioner's Guide in Atmospheric Science*, Second Edition. Edited by Ian T. Jolliffe and David B. Stephenson.
© 2012 John Wiley & Sons, Ltd. Published 2012 by John Wiley & Sons, Ltd.

often issued for categorical predictands with two or more categories. In the case of continuous predictands a forecast probability density function (see Section 2.5) may be produced, for example based on ensembles (see Chapter 8). Probability forecasts are discussed in greater detail in Chapter 7.

Forecasts are made at different temporal and spatial scales. A very short-range forecast may cover the next 12 hours, whereas long-range forecasts may be issued from 30 days to 2 years ahead and be forecasts of the mean value of a variable over a month or an entire season. Climate change predictions are made at decadal and longer timescales. The verification of seasonal and decadal forecasts is considered in Chapter 11. Prediction models often produce forecasts of spatial fields, usually defined by values of a variable at many points on a regular grid. These vary both in their geographical extent and in the distance between grid points within that area. Forecasts of spatial fields are considered in Chapter 6.

Meteorological data are autocorrelated in both space and time. At a given location, the correlation between observations a day apart will usually be greater than that between observations separated by longer time intervals. Similarly, at a given time, the correlation between observations at grid points that are close together will generally be greater than between those that are further apart, although teleconnection patterns such as the North Atlantic Oscillation can lead to correlation between weather patterns in areas that are separated by vast distances.

Both temporal and spatial autocorrelation have implications for forecast verification. Temporal autocorrelation means that for some types of short-range forecast, persistence often performs well when compared to a forecast of the climatological average. A specific user may be interested only in the quality of forecasts at a particular site, but meteorologists are often interested in evaluating the forecasting system in terms of its ability to predict the whole spatial field. The degree of spatial autocorrelation will affect the statistical distribution of the performance measures used. When spatial autocorrelation is present in both the observed and forecast fields it is likely that, if a forecast is fairly accurate at one grid point, it will also be fairly accurate at neighbouring grid points. Similarly, it is likely that if the forecast is not very accurate at one grid point, it will also not be very accurate at neighbouring grid points. Consequently, the significance of a particular value of a performance measure calculated over a spatial field will be quite different from its significance if it was calculated over the same number of independent forecasts.

## 2.3 Exploratory methods

Exploratory methods should be used to examine the forecast and observed data graphically; further information about these techniques can be found in Tukey (1977); see also Wilks (2006b, Chapter 3). For continuous variables boxplots (Figure 2.1) provide a means of examining the location, spread and skewness of the forecasts and the observations. The box covers the *interquartile range* (the central 50% of the data) and the line across the centre of the box marks the *median* (the central observation). The whiskers attached to the box show the range of the data, from minimum to maximum. Boxplots are especially useful when several of them are placed side by side for comparison.

Figure 2.1 shows boxplots of high-temperature forecasts for Oklahoma City made by the National Weather Service Forecast Office at Norman, Oklahoma. Outputs from three different forecasting systems are shown, together with the corresponding observations. These data were used in Brooks and Doswell (1996) and a full description of the forecasting systems can be found in that paper. In Figure 2.1 the median of the observed data is 24°C; 50% of the values lie between 14°C and 31°C; the minimum value is −8°C and the maximum value is 39°C. Sometimes a schematic boxplot is drawn, in which the whiskers extend only as far as the most extreme points inside the fences; outliers beyond this are drawn individually. The fences are at a distance of 1.5 times the interquartile range from the quartiles. Figure 2.2 shows boxplots of this type for forecasts and observations of winter temperature at 850 hPa over France from 1979/80 to 1993/94; these are the data used in the example given in Chapter 5 and are fully described there. These boxplots show that in this example the spread of the forecasts is considerably less than the spread of the observations. Notches may be drawn in each box to provide a measure of the rough significance of differences between the (sample) medians (McGill *et al.*, 1978).

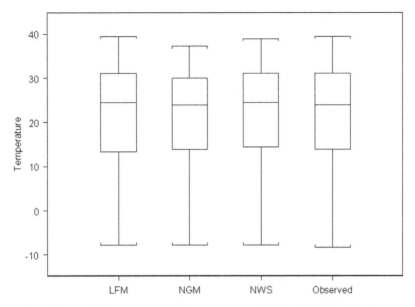

**Figure 2.1** Boxplots of 12–24-h forecasts of high-temperature (°C) for Oklahoma City from three forecasting systems and the corresponding observations

If the notched intervals for two groups of data do not overlap, this suggests that the corresponding population medians are different. Figure 2.3 shows notched boxplots for the observed data used in Figure 2.1 together with some artificial forecasts that were generated by adding a constant value to the actual forecasts. The notched intervals do not overlap, indicating a significant difference in the medians.

Histograms and bar charts provide another useful way of comparing the distributions of the observations and forecasts. A bar chart indicating the frequency of occurrence of each category can

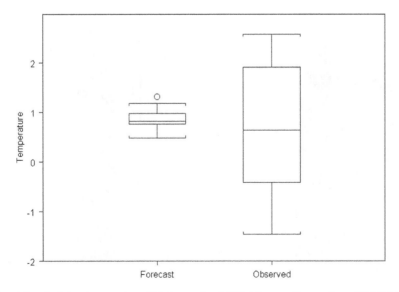

**Figure 2.2** Boxplots of winter temperature (°C) forecasts at 850 hPa over France from 1979/80 to 1993/94 and the corresponding observations

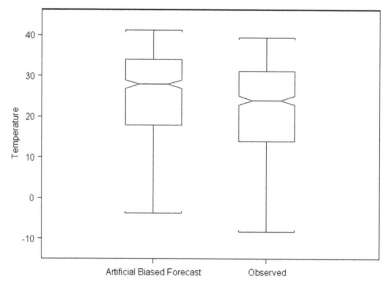

**Figure 2.3** Notched boxplot of artificial biased forecasts of high-temperature (°C) for Oklahoma City and the corresponding observations

be used to compare the distribution of forecasts and observations of categorical variables. The *mode* is the value that occurs most frequently. Bar charts for Finley's tornado data, which were presented in Chapter 1, are shown in Figure 2.4. In the case of continuous variables the values must be grouped into successive class intervals (bins) in order to produce a histogram. Figure 2.5 shows histograms for the observations and one of the sets of forecasts used in Figure 2.1. The appearance of the histogram may be quite sensitive to the choice of bin width and position of bin boundaries. If the bin width is

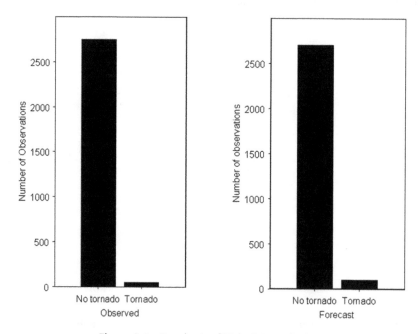

**Figure 2.4**   Bar charts of Finley's tornado data

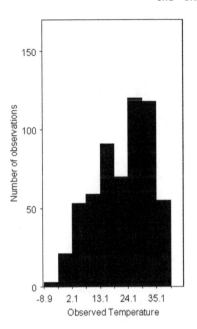
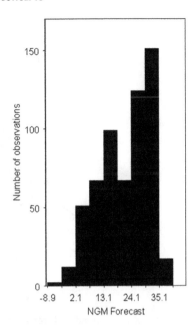

**Figure 2.5** Histograms of observed high-temperatures (°C) and 12–24-h forecasts for Oklahoma City

too small, the histogram reduces to a spike at each data point, but if it is too large, important features of the data may be hidden. Various rules for selecting the number of classes have been proposed, for example by Sturges (1926) and Scott (1979).

Boxplots and histograms can indicate systematic problems with the forecast system. For example, the forecasts may tend to be close to the climatological average with the consequence that the spread of the observations is much greater than the spread of the forecasts. Alternatively, the forecasts may be consistently too large or too small. However, the main concern of forecast verification is to examine the relationship between the forecasts and the observations. For continuous variables this can be done graphically by drawing a scatterplot. Figure 2.6 shows a scatterplot for persistence forecasts of the Oklahoma City high-temperature observations. If the forecasting system were perfect, all the points would lie on a straight line that starts at the origin and has a slope of unity. In Figure 2.6 there is a fair amount of scatter about this line. Figure 2.7, which is the scatterplot for one of the actual sets of forecasts, shows a stronger linear relationship. Figure 2.8 shows the scatterplot for the artificial set of forecasts used in Figure 2.3. There is still a linear relationship but the points do not lie on the line

through the origin. Figure 2.9 shows the scatterplot for another set of forecasts that have been generated artificially, in this case by reducing the spread of the forecasts. The points again lie close to a straight line but the line does not have a slope of unity.

In the case of categorical variables a contingency table can be drawn up showing the frequency of occurrence of each combination of forecast and observed category. Table 1.1 (see Chapter 1), showing Finley's tornado forecasts, is an example of such a table. If the forecasting system were perfect all the entries apart from those on the diagonal of the table would be zero. The relationship between forecasts and observations of continuous or categorical variables may be examined by means of a bivariate histogram or bar chart. Figure 2.10 shows a bivariate histogram for the data used in Figure 2.5.

## 2.4 Numerical descriptive measures

Boxplots and histograms provide a good visual means of examining the distribution of forecasts and observations. However, it is also useful to look at numerical summary statistics. Let $\hat{x}_1 \dots \hat{x}_n$ denote

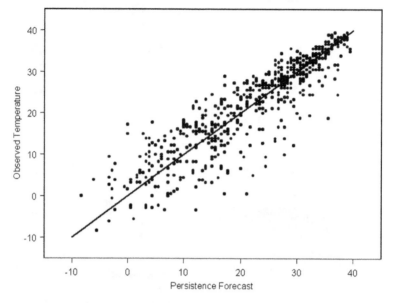

**Figure 2.6**   Scatterplot of observed high-temperatures (°C) against persistence forecasts for Oklahoma City

the set of forecasts and $x_1 \ldots x_n$ denote the corresponding observations. The *sample mean* of the observations is simply the average of all the observed values. It is calculated from the formula

$$\bar{x} = \frac{1}{n} \sum_{i=1}^{n} x_i \qquad (2.1)$$

One aspect of forecast quality is the (unconditional) *bias*, which is the difference between the mean forecast $\bar{\hat{x}}$ and the mean observation $\bar{x}$. It is desirable that the bias should be small. The forecasts in Figure 2.3 have a bias of 4°C.

The median is the central value; half of the observations are less than it and half are greater. For a

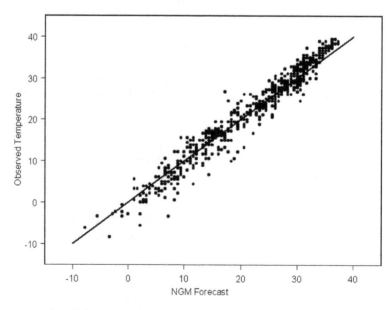

**Figure 2.7**   Scatterplot of observed high-temperatures (°C) against 12–24-h forecasts for Oklahoma City

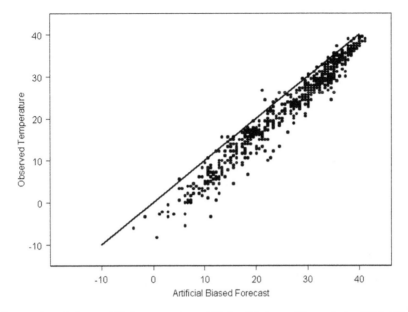

**Figure 2.8**  Scatterplot of observed high-temperatures (°C) for Oklahoma City against artificial biased forecasts

variable which has a reasonably symmetric distribution, the mean and the median will usually be fairly
similar. In the case of the winter 850 hPa temperature observations in Figure 2.2, the mean is 0.63°C
and the median is 0.64°C. Rainfall, in contrast, has a
distribution that is positively *skewed*, which means
that the distribution has a long right-hand tail. Daily

rainfall has a particularly highly skewed distribution
but even monthly averages display some skewness.
For example, Figure 2.11 is a histogram showing
the distribution of monthly precipitation at Oxford,
UK, over the period 1853–2009. Positively skewed
variables have a mean that is higher than the median. In the case of the data in Figure 2.11 the mean

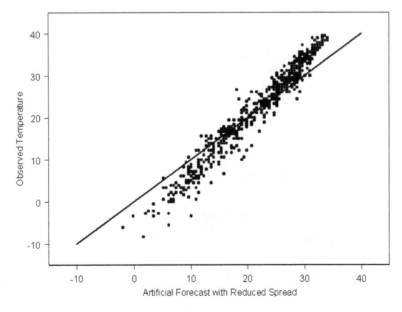

**Figure 2.9**  Scatterplot of observed high-temperatures (°C) for Oklahoma City against artificial forecasts that have
less spread than the observations

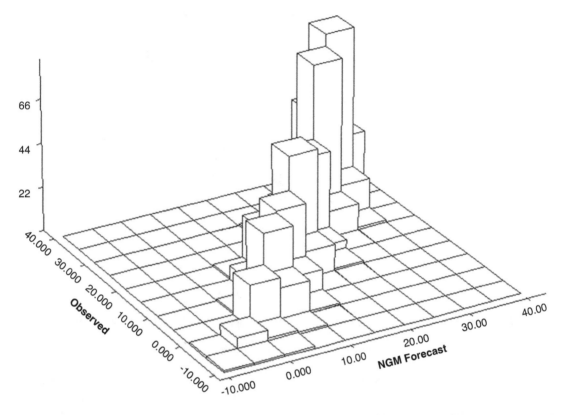

**Figure 2.10**   Bivariate histogram of observed high-temperatures (°C) and 12–24-h forecasts for Oklahoma City

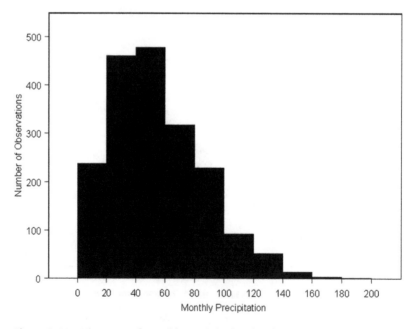

**Figure 2.11**   Histogram of monthly precipitation (mm) at Oxford, UK, 1853–2009

is 55 mm but the median is only 50 mm. Other variables, such as atmospheric pressure, may be negatively skewed, which means that the distribution has a long left-hand tail. The difference between the mean and the median divided by the standard deviation (defined below) provides one measure of the skewness of the distribution. Another measure of the skewness of the observations, which is described in more detail in Chapter 5, is

$$\text{skewness} = \frac{\frac{1}{n}\sum_{i=1}^{n}(x_i - \bar{x})^3}{\left(\frac{1}{n-1}\sum_{i=1}^{n}(x_i - \bar{x})^2\right)^{3/2}} \qquad (2.2)$$

Other definitions of skewness also exist, although for large samples there is little to choose between them (Joanes and Gill, 1998).

If the data come from a normal (Gaussian) distribution (Wilks, 2006b, section 4.4.2) then, provided the sample size is sufficiently large, the histogram should have approximately a symmetric bell-shaped form. For normally distributed data, the sample mean has a number of optimal properties, but in situations where the distribution is asymmetric or otherwise non-normal, other measures such as the median may be more appropriate (Garthwaite *et al.*, 2002, p. 15; DeGroot, 1986, pp. 567–9). Measures that are not sensitive to particular assumptions about the distribution of the data are known as *robust* measures. The mean can be heavily influenced by any extreme values; so use of the median is also preferable if there are outliers. Measures that are not unduly influenced by a few outlying values are known as *resistant* measures. The median is more robust and resistant than the mean, but even it can sometimes display surprising sensitivity to small changes in the data (Jolliffe, 1999).

The mean and median are not the only measures of the location of a data set (Wilks, 2006b, section 3.2) but we now move on to consider the spread of the values. The *sample variance* of the observations is defined as

$$s_x^2 = \frac{1}{n-1}\sum_{i=1}^{n}(x_i - \bar{x})^2 \qquad (2.3)$$

The most commonly used measure of spread is the *standard deviation*, which is the square root of this quantity. The standard deviations for the 850 hPa winter temperature data in Figure 2.2 are 0.2 for the forecasts and 1.3 for the observations. A more robust measure of spread is the interquartile range (IQR), which is the difference between the upper and lower *quartiles*. If the data are sorted into ascending order, the lower quartile and upper quartiles are one-quarter and three-quarters of the way through the data respectively. Like the median, the IQR is a measure that is resistant to the influence of extreme values and it may be a more appropriate measure than the standard deviation when the distribution is asymmetric. The Yule–Kendall index, which is the difference between the upper quartile minus the median and the median minus the lower quartile, divided by the IQR, provides a robust and resistant measure of skewness.

The median is the *quantile* for the proportion 0.5, and the lower and upper quartiles are the quantiles for the proportions 0.25 and 0.75. In general the quantile for the proportion $p$, also known as the 100$p$th *percentile*, is the value that is 100$p$% of the way through the data when they are arranged in ascending order. Other quantiles in addition to the median and the quartiles may also be useful in assessing the statistical characteristics of the distributions of forecasts and observations. For example, Murphy *et al.* (1989) use the difference between the 90th percentile minus the median and the median minus the 10th percentile as a measure of the asymmetry of the distribution.

There are also summary statistics that can be used to describe the relationship between the forecasts and the observations. The *sample covariance* between the forecasts and observations is defined as

$$s_{\hat{x}x} = \frac{1}{n-1}\sum_{i=1}^{n}(x_i - \bar{x})(\hat{x}_i - \bar{\hat{x}}) \qquad (2.4)$$

The *sample correlation coefficient* can be obtained from the sample covariance and the sample variances using the definition

$$r_{\hat{x}x} = \frac{s_{\hat{x}x}}{\sqrt{s_{\hat{x}}^2 s_x^2}} \qquad (2.5)$$

Further discussion of various forms of the correlation coefficient is given in Chapters 5 and 6.

## 2.5 Probability, random variables and expectations

If observations of a categorical variable are made over a sufficiently long period of time then the relative frequency of each event will tend to some limiting value, which is the *probability* of that event. For example, in Table 1.1 the relative frequency of the event 'tornado' is $51/2803 = 0.018$. The probability of a tornado occurring on any given day is therefore estimated to be 0.018. A *random variable*, denoted by $X$, associates a unique numerical value with each mutually exclusive event. For example, $X = 1$ if a tornado occurs and $X = 0$ if there is no tornado. A particular value of the random variable $X$ is denoted by $x$. The *probability function* $p(x)$ of a discrete variable associates a probability with each of the possible values that can be taken by $X$. For example, in the case of the tornado data, the estimated probability function is $p(0) = 0.982$ and $p(1) = 0.018$. The sum of $p(x)$ over all possible values of $x$ must by definition be unity.

In the case of continuous random variables the probability associated with any particular exact value is zero and positive probabilities can only be assigned to a range of values of $X$. The *probability density function* $f(x)$ for a continuous variable has the following properties:

$$f(x) \geq 0 \qquad (2.6)$$

$$\int_a^b f(x)dx = P(a \leq X \leq b) \qquad (2.7)$$

where $P(a \leq X \leq b)$ denotes the probability that $X$ lies in the interval from $a$ to $b$; and

$$\int_{-\infty}^{\infty} f(x)dx = 1 \qquad (2.8)$$

The *expectation* of a random variable $X$ is given by

$$E[X] = \sum_x xp(x) \qquad (2.9)$$

for discrete variables, and by

$$E[X] = \int_{-\infty}^{\infty} xf(x)dx \qquad (2.10)$$

for continuous variables. In both cases $E[X]$ can be viewed as the 'long-run average' value of $X$, so the sample mean provides a natural estimate of $E[X]$.

The *variance* of $X$ can be found from:

$$\text{var}(X) = E[(X - E[X])^2] \qquad (2.11)$$

The sample variance, $s_x^2$, provides an unbiased estimate of $\text{var}(X)$.

## 2.6 Joint, marginal and conditional distributions

In the case of discrete variables, the probability function for the *joint distribution* of the forecasts and observations $p(\hat{x}, x)$ gives the probability that the forecast $\hat{x}$ has a particular value and at the same time the observation $x$ has a particular value. So in the case of the tornado forecasts: $p(1,1) = 0.010$, $p(1,0) = 0.026$, $p(0,1) = 0.008$ and $p(0,0) = 0.956$. The sum of $p(\hat{x}, x)$ over all possible values of $\hat{x}$ and $x$ is by definition unity. In the case of continuous variables, the joint density function $f(\hat{x}, x)$ is a function with the following properties:

$$f(\hat{x}, x) \geq 0 \qquad (2.12)$$

$$\int_a^b \int_c^d f(\hat{x}, x)d\hat{x}dx = P(a \leq X$$
$$\leq b \text{ and } c \leq \hat{X} \leq d) \qquad (2.13)$$

$$\int_{-\infty}^{\infty} \int_{-\infty}^{\infty} f(\hat{x}, x)d\hat{x}dx = 1 \qquad (2.14)$$

The distributions with probability density functions $f(\hat{x})$ and $f(x)$, or probability functions $p(\hat{x})$ and $p(x)$ in the case of discrete random variables, are known as the *marginal distributions* of $\hat{X}$ and $X$ respectively. The marginal probability function $p(\hat{x})$ may be obtained by forming the sum of $p(\hat{x}, x)$

over all possible values of $x$. For example, in the case of the tornado forecasts

$$p(\hat{x}) = \begin{cases} p(1, 1) + p(1, 0) = 0.010 + 0.026 \\ \qquad\qquad = 0.036 \text{ for } \hat{x} = 1 \\ p(0, 1) + p(0, 0) = 0.008 + 0.956 \\ \qquad\qquad = 0.964 \text{ for } \hat{x} = 0 \end{cases}$$

(2.15)

Similarly,

$$p(x) = \sum_{\hat{x}} p(\hat{x}, x) \qquad (2.16)$$

In the case of continuous variables,

$$f(\hat{x}) = \int_x f(\hat{x}, x)\mathrm{d}x \qquad (2.17)$$

and

$$f(x) = \int_{\hat{x}} f(\hat{x}, x)\mathrm{d}\hat{x} \qquad (2.18)$$

The *conditional distribution*, which has the probability function $p(x|\hat{x})$, gives the probability that the observation will assume a particular value $x$ when a fixed value $\hat{x}$ has been forecast. The conditional probability function is given by the formula

$$p(x|\hat{x}) = \frac{p(\hat{x}, x)}{p(\hat{x})} \qquad (2.19)$$

A corresponding formula applies to continuous variables, with probability density functions replacing probability functions. In the case of the tornado data

$$p(x|\hat{X} = 1) = \begin{cases} 0.28 \text{ for } x = 1 \\ 0.72 \text{ for } x = 0 \end{cases} \qquad (2.20)$$

and

$$p(x|\hat{X} = 0) = \begin{cases} 0.01 \text{ for } x = 1 \\ 0.99 \text{ for } x = 0 \end{cases} \qquad (2.21)$$

Similarly, the probability function for the various forecast values, given that a fixed value has been observed, is given by the conditional probability function $p(\hat{x}|x)$. This function satisfies the equation:

$$p(\hat{x}|x) = \frac{p(\hat{x}, x)}{p(x)} \qquad (2.22)$$

So for the tornado data

$$p(\hat{x}|X = 1) = \begin{cases} 0.55 \text{ for } \hat{x} = 1 \\ 0.45 \text{ for } \hat{x} = 0 \end{cases} \qquad (2.23)$$

and

$$p(\hat{x}|X = 0) = \begin{cases} 0.03 \text{ for } \hat{x} = 1 \\ 0.97 \text{ for } \hat{x} = 0 \end{cases} \qquad (2.24)$$

The *conditional expectation* is the mean of the conditional distribution. In the case of discrete variables it is defined by

$$E[X|\hat{x}] = \sum_x x p(x|\hat{x}) \qquad (2.25)$$

which is a function of $\hat{x}$ alone, and in the case of continuous variables by

$$E[X|\hat{x}] = \int_x x f(x|\hat{x})\mathrm{d}x \qquad (2.26)$$

If $p(x|\hat{x}) = p(x)$ or $f(x|\hat{x}) = f(x)$, then the forecasts and observations are statistically *independent*. This would occur, for example, if forecasts were made at random, or if the climatological average value was always forecast. Forecasts that are statistically independent of the observations may perhaps be regarded as the least useful kind of forecasts. If the forecasts are taken at face value, then it is clearly possible to have worse forecasts. In the extreme case, it would be possible to have a situation in which rain was always observed when the forecast was no rain and it was always dry when rain was forecast. However, such forecasts would actually be very useful if a user was aware of this and inverted (recalibrated) the forecasts accordingly.

## 2.7  Accuracy, association and skill

A score function or scoring rule is a function of the forecast and observed values that is used to assess the quality of the forecasts. Such verification measures often assess the accuracy or association of the forecasts and observations. Accuracy is a measure of the correspondence between individual pairs of forecasts and observations, while association is the overall strength of the relationship between individual pairs of forecasts and observations. The correlation coefficient is thus a measure of linear association, whereas mean absolute error and mean squared error, which will be discussed in Chapter 5, are measures of accuracy.

As discussed in Chapter 1, skill scores are used to compare the performance of the forecasts with that of a reference forecast such as climatology or persistence. Skill scores are often in the form of an index that takes the value 1 for a perfect forecast and 0 for the reference forecast. Such an index can be constructed in the following way:

$$\text{Skill score} = \frac{\text{Score} - \text{Sref}}{\text{Sperf} - \text{Sref}} \quad (2.27)$$

where Sref is the score for a reference forecast and Sperf is the score for a perfect forecast.

The choice of reference forecast will depend on the temporal scale. As already noted, persistence may be an appropriate choice for short-range forecasts, whereas climatology may be more appropriate for longer-range forecasts.

## 2.8  Properties of verification measures

The process of determining whether verification measures possess certain properties is called *metaverification* (Murphy, 1996a). Here we introduce the best-known properties; there is some further discussion of metaverification, including some additional measures, in Section 3.4.

Murphy (1993) identified *consistency* as one of the characteristics of a good forecast. A forecast is consistent if it corresponds with the forecaster's judgement. Murphy (1993) argues that since forecasters' judgements necessarily contain an element of uncertainty, this uncertainty must be reflected accurately in the forecasts, which means that they must be expressed in probabilistic terms, although it is possible that words rather than numbers may be used to express this uncertainty. Some scoring systems encourage forecasters to be inconsistent or to 'hedge' (Murphy and Epstein, 1967b). For example, with some verification measures a better score is obtained on average by issuing a forecast that is closer to the climatological average than the forecaster's best judgement. Murphy and Daan (1985) apply the term *consistent* to scoring rules for a particular directive, rather than to forecasts. For example, a scoring rule is consistent for the directive 'forecast the expected value of the variable' if the best possible score is obtained when the forecaster forecasts the mean of his or her subjective probability distribution.

A *proper* scoring rule for probabilistic forecasts is one that is defined in such a way that forecasters are rewarded with the best expected scores if their forecast probabilities correspond with their true beliefs. A scoring rule is *strictly proper* when the best scores are obtained if and only if the forecasts correspond with the forecaster's judgement. An example of a strictly proper scoring rule is the Brier score, described in Chapter 7.

A deterministic forecast may correspond to the mean, median or mode of the forecaster's subjective probability distribution. The full distribution may not be known or may be too complicated to communicate. The term 'proper scoring rule' is generally only applied to forecasts that specify the full probability distribution. However, in the statistical literature the term has occasionally been applied to the truthful elicitation of a specific parameter of the distribution. Savage (1971) discussed proper scoring rules for eliciting expectations, while Cervera and Muñoz (1996) and Gneiting and Raftery (2007) wrote about proper scoring rules for quantiles. These are equivalent to consistent scoring rules for directives stating that these properties should be forecast. A score that is optimized when the mean of the forecaster's subjective probability distribution is truthfully reported is certainly a proper scoring rule, but it is not a strictly proper one because every other probability distribution with the same mean will achieve the same score (Bröcker and Smith, 2007b). So the distinction between proper scoring rules for the full probability distribution and consistent ones for particular parameters is helpful. However, the terminology used to make the distinction differs. For example, Lambert *et al.* (2008) distinguish between scoring rules for full probability distributions and score functions for distribution properties. Lambert and Shoham (2009) describe (strictly) proper payoffs for multiple choice questions, of which categorical forecasts are an example.

Jolliffe (2008) notes that the definition of consistency in Murphy and Daan (1985) is circular in the sense that rather than choosing a directive first and then finding a consistent scoring rule, it is equally plausible to choose a scoring rule and then look for a directive that is consistent with it. It follows that although a scoring rule may be chosen because it is consistent with a particular directive, it is sufficient in theory for the forecaster to be informed of the scoring rule and not the directive, as a forecaster seeking to optimize his or her score will automatically follow the directive. Not all properties are elicitable (Lambert *et al.*, 2008), which means that not every directive will have a scoring rule that is consistent for it.

An *equitable* (Gandin and Murphy, 1992) verification measure is one for which all unskilful forecasts, such as constant forecasts and random forecasts, receive the same expected score over a long sequence of forecasts. Thus the definition implies a frequentist interpretation of probability, whereas the definition of propriety implies a subjective one. The concept of equitability is applied to deterministic categorical forecasts. Jolliffe and Stephenson (2008) show that scoring rules for probabilistic forecasts can never be both equitable and proper, assuming that the verification measure for a set of forecasts is equal to the sum or average of the scores for each individual forecast. Equitability is discussed in greater detail in Chapter 3.

A score for a probabilistic forecast is *local* if it depends only on the probability assigned to the outcome that is actually observed. However, except in the trivial case of binary predictands, this property is incompatible with the property of being *sensitive-to-distance* (Jose *et al.*, 2009). It is widely accepted that locality may not be a desirable property, since for ordinal predictands it is intuitively appealing

to credit forecasts that give higher probabilities to categories that are close to the one observed. However, in the case of a non-local score, a forecast with a higher probability assigned to the observed category will not necessarily achieve a better score than one with a lower probability assigned to it (Mason, 2008).

## 2.9 Verification as a regression problem

It is possible to interpret verification in terms of simple linear *regression* models in which the forecasts are regressed on the observations and vice versa (Murphy *et al.*, 1989). The book by Draper and Smith (1998) provides a comprehensive review of regression models. In the case in which the observations are regressed on the forecasts, the linear regression model is

$$x_i = \alpha + \beta \hat{x}_i + \varepsilon_i \qquad (2.28)$$

where $\varepsilon_i$, $i = 1 \ldots n$ are error terms. It is assumed that $E[\varepsilon_i] = 0$ for all $i = 1 \ldots n$ and that the errors are uncorrelated. The regression equation can be rewritten as

$$E[X|\hat{x}] = \alpha + \beta \hat{x} \qquad (2.29)$$

Estimates $\hat{\alpha}$ and $\hat{\beta}$ of the parameters $\alpha$ and $\beta$ can be obtained by the method of least squares. These estimates are

$$\hat{\alpha} = \bar{x} - \hat{\beta}\bar{\hat{x}} \qquad (2.30)$$

and

$$\hat{\beta} = \frac{s_x}{s_{\hat{x}}} r_{\hat{x}x} \qquad (2.31)$$

where $s_x$ and $s_{\hat{x}}$ are the sample standard deviations of the observations and forecasts respectively (see Equation (2.3) and $r_{\hat{x}x}$ is the sample correlation coefficient between the forecasts and the observations (see Equation 2.5). It is desirable that the conditional bias of the observations given the forecasts $(E[X|\hat{x}] - \hat{x})$ should be zero (*unbiased*). This will

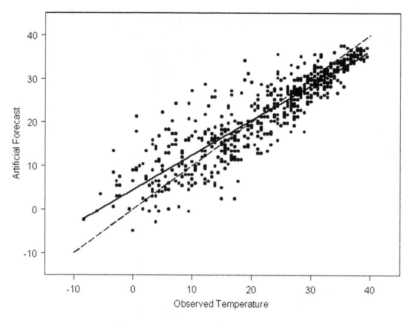

**Figure 2.12** Scatterplot of artificial forecasts against observed high-temperature (°C) at Oklahoma City. The solid line shows the regression line obtained by regressing the forecasts on the observations. The dashed line is the 45° line and the forecasts have been generated in such a way that this is the regression line obtained by regressing the observations on the forecasts.

only be satisfied if $\alpha = 0$ and $\beta = 1$, which means that the regression line has an intercept of zero and a slope of unity. So ideally the regression line will coincide with the 45° line. Figure 2.12 shows an artificial set of Oklahoma City high-temperature forecasts for which this is the case. Note that in the literature on forecast verification, the phrase *conditional bias of the forecasts* is used to refer to both the conditional bias of the observations given the forecasts, $E[X|\hat{x}] - \hat{x}$ (type 1 conditional bias), and to $E[\hat{X}|x] - x$ (type 2 conditional bias). Since the correlation coefficient between the forecasts and the observations is always less than or equal to unity, it follows from Equation 2.31 that the condition $\hat{\beta} = 1$ can only be satisfied if the standard deviation of the forecasts is less than or equal to the standard deviation of the observations. The points will all lie exactly on the fitted straight line only if $r_{\hat{x}x}^2 = 1$.

In the case in which the forecasts are regressed on the observations, the linear regression model can be written as

$$E[\hat{X}|x] = \gamma + \delta x \qquad (2.32)$$

and estimates $\hat{\gamma}$ and $\hat{\delta}$ of the parameters $\gamma$ and $\delta$ are given by

$$\hat{\gamma} = \bar{\hat{x}} - \hat{\delta}\bar{x} \qquad (2.33)$$

and

$$\hat{\delta} = \frac{s_{\hat{x}}}{s_x} r_{\hat{x}x} \qquad (2.34)$$

In situations where $\hat{\alpha} = 0$ and $\hat{\beta} = 1$ (from which we would infer that the forecasts are conditionally unbiased), $s_{\hat{x}} < s_x$ unless the forecasts and observations are perfectly correlated (i.e. $r_{\hat{x}x} = 1$), and it follows that $\hat{\delta} < 1$ unless the forecasts are perfect. In Figure 2.12 the regression line obtained by regressing the forecasts on the observations does not have a slope of unity even though the regression line obtained by regressing the observations on the forecasts does have. Thus forecasts that are conditionally unbiased from the perspective of regressing the observations on the forecasts will, unless they are perfect, be conditionally biased from the perspective of regressing the forecasts on the observations.

## 2.10 The Murphy–Winkler framework

Murphy and Winkler (1987) outlined a general framework for forecast verification based on the joint distribution of the forecasts and observations. The traditional *measures-oriented* approach to forecast verification involves the reduction of the information from a set of forecasts and observations into a single verification measure or perhaps a small number of performance measures. These measures are usually concerned with some overall aspect of forecast quality such as accuracy, association or skill. The alternative approach involving the use of the joint distribution of forecasts and observations is known as *distributions-oriented* verification or *diagnostic* verification. Forecast verification usually involves posterior evaluation of a sample of past forecasts and observations. In this context the joint distribution $p(\hat{x}, x)$ is usually interpreted as being a discrete empirical relative frequency distribution. In practice even a variable that can theoretically be measured on a continuous scale is only recorded with a certain precision. For example, temperature is usually recorded to the nearest tenth of a degree Celsius. This means that only a certain number of distinct values of the forecasts and observations will be found in the verification data set. So provided that a sufficiently large data set is available, it is possible to examine the empirical relative frequency distribution even in the case of variables such as temperature that are theoretically continuous.

With a sufficiently large data set, graphical techniques such as boxplots and histograms and simple summary statistics can be applied to the conditional as well as the marginal distributions. Summary statistics such as the median and the interquartile range of the conditional distributions can help to identify particular values of the forecasts and observations for which the forecasting system performs especially well or especially badly. This may give insights into ways in which forecasts could be improved that would not be available through traditional overall measures of forecast performance.

It follows from Equations 2.19 and 2.22 that the joint distribution of the forecasts and the observations can be factored into a conditional and a marginal distribution in two different ways. The factorization

$$p(\hat{x}, x) = p(x|\hat{x})p(\hat{x}) \qquad (2.35)$$

is known as the *calibration-refinement* factorization. A set of deterministic forecasts is said to be perfectly *calibrated* or *reliable* if $E[X|\hat{X} = \hat{x}] = \hat{x}$ for all $\hat{x}$. The concept of a set of forecasts being completely reliable is therefore equivalent to the observations given the forecasts being conditionally unbiased. If a set of forecasts is conditionally unbiased for all forecast values, it must also be unconditionally unbiased (i.e. $E[\hat{X}] = E[X]$). Probabilistic forecasts of a binary variable are perfectly calibrated if $E[X|\hat{p}(1)] = \hat{p}(1)$ for all $\hat{p}(1)$, where $\hat{p}(1)$ is the forecast probability that $X = 1$. An overall measure of reliability is

$$\text{REL} = E_{\hat{X}}[(\hat{X} - E[X|\hat{X}])^2] \qquad (2.36)$$

where the notation $E_{\hat{X}}[.]$ means that the expectation is calculated with respect to the marginal distribution of the forecasts. A similar definition applies to probabilistic forecasts of a binary variable with $\hat{X}$ replaced by the random variable $\hat{P}(1)$.

The marginal distribution $p(\hat{x})$ indicates how often different forecast values occur. If the same forecast is always issued, forecasts are said not to be *sharp*. Thus a forecaster who always forecasts climatological average is not sharp. Sharpness is difficult to define in the case of deterministic forecasts (Murphy *et al.*, 1989) but for perfect forecasts it must be the case that $p(\hat{x})$ is equal to the marginal distribution of the observations $p(x)$. Murphy and Epstein (1967a), citing work by Bross (1953, pp. 48–52), define probabilistic forecasts as being sharp if the predicted probabilities are all either zero or unity. They suggest using the Shannon–Weaver information quantity

$$I = -\frac{1}{n}\sum_{i=1}^{n}\sum_{x=0}^{K-1}\hat{p}_i(x)\ln(\hat{p}_i(x)) \qquad (2.37)$$

as a measure of the sharpness of probabilistic forecasts of a categorical variable, where $K$ is the number of categories and $n$ is the number of forecasts. This index has a minimum value of zero (maximum sharpness) when all the values of $\hat{p}(x)$ are zero or unity and a maximum value of $\ln(K)$ when all the values of $\hat{p}(x)$ are equal to $1/K$. The use of 'information' in forecast verification dates back to Holloway and Woodbury (1955). In the case of a binary variable, an alternative measure (Daan, 1984) is

$$S = \frac{1}{n} \sum_{i=1}^{n} (\hat{p}_i(1)(1 - \hat{p}_i(1))) \qquad (2.38)$$

This function has a minimum value of zero (maximum sharpness) when all the values of $\hat{p}(1)$ are zero or unity, and a maximum value of 0.25 when all the values of $\hat{p}(1)$ are 0.5. Murphy and Winkler (1992) suggest using the variance of the forecasts $\text{var}(\hat{P}(1))$ as a measure of the sharpness of probabilistic forecasts of a binary variable, with larger values of the variance indicating greater sharpness. However, this leads to a contradiction. If the variance of the forecasts is used, a forecaster who always forecasts 'no rain' is not at all sharp because the forecast probabilities have a variance of zero, whereas according to the definition of Murphy and Epstein (1967a) he or she is perfectly sharp, since only forecasts of zero or unity are used. However, for perfectly calibrated forecasts, the measures of sharpness are consistent with one another, as in this case:

$$E[\hat{P}(1)(1 - \hat{P}(1))] = -E[(\hat{P}(1))^2] + E[\hat{P}(1)]$$
$$= -E[(\hat{P}(1))^2] + E[X]$$
$$(2.39)$$

and

$$\text{var}(\hat{P}(1)) = E[(\hat{P}(1))^2] - (E[\hat{P}(1)])^2$$
$$= E[(\hat{P}(1))^2] - (E[X])^2$$
$$(2.40)$$

The terms involving $E[X]$ in Equations 2.39 and 2.40 do not depend on the forecast and so can be ignored.

DeGroot and Fienberg (1982, 1983) introduced the concept of *refinement* as a means of comparing forecasters who are perfectly calibrated. Denoting two perfectly calibrated forecasters by A and B, A is at least as refined as B if, knowing A's predictions, it is possible to simulate B's predictions by

means of an auxiliary randomization. The mathematics of this definition is beyond the scope of this book. However, the least refined of all perfectly calibrated forecasts are those in which the forecast probabilities are simply the climatological probabilities, and the most refined are those in which the forecast probabilities are always zero or unity. If the forecasts are perfectly calibrated and the forecast probabilities are all zero or unity, the forecasts must be perfect. According to the definition given by DeGroot and Fienberg, it is not always possible to compare forecasters in terms of refinement; sometimes it is the case that neither is forecaster A at least as refined as forecaster B, nor is forecaster B at least as refined as forecaster A. So, strictly speaking, the concept of refinement is not the same as that of sharpness, although many authors have treated the terms as synonymous. The extension of this definition of refinement to forecasts that are not perfectly calibrated leads to the concept of *sufficiency* (DeGroot and Fienberg, 1983; Ehrendorfer and Murphy, 1988). A's forecasts are sufficient for B's if B's forecasts can be obtained from A's by a stochastic transformation.

Sanders (1963) used the term 'sharpness' for the concept termed '*resolution*' in an earlier paper (Sanders, 1958). His definition of the resolution of probabilistic forecasts of a binary variable is

$$\text{RES}_{\text{Sanders}} = E_{\hat{P}(1)}[E[X|\hat{P}(1)](1 - E[X|\hat{P}(1)])] \qquad (2.41)$$

Small values of this quantity are preferable to large ones. Another measure of resolution (Murphy and Winkler, 1987) is:

$$\text{RES}_{\text{Murphy}} = E_{\hat{X}}[(E[X|\hat{X}] - E[X])^2]$$
$$= \text{var}_{\hat{X}}(E[X|\hat{X}]) \qquad (2.42)$$

For probabilistic forecasts of a binary variable, $\hat{X}$ should be replaced by $\hat{P}(1)$ in this definition. Large values of this quantity are preferable to small ones because they indicate that on average different forecasts are followed by different observations. Resolution is not equivalent to the definitions of sharpness given by Murphy and Epstein (1967a) or Murphy and Winkler (1992) since it involves the distribution of the observations as well as that of the forecasts. However, it should be noted that for perfectly calibrated forecasts of a binary variable they become identical concepts. For perfectly calibrated forecasts $E[X|\hat{p}(1)] = \hat{p}(1)$ and $E[\hat{P}(1)] = E[X]$; so Equation 2.41 becomes

$$\text{RES}_{\text{Sanders}} = E[\hat{P}(1)(1 - \hat{P}(1))] \qquad (2.43)$$

and Equation 2.42 becomes

$$\text{RES}_{\text{Murphy}} = E[(\hat{P}(1) - E[\hat{P}(1)])^2] = \text{var}(\hat{P}(1)) \qquad (2.44)$$

Both calibration and sharpness are important. Suppose the climatological average probability of rain at a particular place is 0.3. A forecaster who always forecast a 30% chance of rain would be perfectly calibrated but not sharp. A forecaster who forecast 100% chance of rain on 30% of days chosen at random and 0% chance on the other 70% of days so that $E[X|\hat{P}(1) = 1] = 0.3$ and $E[X|\hat{P}(1) = 0] = 0.3$ would be sharp but not at all calibrated. Both examples show no resolution. In the first case the expected value of $X$ following a forecast of *30% chance of rain* is 0.3, which is identical to the unconditional expectation of $X$. In the second case the expected values of $X$ following forecasts of either *rain* or *no rain* are both equal to the unconditional expectation of 0.3.

The second factorization

$$p(\hat{x}, x) = p(\hat{x}|x)p(x) \qquad (2.45)$$

is known as the *likelihood-base rate factorization*. For a given forecast $\hat{x}$, the conditional probabilities $p(\hat{x}|x)$ are known as the *likelihoods* associated with the forecast. If $p(\hat{x}|x)$ is zero for all values of $x$ except one, the forecast is perfectly *discriminatory*. When $p(\hat{x}|x)$ is the same for all values of $x$, the forecast is not at all discriminatory. Two measures of discrimination are (Murphy, 1993):

$$\text{DIS1} = E_X[(E[\hat{X}|X] - X)^2] \qquad (2.46)$$

and

$$\text{DIS2} = E_X[(E[\hat{X}|X] - E[\hat{X}])^2] \qquad (2.47)$$

For probabilistic forecasts of a binary variable, $\hat{X}$ should again be replaced by $\hat{P}(1)$ in these definitions. Good discrimination implies that DIS1 is small and DIS2 is large. Murphy et al., (1989) used a measure of discrimination based on the likelihood ratio

$$\text{LR}(\hat{x}; x_i, x_j) = \frac{p(\hat{x}|x_i)}{p(\hat{x}|x_j)} \qquad (2.48)$$

The discrimination between the observations $x_i$ and $x_j$ provided by the forecast $\hat{x}$ is the maximum of $\text{LR}(\hat{x}; x_i, x_j)$ and $1/\text{LR}(\hat{x}; x_i, x_j)$.

The marginal distribution $p(x)$ is sometimes known as the *uncertainty* or *base rate*. It is a characteristic of the forecasting situation rather than of the forecast system. Forecasting situations involving a fairly peaked distribution are characterized by relatively little uncertainty and are less difficult situations in which to forecast than ones in which the distribution of possible values is fairly uniform. A binary variable is easier to forecast if the climatological probability is close to zero or unity than if it is close to 0.5. A measure of whether a continuous distribution is peaked or not is the *kurtosis*, which is defined as:

$$\text{kurtosis} = \frac{E[(X - E[X])^4]}{(E[(X - E[X])^2])^2} - 3 \qquad (2.49)$$

For a normal distribution the kurtosis is zero, whereas for a uniform distribution it is $-1.2$.

By equating the two factorizations we find that

$$p(x|\hat{x})p(\hat{x}) = p(\hat{x}|x)p(x) \qquad (2.50)$$

This equation can be rewritten in the form

$$p(x|\hat{x}) = \frac{p(x)p(\hat{x}|x)}{p(\hat{x})} \qquad (2.51)$$

which is known as *Bayes' theorem*.

If forecasts of a binary variable are perfectly calibrated and completely sharp, then the forecasts must be perfectly discriminatory. If the forecasts are completely sharp, $\hat{p}(x)$ takes only the values zero and unity. In this situation

$$DIS1 = P(X = 1)(P(\hat{P}(1) = 1|X = 1) - 1)^2$$
$$+ P(X = 0)(P(\hat{P}(1) = 1|X = 0) - 0)^2$$
(2.52)

It follows from Equation 2.51 that

$$P(\hat{P}(1) = 1|X = 1)$$
$$= \frac{P(X = 1|\hat{P}(1) = 1)P(\hat{P}(1) = 1)}{P(X = 1)}$$
(2.53)

If the forecasts are perfectly calibrated

$$P(X = 1|\hat{P}(1) = 1) = 1 \qquad (2.54)$$

and

$$P(\hat{P}(1) = 1) = P(X = 1) \qquad (2.55)$$

so

$$P(\hat{P}(1) = 1|X = 1) = 1 \qquad (2.56)$$

Similarly,

$$P(\hat{P}(1) = 1|X = 0)$$
$$= \frac{P(X = 0|\hat{P}(1) = 1)P(\hat{P}(1) = 1)}{P(X = 0)}$$
$$= \frac{(1 - P(X = 1|\hat{P}(1) = 1))P(\hat{P}(1) = 1)}{P(X = 0)}$$
$$= \frac{(1 - 1)P(\hat{P}(1) = 1)}{P(X = 0)}$$
$$= 0$$
(2.57)

Thus

$$DIS1 = P(X = 1)(1 - 1)^2$$
$$+ P(X = 0)(0 - 0)^2 = 0 \quad (2.58)$$

However, the converse is not true. It is possible for forecasts to be perfectly discriminatory even if they are not calibrated. For example, if forecasts of rain are always followed by no rain and forecasts of no rain are always followed by rain, the forecasts are perfectly discriminatory but not calibrated. If users are aware of the base rates and the likelihoods and use the forecasts appropriately, the fact that the forecasts are not well calibrated is irrelevant as long as they are perfectly discriminatory. However, calibration is relevant if the forecasts are taken at face value. The likelihood-base rate factorization therefore gives information about the potential skill of the forecasts, whereas the calibration-refinement factorization gives information about the actual skill. The base rate $p(x)$ represents the probability of an event occurring before the forecast is issued, and Bayes' theorem shows how this probability should be updated when a particular forecast is issued. For example, in the case of Finley's tornado data, a forecast of *tornado* could be interpreted as *28% chance of a tornado* and a forecast of *no tornado* could be interpreted as *1% chance of a tornado* (see Equations 2.20 and 2.21).

## 2.11 Dimensionality of the verification problem

One difficulty that arises from the use of the distributions-oriented approach is the high *dimensionality* of many typical forecast verification problems (Murphy, 1991a). Dimensionality is defined as the number of probabilities that must be specified to reconstruct the basic distribution of forecasts and observations. It is therefore one fewer than the total number of distinct combinations of forecasts and observations. Problems involving probabilistic forecasts or non-probabilistic forecasts of predictands that can take a large number of possible values are of particularly high dimensionality. Of even higher dimensionality are *comparative verification* problems. Whereas *absolute verification* is concerned with the performance of an individual forecasting system, comparative verification is concerned with comparing two or more forecasting systems. The situation in which these produce forecasts under identical conditions is known as *matched comparative verification*. *Unmatched comparative verification* refers to the situation in which they produce forecasts under different conditions. Whereas absolute verification is based on the joint distribution of two variables, which can be factored into conditional and marginal distributions in two different ways, matched comparative verification is based on a three-variable distribution, which has six distinct factorizations. Unmatched comparative verification is based on a four-variable distribution, which has 24 different factorizations. The number of distinct factorizations represents the *complexity* of the verification problem (Murphy, 1991a).

In many cases the dimensionality of the verification problem will be too great for the size of the data set available. One approach to reducing dimensionality, which was used by Brooks and Doswell (1996), is to create a categorical variable with a smaller number of categories than the number of distinct values recorded, by dividing the values into bins. An alternative approach to reducing dimensionality, which is discussed by Murphy (1991a), is to fit parametric statistical models to the conditional or unconditional distributions. The evaluation of forecast quality is then based on the parameters of these distributions. For example, fitting a bivariate normal distribution to forecasts and observations of a continuous variable would lead to five parameters: the means and variances of the forecasts and observations respectively and the correlation between them (Katz *et al.*, 1982).

# 3

# Deterministic forecasts of binary events

**Robin J. Hogan[1] and Ian B. Mason[2]**

[1]*Department of Meteorology, University of Reading, Reading, UK*
[2]*Retired (formerly Canberra Meteorological Office, Canberra, Australia)*

## 3.1 Introduction

Many meteorological phenomena can be regarded as simple binary (dichotomous) events, and forecasts or warnings for these events are often issued as unqualified statements that they will or will not take place. Rain, floods, tornados, frosts, fogs, etc. either do or do not occur, and appropriate forecasts or warnings either are or are not issued. These kinds of predictions are sometimes referred to as yes/no forecasts, and represent the simplest type of forecasting and decision-making situation. The $2 \times 2$ possible outcomes (contingencies) for an event are shown in Table 3.1. For a sequence of binary forecasts, we seek measures of performance that can be formulated as a function of the number of *hits* ($a$), *false alarms* ($b$), *misses* ($c$) and *correct rejections* ($d$).

The search for good measures of the quality of deterministic binary forecasts has a long history, dating at least to 1884 when Sergeant Finley of the US Army Signal Corps published the results of some experimental tornado forecasts (Finley, 1884). Murphy (1996a) has given a detailed history of the so-called 'Finley affair', briefly outlined in Chapter 1. Many of the basic issues in verification were first raised in this episode, and Finley's forecasts are often used as an example of deterministic binary forecasts. The $2 \times 2$ contingency table of forecasts and observations for the whole period of Finley's experimental programme is shown as Table 1.1 of this book.

Finley measured the performance of his forecasts using *percent correct*, the proportion of correct forecasts of either kind expressed as a percentage, i.e. $100 \times (a + d)/n$. By this measure, his forecasts were 96.6% correct. Gilbert (1884) promptly pointed out that there would have been more correct forecasts if Finley had simply forecast 'no tornado' every time, giving 98.2% correct. This was the first of many comments in the meteorological literature and elsewhere on the inadequacies of percent correct as a measure of forecasting performance, and more broadly on the need for measures that do not reward such behaviour. Gilbert (1884) proposed two new measures, one of which is most commonly known now as the *Critical Success Index* (Donaldson *et al.*, 1975). Gilbert's second measure was studied by Schaefer (1990), who recognized the historical precedence by naming it the *Gilbert Skill Score*. The well-known philosopher C.S. Peirce (1884) also took an interest in Finley's forecasts and proposed another measure of

**Table 3.1**  Schematic contingency table for deterministic forecasts of a sequence of $n$ binary events. The numbers of observations/forecasts in each category are represented by $a$, $b$, $c$ and $d$

|              | Event observed |          |           |
| ------------ | -------------- | -------- | --------- |
| Event forecast | Yes | No | Total |
| Yes   | $a$ (Hits)   | $b$ (False alarms)         | $a + b$ |
| No    | $c$ (Misses) | $d$ (Correct rejections)   | $c + d$ |
| Total | $a + c$      | $b + d$                    | $a + b + c + d = n$ |

forecasting skill, which we shall refer to as the *Peirce Skill Score*. A third paper stimulated by Finley's forecasts was published by Doolittle (1885). He proposed an association measure similar to the correlation coefficient for the $2 \times 2$ case, which has been used as an accuracy index for weather forecasts (Pickup, 1982). Doolittle (1888) later proposed a $2 \times 2$ version of the score, now known as the *Heidke Skill Score* (Heidke, 1926). A number of other papers followed during the 1880s and 1890s, reviewed by Murphy (1996a) and which he referred to as the 'aftermath' of the first three methodological papers mentioned above.

It is probably fair to say that there was very little change in verification practice for deterministic binary forecasts until the 1980s, with the introduction of methods from *signal detection theory* by Mason (1980, 1982a,b, 1989), and the development of a general framework for forecast verification by Murphy and Winkler (1987). The apparent simplicity of the $2 \times 2$ contingency table means that it is common to reduce the verification of a deterministic forecast of a continuous variable (e.g. rainfall) to a binary verification problem, by evaluating the performance of the forecast system in predicting exceedance of a range of chosen thresholds. In this case, larger thresholds lead to fewer hits but also fewer false alarms. Likewise, the evaluation of probabilistic forecasts of a binary event can be more easily evaluated by considering the set of deterministic binary forecasts obtained by choosing a range of probability decision thresholds (Mason, 1979). For these kinds of multiple binary forecast problems, methods from signal detection theory offer two broad advantages. Firstly, they provide a means of assessing the performance of a forecasting sys-tem that distinguishes between the intrinsic discrim-ination capacity and the decision threshold of the system. The main analysis tool that accomplishes this is the *Relative* (or *Receiver*) *Operating Char-acteristic* (ROC). Secondly, signal detection theory provides a framework within which other methods of assessing binary forecasting performance can be analysed and evaluated.

Since the 1980s, there has been considerable work to identify and quantify the *desirable proper-ties* that binary verification measures should have. For example, Gandin and Murphy (1992) provided the first definition and discussion of *equitability*, the property that all random forecasting systems, in-cluding those that always predict occurrence or non-occurrence, should be awarded the same expected score. Hogan *et al.* (2010) showed that several mea-sures reported as equitable are in fact not equitable (including, ironically, the *Equitable Threat Score*, a common alternative name for the Gilbert Skill Score), but went on to provide a recipe for gen-erating truly equitable measures. Stephenson *et al.* (2008a) tackled the problem in verifying forecasts of rare events that almost all measures tend to a meaningless limit (usually zero) as the rate of oc-currence falls to zero. Their *Extreme Dependency Score* and its successors do not have this problem and so open the way for the skill of forecasts of extreme events to be reliably verified and their per-formances compared (see also Chapter 10). Work such as this has led to a large number of available verification measures, which can be bewildering for the newcomer. Indeed, this complexity is surprising given the apparent simplicity of a $2 \times 2$ contingency table. The aim of this chapter is to provide the the-oretical framework to allow these measures to be judged according to a range of different properties, and hence decide which measure is appropriate for a particular verification problem.

The remainder of this chapter is in four parts. Section 3.2 presents some underpinning theory based on Murphy and Winkler's (1987) general joint probability framework and its relationship to diagrams that can be used to visualize verification measures, as well as introducing several common verification measures. Section 3.3 outlines signal detection theory (SDT) and the use of SDT methods such as the ROC for the verification of binary forecasts. Section 3.4 describes the various desirable properties of binary verification measures. Section 3.5 then presents the most widely used performance measures and their properties, as well as demonstrating how confidence intervals may be estimated and how *optimal threshold probabilities* may be derived. Probabilistic forecasts of binary events are covered in more detail in Chapters 7 and 8.

## 3.2 Theoretical considerations

This section presents some of the basic theory behind forecast verification relevant to the verification of deterministic binary forecasts. A more general discussion is given in Sections 2.8 and 2.10 (see Chapter 2).

### 3.2.1 Some basic descriptive statistics

In order to build the theoretical framework for analysing the behaviour of performance measures, it is first necessary to define a number of basic verification measures. Murphy (1997) distinguishes be-

tween *verification measures* and *performance measures*. A verification measure is any function of the forecasts, the observations, or their relationship, and includes for example the probability of the event being observed (the *base rate*), even though this is not concerned with the correspondence between forecasts and observations. Performance measures constitute a subset of verification measures that assess some aspect of the correspondence between forecasts and observations, either on an individual or collective basis, for example Finley's *percent correct*.

Table 3.1 gives the cell counts for each of the four possible combinations of forecast and observed event represented by $a, b, c$ and $d$. Everitt (1992) and Agresti (2007) reviewed the statistical methods that can be used to analyse categorical data presented in contingency tables. The sum $n = a + b + c + d$ is known as the *sample size* and is crucial for quantifying the sampling uncertainty in verification statistics. The counts in Table 3.1 may be converted to relative frequencies by dividing by $n$, and then interpreted as sample estimates of joint probabilities. It is good practice to always quote cell counts rather than derived relative frequencies since counts provide useful information about the number of events and are less prone to misinterpretation (Hoffrage *et al.*, 2000). Table 3.2 shows the joint probabilities estimated from the counts in Table 3.1; for example, $p(\hat{x} = 1, x = 0) = b/n$ is the estimated probability of joint occurrence of a forecast event ($\hat{x} = 1$) *and* an observed non-event ($x = 0$).

We first consider statistics that are verification measures in Murphy's (1997) sense, being functions of the forecasts and observations, but are not performance measures since they are not concerned

**Table 3.2** Schematic contingency table for deterministic forecasts of a binary event in terms of joint and marginal probabilities

| Event forecast | Event observed | | |
| --- | --- | --- | --- |
| | Yes ($x = 1$) | No ($x = 0$) | Total |
| Yes ($\hat{x} = 1$) | $p(\hat{x} = 1, x = 1)$ | $p(\hat{x} = 1, x = 0)$ | $p(\hat{x} = 1)$ |
| No ($\hat{x} = 0$) | $p(\hat{x} = 0, x = 1)$ | $p(\hat{x} = 0, x = 0)$ | $p(\hat{x} = 0)$ |
| Total | $p(x = 1)$ | $p(x = 0)$ | 1.0 |

directly with the correspondence between forecasts and observations. The *base rate, s,* is a sample estimate of the marginal probability of the event occurring, and is given by

$$s = (a + c)/n = \hat{p}(x = 1) \qquad (3.1)$$

It is purely a characteristic of the observations rather than of the forecasting system, and is also known as the *sample climate* or the *climatological probability* of the weather event. It strictly should have no direct relevance to assessment of forecasting skill, because the forecasting system has no control over the rate of occurrence of the observed events. However, many performance measures do depend on *s*, and are therefore (often unduly) sensitive to natural variations in observed weather and climate. The sampling distribution of *s* is that of a simple binary proportion with confidence interval given by Equation 3.23.

The *forecast rate, r,* is a sample estimate of the marginal probability of the event being forecast, and is given by

$$r = (a + b)/n = \hat{p}(\hat{x} = 1) \qquad (3.2)$$

As with *s*, the 95% confidence interval for the population probability corresponding to *r* is given by Equation 3.23.

The frequency bias, *B*, is the ratio of the number of forecasts of occurrence to the number of actual occurrences:

$$B = \frac{r}{s} = \frac{a + b}{a + c} \qquad (3.3)$$

It is referred to simply as *bias* when there is no risk of confusion with other meanings of the term (see Glossary). It is important to distinguish between *bias* and *skill*. When a free-running climate model is compared to observations, it is only possible to assess the bias of the model, because the model is not attempting to simulate the individual weather systems that actually occurred. Even if such a model were unbiased, it would be expected to have no skill, since there would be no correlation between the timing of events in the model and the observations. Therefore, bias alone conveys no information about skill. The contrast between these two quantities is exploited in Section 3.2.4 where a *skill-bias*

*diagram* is used to visualize a range of performance measures.

It is often stated that a bias of 1 is desirable; in other words, the event of interest should be forecast at the same rate as it is observed to occur. This allows certain types of forecast to be 'calibrated'. For example, if a deterministic forecast of tornado occurrence is issued by applying a decision threshold to a probability forecast, then any value of bias can be attained by changing the decision threshold. However, a bias of 1 is only desirable for users of the forecasts whose economically optimal threshold probability corresponds to a bias of 1 (in that particular forecasting system and climate), but not necessarily for all possible users. Users of tornado forecasts might tolerate false alarms but not misses, and therefore their favoured threshold probability corresponds to a bias of greater than 1.

### 3.2.2  A general framework for verification: the distributions-oriented approach

The conventional approach to assessment of skill in forecasting binary events prior to the mid-1980s consisted of calculating values for one or more summary measures of the correspondence between forecasts and observations, and then drawing conclusions about forecasting performance on the basis of these scores. This process is known as the 'measures-oriented' approach to verification and led to the development of a surprisingly large range of measures, a selection of which is shown in Table 3.3. The properties of these will be reviewed in full in Section 3.5. While they were generally plausible, there was little in the way of an agreed background of basic theory to guide selection of appropriate measures for particular verification problems, or to systematically discuss the properties of particular measures. With the aim of remedying this situation, Murphy and Winkler (1987) proposed a general framework for verification based on the joint probability distribution of forecasts and observations (see Table 3.2), and they defined forecast verification as the process of assessing the statistical characteristics of this joint distribution. This approach is known as 'distributions-oriented' verification, or sometimes 'diagnostic' verification. See Section 2.10 for a discussion of this probabilistic framework.

**Table 3.3**  Summary of verification measures discussed in this chapter

| Name of measure | Definition | Range | References |
|---|---|---|---|
| **Basic descriptive measures** | | | |
| Base rate, $s$ | $s = (a+c)/n$ | $[0,1]$ | Donaldson et al. (1975) |
| Forecast rate, $r$ | $r = (a+b)/n = (1-s)F + sH$ | $[0,1]$ | Swets (1986a) |
| Frequency bias, $B$ | $B = (a+b)/(a+c) = (1-s)F/s + H$ | $[0,\infty]$ | Swets (1986a) |
| **Performance measures** | | | |
| Hit rate, $H$ | $H = a/(a+c)$ | $[0,1]$ | Donaldson et al. (1975) |
| False alarm rate, $F$ | $F = b/(b+d)$ | $[0,1]$ | Finley (1884) |
| False alarm ratio, FAR | $FAR = b/(a+b) = \left[1 + \left(\dfrac{s}{1-s}\right)\dfrac{H}{F}\right]^{-1}$ | $[0,1]$ | Gilbert (1884); Donaldson et al. (1975) |
| Proportion Correct, PC | $PC = (a+d)/n = (1-s)(1-F) + sH$ | $[0,1]$ | Gilbert (1884); Schaefer (1990) |
| Critical Success Index, CSI | $CSI = \dfrac{a}{a+b+c} = \dfrac{H}{1 + F(1-s)/s}$ | $[0,1]$ | Doolittle (1888); Heidke (1926); Murphy and Daan (1985) |
| Gilbert Skill Score, GSS | $GSS = \dfrac{a-a_r}{a+b+c-a_r} = \dfrac{H-F}{(1-sH)/(1-s)+F(1-s)/s}$, where $a_r = (a+b)(a+c)/n$ is the expected $a$ for a random forecast with the same $r$ and $s$ | $[-1/3,1]$ | Peirce (1884); Hanssen and Kuipers (1965); Murphy and Daan (1985) |
| Heidke Skill Score, HSS | $HSS = \dfrac{a+d-a_r-d_r}{n-a_r-d_r} = \dfrac{2s(1-2s)(H-F)}{s+s(1-2s)H+(1-s)(1-2s)F}$, where $d_r = (b+d)(c+d)/n$ | $[-1,1]$ | |
| Peirce Skill Score, PSS | $PSS = \dfrac{ad-bc}{(b+d)(a+c)} = H - F$ | $[-1,1]$ | |

*(Continued)*

**Table 3.3**    (Continued)

| Name of measure | Definition | Range | References |
|---|---|---|---|
| Clayton Skill Score, CSS | $CSS = \dfrac{a}{a+b} - \dfrac{c}{c+d} = \left[1 + \dfrac{(1-s)F}{sH}\right]^{-1} - \left[1 + \dfrac{(1-s)(1-F)}{s(1-H)}\right]^{-1}$ | [−1,1] | Clayton (1934); Murphy (1996a); Wandishin and Brooks (2002) |
| Doolittle Skill Score, DSS | $DSS = \dfrac{ad-bc}{\sqrt{(a+b)(c+d)(a+c)(b+d)}}$ $= [H(1-F) - (1-H)F]\left[\left(\dfrac{H}{1-s} + \dfrac{F}{s}\right)\left(\dfrac{1-H}{1-s} + \dfrac{1-F}{s}\right)\right]^{-1/2}$ | [−1,1] | Doolittle (1885); Pickup (1982) |
| Log of Odds Ratio, LOR | $\theta = ad/(bc) = [H(1-F)]/[(1-H)F]; LOR = \ln(\theta)$ | $(-\infty, +\infty)$ | Stephenson (2000) |
| Odds Ratio Skill Score, ORSS; Yule's Q | $Q = \dfrac{ad-bc}{ad+bc} = \dfrac{\theta-1}{\theta+1} = \dfrac{H-F}{H(1-F)+F(1-H)}$ | [−1,1] | Yule (1900); Stephenson (2000) |
| Discrimination distance, $d'$ | $d' = \Phi^{-1}(H) - \Phi^{-1}(F)$ | $(-\infty, +\infty)$ | Tanner and Birdsall (1958); Swets (1986a) |
| ROC Skill Score, ROCSS | In the $2 \times 2$ case only, $A_z = \Phi(d'/\sqrt{2})$; $ROCSS = 2A_z - 1$ | [−1,1] | Swets (1986a) |
| Extreme Dependency Score, EDS | $EDS = 2\ln[(a+c)/n]/\ln(a/n) - 1 = 2\ln(s)/\ln(Hs) - 1$ | [−1,1] | Stephenson et al. (2008a) |
| Symmetric Extreme Dependency Score, SEDS | $SEDS = \ln(a_r/a)/\ln(a/n) = \ln[(1-s)F/H + s]/\ln(Hs)$ | [−1,1] | Hogan et al. (2009) |
| Symmetric Extremal Dependence Index, SEDI | $SEDI = \dfrac{\ln F - \ln H + \ln(1-H) - \ln(1-F)}{\ln F + \ln H + \ln(1-H) + \ln(1-F)}$ | [−1,1] | Ferro and Stephenson (2011) |
| Equitably Transformed SEDI | $SEDI_{ET} = \dfrac{SEDI - E[SEDI|r,s]}{1 - E[SEDI|r,s]}$ | [variable,1] | Hogan et al. (2010) |

The *dimensionality* of the joint probability distribution is defined by Murphy (1991a) as the number of probabilities (or parameters) that must be specified to reconstruct it. In general, the dimensionality of a verification problem is equal to $MN - 1$, where $M$ is the number of different forecast categories or values available to be used, and $N$ is the number of different observed categories or values possible. For a $2 \times 2$ contingency table, the dimensionality is three – the fourth degree of freedom being fixed by the constraint that the joint probabilities sum to unity. This implies that a full description of forecast quality in the $2 \times 2$ case requires only three parameters that contain all the information needed to reconstruct the joint distribution. Therefore, despite the many different verification measures for binary forecasts, there are only three independent dimensions and so the different measures are strongly interrelated with one another.

The joint distribution can be factored in two different ways into conditional and marginal probabilities that reveal different aspects of forecast quality. The calibration-refinement factorization is given by

$$p(\hat{x}, x) = p(x|\hat{x})p(\hat{x}). \qquad (3.4)$$

For example, the probability of a hit $p(\hat{x} = 1, x = 1) = a/n$ is equal to $p(x = 1 \mid \hat{x} = 1)p(\hat{x} = 1) = [a/(a + b)] \times [(a + b)/n]$. The refinement term $p(\hat{x})$ in the calibration-refinement factorization is the marginal distribution of the forecasts, which in the case of binary forecasts depends on the threshold for issuing a forecast of occurrence. The calibration term in the factorization, $p(x|\hat{x})$, is the quantity usually of most interest to users of forecasts, who wish to know the probability of the weather event given that it was or was not forecast. The second way of factoring the joint distribution is known as the likelihood-base rate factorization and is given by

$$p(\hat{x}, x) = p(\hat{x}|x)p(x) \qquad (3.5)$$

For example, the probability of a hit $p(\hat{x} = 1, x = 1) = a/n$ is equal to $p(\hat{x} = 1 \mid x = 1)p(x = 1) = [a/(a + c)] \times [(a + c)/n] = a/n$. The term $p(x)$ is simply the *base rate* given by Equation 3.1. The likelihood term $p(\hat{x}|x)$, considered to be a function of $x$, is the conditional probability of the forecast $\hat{x}$ given the observation $x$. The calibration-refinement and likelihood-base rate factorizations are related

by Bayes' theorem, obtained by equating the right-hand sides of Equations 3.4 and 3.5:

$$p(x|\hat{x}) = p(\hat{x}|x)p(x)/p(\hat{x}) \qquad (3.6)$$

### 3.2.3 Performance measures in terms of factorizations of the joint distribution

To demonstrate that only three parameters contain all information necessary to reconstruct the joint distribution $p(\hat{x}, x)$, consider first the likelihood and base rate components of the likelihood-base rate factorization. Obviously one parameter is the base rate, given by Equation 3.1. The second parameter is the *hit rate*, $H$, which is the proportion of occurrences that were correctly forecast and is given by

$$H = a/(a + c) = \hat{p}(\hat{x} = 1 \mid x = 1) \qquad (3.7)$$

The third is the *false alarm rate*, $F$, and is the proportion of non-occurrences that were incorrectly forecast. It is given by

$$F = b/(b + d) = \hat{p}(\hat{x} = 1 \mid x = 0) \qquad (3.8)$$

Inverting these relationships, the elements of the contingency table can be expressed in terms of $H$, $F$ and $s$ as follows:

$$a/n = sH \qquad (3.9)$$
$$b/n = (1 - s)F \qquad (3.10)$$
$$c/n = s(1 - H) \qquad (3.11)$$
$$d/n = (1 - s)(1 - F) \qquad (3.12)$$

Thus, all verification measures written in terms of $a, b, c$ and $d$ can be expressed instead in terms of $H, F, s$ and $n$. In practice, $n$ cancels in the definition of all the measures in Table 3.3, and thus we are able to express them all also in terms of $H, F$ and $s$. This property is useful for a number of reasons. Firstly, it highlights that since a forecaster has no control of the base rate, then for a given set of observed events in reality (i.e. a given $s$), the three-dimensional verification problem has reduced to a two-dimensional one defined by $H$ and $F$. As shown in Section 3.2.4, this enables the scores that would be awarded for all possible forecasts to be plotted on a graph of $H$ versus $F$, enabling the merits and disadvantages of various measures to be more easily visualized and diagnosed. A further use for expressing measures in terms of $H$, $F$ and $s$ is that it enables confidence intervals to be estimated, since $H$ and $F$ can be treated as random, binomially distributed variables. This is explained in Section 3.5.2.

A few comments should be made about $H$ and $F$. The hit rate, $H$, is also known as the *Probability of Detection* (POD) (Donaldson *et al.*, 1975), and in medical statistics it is referred to as *sensitivity*. The false alarm rate, $F$, is sometimes also called *Probability of False Detection* (POFD). The quantity $1 - F$ is referred to as *specificity* in medical statistics, and as *confidence* by the UK Met Office (see Chapter 10), and is an estimate of the conditional probability of correct rejections given that the event did not occur. As will be explained in Section 3.5.1, neither $H$ nor $F$ is suitable as a performance measure on its own.

An alternative way to formulate verification measures is based on the calibration-refinement factorization. Doswell *et al.* (1990) defined the *frequency of hits* as

$$\text{FOH} = a/(a + b) = \hat{p}(x = 1 \mid \hat{x} = 1)$$

and the *detection failure* ratio as

$$\text{DFR} = c/(c + d) = \hat{p}(x = 1 \mid \hat{x} = 0)$$

so we may follow the same procedure as above to describe the elements of the contingency table in terms of the three quantities FOH, DFR and the forecast rate, $r$, given by Equation 3.2. However, since the forecaster is able to influence all three of these variables, we no longer have the ability to compress a three-dimensional verification problem into two dimensions, and so this formulation is generally much less useful. Note that FOH and DFR were referred to as *correct alarm ratio* and *miss ratio,* respectively, by Mason and Graham (2002).

Some other combinations of three variables can be used to reconstruct the contingency table. A useful one is the base rate, $s$, the forecast rate, $r$ (or alternatively the frequency bias $B = r/s$) and a measure of skill such as the *Peirce Skill Score* (PSS = $H - F$). In this chapter we use the term *skill* to mean the ability to correctly forecast occurrence and/or non-occurrence more often than would be expected by chance; thus a free-running

climate model would be expected to have no skill in predicting the timing of individual weather events, even though it might have no bias. This combination of three variables therefore highlights that given a particular base rate, all possible sets of forecasts can be characterized purely by their *bias* and their *skill*. In the next section this will be used to introduce the *skill-bias* diagram for visualizing performance measures.

### 3.2.4  Diagrams for visualizing performance measures

In the previous section it was shown that for a given base rate $s$ and number of forecasts $n$, all possible sets of forecasts can be described by just two numbers. This enables verification measures to be easily plotted, which is useful for judging their merits and drawbacks. The most widely used of such graphs is a plot of hit rate, $H$, against false alarm rate, $F$, as shown in Figure 3.1a. The traditional use of this diagram is for assessing probabilistic forecasts of binary events, for which it is referred to as a *Relative Operating Characteristic* (ROC) diagram; this application is described in Section 3.3. On these axes, a perfect set of forecasts lies at the point $(F, H) = (0, 1)$ in the top left-hand corner, while the worst possible set of forecasts (every event was missed and for every non-event, an event was forecast) lies at $(1, 0)$. The point $(0, 0)$ in the lower left-hand corner corresponds to never forecasting occurrence, while the point $(1, 1)$ corresponds to constantly forecasting occurrence. The diagonal $H = F$ line between these two points represents zero skill. Specifically, forecasts produced by a random number generator that predicts occurrence with rate $r$ would have expected values of $H$ and $F$ that are equal. This can be demonstrated by noting that the expected value of $a$ from a random forecasting system (which we shall denote $a_r$), is proportional to the product of the marginal probabilities of the event occurring and the event being forecast:

$$a = a_r = np(\hat{x} = 1)p(x = 1) = nrs$$
$$= (a + b)(a + c)/n \tag{3.13}$$

Substitution of Equations 3.9, 3.10 and 3.11 into Equation 3.13 then leads to $H = F$. Thus,

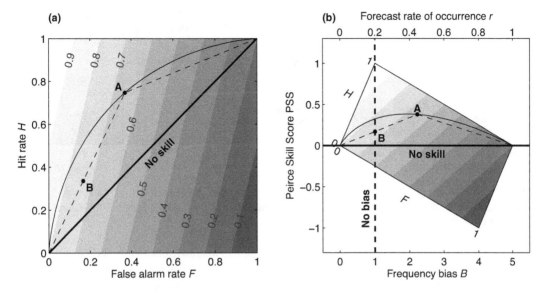

**Figure 3.1** Diagrams for visualizing verification measures. (a) Relative operating characteristic (ROC) diagram, where the shaded contours represent the measure *Proportion Correct* for a base rate of 0.2, the dashed line represents the trajectory taken in hedging forecast system A by randomly reassigning forecasts of occurrence to non-occurrence (or vice versa), and the thin solid line shows the fitted ROC curve to point A. (b) The equivalent skill-bias diagram for a base rate of 0.2; note that the axes of the tilted rectangle are simply hit rate *H* and false-alarm rate *F*. Point B in each panel represents a forecast system that has been hedged towards climatology from its original point A

forecasting systems with some degree of skill lie in the upper-left half of the diagram.

To demonstrate how the possible sets of forecasts represented on these axes would be rewarded by a particular verification measure, the shaded contours in Figure 3.1a depict the values of *Proportion Correct*, $PC = (a + d)/n$, for the case when base rate $s = 0.2$. As shown in Table 3.3, this measure is a function of $s$ as well as $H$ and $F$, so these contours change when $s$ is changed. PC is essentially the same as *percent correct* used by Finley (1884). The shortcoming of PC identified by Gilbert (1884) is immediately apparent: a forecaster at point A would be awarded a score of 0.66, but could increase his or her score simply by never forecasting occurrence, which would move the forecasts to the lower-left point on the graph, where the score awarded would be 0.8. This is an example of *hedging* a forecast, and Section 3.4.1 discusses how to judge whether measures are difficult to hedge.

An alternative way of plotting the information in a ROC is to rotate it somewhat such that the axes correspond to the Peirce Skill Score and the frequency bias. We refer to this as the *skill-bias*

diagram; an example is shown in Figure 3.1b. These are perhaps more natural axes for comparing verification measures, since we can more easily see how different measures reward skill and penalize bias (note that lines of constant bias could be drawn on ROC axes, but their position and angle depends on the base rate). The valid region of the skill-bias diagram in Figure 3.1b is indicated by the shaded rectangle bounded by $0 \leq H \leq 1$ and $0 \leq F \leq 1$ (as in Figure 3.1a). The choice of PSS as an ordinate as opposed to any other measure of skill is simply to avoid warping the shape of the diagram from a simple rectangle, not because we regard PSS as the best measure of skill. It should be noted that the location of the rectangle in a skill-bias diagram is dependent on the base rate, which means that if forecasting systems with different base rates are plotted on the same diagram, their location on the diagram does not uniquely define the value of any performance measure other than PSS and bias. The equivalent limitation of the ROC diagram is that if forecasting systems with different base rates are plotted together, their location on the diagram does not uniquely define either their bias or the

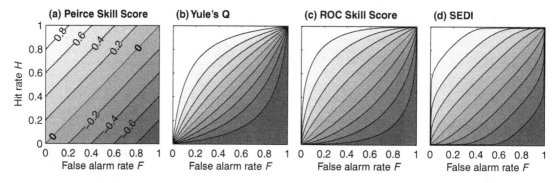

**Figure 3.2** Four base-rate independent performance measures plotted on ROC axes. All share the same colour scale as the Peirce Skill Score in panel (a). SEDI is the Symmetric Extremal Dependence Index. A full colour version of this figure can be found in the colour plate section

values of any 'base-rate dependent' performance measures (see Section 3.4.1). This is simply because, as discussed in Section 3.2.2, if base rate is allowed to vary then we have a three-dimensional verification problem that cannot be collapsed uniquely onto a two-dimensional diagram. Roebber (2009) presented an alternative diagram in which bias was

uniquely defined, but which lacked a unique 'no skill' line to enable forecasts to be compared easily with the equivalent random forecast.

Further examples of ROC and skill-bias diagrams for specific measures are shown in Figure 3.2 and Figure 3.3. In Section 3.2.5, the skill-bias diagram will be used to compare real forecasting systems,

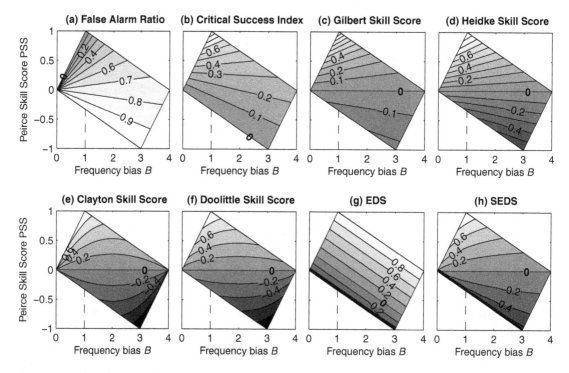

**Figure 3.3** Eight base-rate-dependent performance measures plotted on skill-bias axes for a base rate of 0.25. The vertical dashed lines indicate no bias. EDS, Extreme Dependency Score; SEDS, Symmetric Extreme Dependency Score. A full colour version of this figure can be found in the colour plate section

and in Section 3.4 these diagrams will be used to illustrate the various desirable properties of performance measures.

### 3.2.5 Case study: verification of cloud-fraction forecasts

We now illustrate some of the ideas presented in this section using real data. Most weather forecast models carry a *cloud fraction* variable in each vertical grid-box that simply quantifies the fraction of the grid-box containing cloud between zero (clear) and one (overcast). Hogan *et al.* (2009) used observations of cloud fraction from continuously operating radar and lidar instruments at several European sites to verify the hourly forecasts made by seven models; here we present some of their data on a skill-bias diagram.

Consider first the 12–35-hour forecasts of the European Centre for Medium Range Weather Forecasts (ECMWF) over three sites in 2004. To convert the cloud fraction observations and ECMWF forecasts into a contingency table, we apply a threshold of 0.05 such that a cloud 'event' is deemed to have occurred whenever cloud occupies more than 5% of a grid-box. This yields a contingency table of $(a, b, c, d) = (24097, 17195, 12095, 94918)$. While these cell counts may be used to calculate performance measures, they will yield poor estimates of the corresponding confidence intervals, since in reality each forecast is not independent. Hogan *et al.* (2009) estimated that only every 12th observation-forecast pair was independent, yielding an effective contingency table of approximately $(a, b, c, d) = (2008, 1433, 1008, 7910)$. From this we calculate a base rate of $s = 0.244$, a frequency bias of $B = 1.141 \pm 0.028$ and a Peirce Skill Score of PSS $= 0.512 \pm 0.018$, where the ranges indicate 95% confidence intervals calculated as described in Section 3.5.2. Without dividing the cell counts by 12, the confidence intervals would be a factor of $\sqrt{12}$ smaller.

The filled black ellipse in Figure 3.4 presents the ECMWF forecasts on a skill-bias diagram (for $s = 0.244$), where the horizontal and vertical extents of the ellipse, which are very similar in this example, correspond to the 95% confidence intervals in $B$ and PSS, respectively. Note that the area

of the ellipse contains only 85% of the joint probability distribution. As will be explored in Sections 3.4 and 3.5, there are many alternative measures of skill to PSS with different properties. To illustrate just one, the contours in Figure 3.4 depict the values of the Symmetric Extremal Dependence Index (SEDI; see Section 3.5.1), which for the ECMWF model has a value SEDI $= 0.677 \pm 0.019$. Nothing should be inferred from the fact that this is greater than the PSS awarded to the same set of forecasts; one should only compare values of the same performance measure applied to different sets of data. The other ellipses show the same analysis performed on forecasts by other models using the same observations. We can use this diagram to get a first indication of whether one model is significantly better than another in terms of either bias or skill, but it is not sufficient to simply check whether the confidence intervals overlap. Consider the German Weather Service Lokal model (DWD-LM) 0–11-hour forecasts, which is awarded a SEDI of $0.662 \pm 0.020$. Under the assumption that the two verification data sets are independent, the SEDI awarded to ECMWF is $0.015 \pm 0.028$ higher; therefore we may say that by this measure of skill, neither model is significantly better than the other. In reality the two data sets are not independent since they use the same observed weather, and the errors are likely to be correlated; if one model simulates a particular cloud type badly it is likely that another one will do likewise since they make many of the same assumptions in their parameterizations. This has the effect of *reducing* the error in the difference in skill between the two models. Section 3.5.2 outlines how such correlations can be accounted for in order to determine rigorously whether one model is significantly more skilful than another.

We can use Figure 3.4 to chart how skill and bias degrade with increasing forecast lead time. It can be seen that the two versions of the UK Met Office (UKMO) model both exhibit a steady reduction in skill with lead time, but with no significant change in bias. By contrast, the DWD-LM forecasts show an increase in bias during the first day of each forecast, after which the bias is steady. We could interpret this as being due to the model being initialized with an unbiased cloud and humidity distribution, but after 24 hours the terms governing the water cycle in the model cause it to evolve towards its preferred state

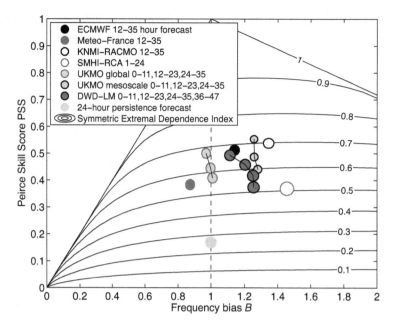

**Figure 3.4**   The ellipses indicate the performance of a number of forecast models in predicting 'cloud fraction' greater than 0.05 over three radar-lidar cloud observing stations in the UK, France and The Netherlands in 2004, plotted on a skill-bias diagram. The horizontal and vertical extents of the ellipses indicate the 95% confidence intervals in bias and Peirce Skill Score, respectively. The properties of the various models were described by Hogan *et al.* (2009). The numbers in the key indicate the forecast lead times being evaluated in hours, and where several ellipses are joined by lines, more than one range of lead times are assessed. '24-hour persistence forecast' refers to forecasting the cloud fraction that was observed 24 hours previously. The contours indicate values of the Symmetric Extremal Dependence Index (SEDI). The vertical dashed line indicates no bias

of slightly too much cloud. Figure 3.4 also depicts the skill that would be attributed to a 'persistence' forecast made by forecasting the cloud fraction that was observed exactly 24 hours before. Naturally, this set of forecasts is unbiased, but it is also significantly better than a random forecast (which would be awarded skill scores of PSS = SEDI = 0). This is due to the fact that weather is inherently autocorrelated, as well as the presence of a diurnal cycle in certain cloud types.

## 3.3   Signal detection theory and the ROC

Underlying virtually all deterministic forecasts of binary events are continuous variables, be they prognostic variables in a numerical forecast model or subjective probabilities being weighed up in a forecaster's head. Signal detection theory (SDT) de-

scribes the process of converting from a continuous variable to a decision as to whether to forecast occurrence or non-occurrence. This provides the framework to elucidate the inherent capability of a system to distinguish one state from another, and thereby can inform the design of verification measures. SDT was originally developed by radar engineers to make the decision as to whether a noisy channel contains signal (e.g. from an aircraft) or noise alone. It has been applied by psychologists as a model of human discrimination processes (Swets, 1973) as well as in numerous other fields.

This section gives a necessarily brief outline of signal detection theory and ROC analysis in evaluation of binary weather forecasts. There is a very extensive literature on these methods in experimental psychology and medical diagnosis, and increasing use in other fields including meteorology. Some references in meteorology include Mason (1980, 1982a, 1989), Levi (1985), McCoy (1986), Stanski et al. (1989), Harvey et al. (1992), Buizza et al. (1999; see also comments by Wilson 2000) and Mason and Graham (1999). Useful overviews and references to the literature outside meteorology can be found in Swets and Pickett (1982), Centor (1991), Swets (1996), and Krzanowski and Hand (2009).

### 3.3.1 The signal detection model

Signal detection theory divides the process of making a deterministic binary forecast into 'discrimination', the production of a scalar quantity pro-portional to the current strength of the evidence (which is subject to error), and the 'decision' as to whether the evidence is strong enough to fore-cast occurrence. Higher values of the 'weight of evidence' variable, which we denote as $W$, corre-spond to a higher probability of occurrence. The decision to forecast occurrence or non-occurrence of the event is then made on the basis of a prede-termined threshold, denoted $w$, so that the event is forecast if $W > w$ and not forecast otherwise. It is assumed that $W$ has a probability density $f_0(W)$ before non-occurrences and $f_1(W)$ before occur-rences. The basic model for the forecasting process is then as shown in Figure 3.5a.

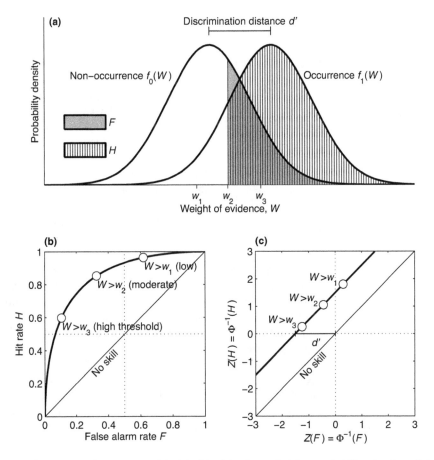

**Figure 3.5** The basic signal detection model. (a) $f_0(W)$ is the underlying probability density of the *weight of evidence* variable, $W$, before non-occurrences ('noise alone'), $f_1(W)$ the density before occurrences ('signal+noise'), and the areas $H$ and $F$ are hit and false-alarm probabilities by assigning a decision threshold $w_2$. The distance between the peaks, divided by their standard deviation, is the discrimination distance $d'$. (b) ROC generated by the model shown in panel (a), where the three points correspond to a low decision threshold ($w_1$), a moderate threshold ($w_2$) and a high threshold ($w_3$). (c) The same ROC but on axes transformed to the standard normal deviates of $H$ and $F$

To make this concept more concrete, consider the case of a statistical forecasting system that attempts to predict the onset or otherwise of El Niño several months in advance, on the basis of the current sea-surface temperature distribution. The weight of evidence, $W$, in this case could be an index summarizing the strength of spatial patterns in the sea-surface temperature that are believed to be a predictor of El Niño, and an El Niño event is forecast if $W > w$. The distribution $f_0(W)$ is the probability density of $W$ when El Niño does not subsequently occur, while $f_1(W)$ is the probability density of $W$ when it does.

The conditional probability of a hit, $H$, is the probability that the weight of evidence exceeds the threshold $w$ if the event occurs, so that

$$H = \int_w^\infty f_1(W)\,dW \qquad (3.14)$$

Similarly the conditional false alarm rate, $F$, is the probability that the weight of evidence exceeds $w$, when the event does not occur, so that

$$F = \int_w^\infty f_0(W)\,dW \qquad (3.15)$$

In order to proceed, it is necessary to make assumptions about the functional form of the distributions. The usual assumption, which is well supported by studies in many fields (e.g. Mason, 1982a; Swets, 1986b), is that the distributions $f_0$ and $f_1$ are normal (Gaussian) with standard deviations of $\sigma_0$ and $\sigma_1$, or that they can be transformed to normality by using a monotonic (possibly non-linear) transformation of the $W$-axis. For practical purposes, normal distributions are found to give a very good fit to empirical data, although other distributions may be more appropriate in certain circumstances. Without loss of generality, it can be assumed that the $W$-axis is scaled so that $f_0$ has a mean of zero and standard deviation of one. The separation of the means of $f_0$ and $f_1$ and the ratio of the standard deviation of $f_0$ to that of $f_1$, $\sigma_0/\sigma_1$, can then be estimated from verification data (Mason, 1982a).

## 3.3.2    The relative operating characteristic (ROC)

The relative (or 'receiver') operating characteristic is the graph of hit rate against false alarm rate that is mapped out as the decision threshold $w$ varies in Figure 3.5a. Suppose a particular decision threshold $w_2$ resulted in a set of binary forecasts located at the middle point on the solid curved line in Figure 3.5b. Reducing the threshold to $w_1$ would result in more hits but more false alarms, and the system would move to the right along the curved line in Figure 3.5b. Conversely, increasing the threshold would reduce both $H$ and $F$, moving the system to the left in Figure 3.5b. In the limit when $w$ is $-\infty$, both $H$ and $F$ are 1, so the ROC point lies at the upper right corner.

The location of the whole curve in the unit square is determined by the intrinsic discrimination capacity of the forecasting system. An empirical ROC can be plotted from forecasts issued as numerical probabilities by stepping a decision threshold through the forecasts, each threshold generating a $2 \times 2$ contingency table and values for $H$ and $F$ (e.g. Mason, 1982a; Stanski et al., 1989). All empirical ROCs have a form similar to the solid black line in Figure 3.1a, necessarily passing through $(0,0)$ and $(1,1)$, and elsewhere interior to the ROC unit square. Perfect discrimination would be represented by a ROC that rises from $(0,0)$ along the $H$-axis to $(0,1)$, then straight to $(1,1)$, while a forecasting system with no skill would form a line close to the $H = F$ diagonal.

When the underlying distributions $f_0(W)$ and $f_1(W)$ are normal, or can be transformed to normality by a monotonic transformation, then we may reveal their properties by presenting the ROC on axes transformed to standard normal deviates corresponding to $H$ and $F$. The new axes are labelled as 'Z-scores' given by $Z(H) = -\Phi^{-1}(1 - H) = \Phi^{-1}(H)$ and $Z(F) = \Phi^{-1}(F)$, where $\Phi^{-1}$ is the inverse of the standard normal distribution function, given by

$$\Phi(z) = \frac{1}{\sqrt{2\pi}} \int_{-\infty}^z \exp\left(-\frac{x^2}{2}\right) dx \qquad (3.16)$$

Mason (1982a) showed that empirical data for weather forecasts are closely linear when plotted on axes transformed in this way, supporting the use of the signal detection model with assumed normal distributions for $f_0(W)$ and $f_1(W)$. The separation of the means can be estimated as the intercept on the horizontal axis, and the ratio of the standard deviations ($\sigma_0/\sigma_1$) as the slope of the line. Details of this procedure are beyond the scope of this chapter, but can be found in many texts, for example Swets and Pickett (1982). When the ROC is presented in this way, it is sometimes referred to as a 'bi-normal' plot. Figure 3.5c shows the ROC in Figure 3.5b plotted on bi-normal axes.

When only a single point is available, as is the case with binary forecasts not explicitly derived from an underlying probability forecast, it is not possible to determine both the slope and intercept of the ROC on bi-normal axes. However, we may compute a *fitted ROC curve* to this point assuming that the variances of the underlying distributions are equal. This is done by defining the *discrimination distance* ($d'$) to be the distance between the centres of the two distributions, normalized by their standard deviation. This is given by

$$d' = \Phi^{-1}(H) - \Phi^{-1}(F) \qquad (3.17)$$

The thin solid line in Figure 3.1a shows the ROC fitted to point A, generated by using Equation 3.17 to trace a curve of $H$ against $F$ with the same value of $d'$ as calculated at A. The equivalent curve is shown on skill-bias axes in Figure 3.1b. It is useful because it provides an estimate of where the forecasting system would move to if it were somehow revised or recalibrated without substantial change to the underlying workings of the system. This curve is higher than the dashed line (corresponding to a naive recalibration strategy of randomly changing forecasts of occurrence to non-occurrence or vice versa) because in reality most forecasting systems *do* know which forecasts are closest to the decision threshold, and so which should be changed if a recalibration is performed. This curve can therefore be used to judge performance measures, since it can be regarded as a line of equal underlying skill regardless of bias. If we do not wish a performance measure to penalize bias at all (e.g. in the case that bias is to be reported separately), then we might regard it desirable for isopleths of a performance measure when plotted on a ROC diagram to be close to these curves.

It should be noted that real forecasting systems plotted on axes of $Z(H)$ versus $Z(F)$ often have a non-unit slope, typically ranging from about 0.7 to 1.5. This is due to the underlying distributions $f_0(W)$ and $f_1(W)$ having unequal width. The effect of this is that on standard $H$ versus $F$ axes, the ROC is not symmetrical about the negative diagonal, so may deviate substantially from the ROCs shown in Figure 3.1a and Figure 3.5b. Nonetheless, performance measures based on fitted ROC curves will be described in Section 3.5.1.

## 3.4 Metaverification: criteria for assessing performance measures

Murphy (1996a) coined the term *metaverification* to refer to the process of evaluating performance measures and the development of assessment criteria for such measures. In this section we discuss first the *desirable properties* that performance measures should ideally have. Then we discuss other properties that have been discussed in the literature and which may arguably be useful for some applications but not necessarily all. Where appropriate, the diagrams in Figure 3.1 are used to illustrate how one can diagnose whether a measure has a particular property. Table 3.4 summarizes which of the performance measures discussed in this chapter have which properties.

### 3.4.1 Desirable properties

*Equitability and asymptotic equitability*
An equitable measure is one that awards all random forecasting systems, including those that always forecast the same value, the same expected score (Gandin and Murphy, 1992). This is desirable because it provides a single no-skill baseline against which a forecaster can be said to have some measure of skill. It was also shown by Hogan *et al.* (2010) that all inequitable measures can be hedged in some situations. We first consider the limit of a very large sample size ($n \to \infty$), in which case the

**Table 3.4** Summary of the properties described in Section 3.3 for a number of different performance measures. The symbols used for each measure are defined in Table 3.3

| Measure | Desirable properties | | | | | | | | Other properties | | |
|---|---|---|---|---|---|---|---|---|---|---|---|
| | Truly equitable | Asymptotically equitable | Not trivial to hedge | Base-rate independent | Non-degenerate | Bounded | Linear | Regular | Biased forecasts can get perfect score | Transpose symmetric | Complement symmetric |
| $H, F$ | N | N | N | Y | N | Y | Y | N | Y | N | N |
| FAR | N | N | N | N | N | Y | Y | N | Y | N | N |
| PC | N | N | N | N | N | Y | Y | N | N | Y | Y |
| CSI | N | Y | Y | N | N | Y | N | N | N | Y | N |
| GSS | Y | Y | Y | N | N | Y | N | N | N | Y | N |
| HSS | Y | Y | Y | N | N | Y | Y | N | N | Y | Y |
| PSS | Y | Y | Y | Y | N | Y | Y | N | N | N | Y |
| CSS | Y | Y | Y | N | N | Y | Y | N | N | N | Y |
| DSS | N | Y | Y | N | N | Y | Y | N | N | Y | Y |
| $\theta$, LOR | N | Y | Y | Y | N | N | Y | Y | N | Y | Y |
| Q/ORSS | N | Y | Y | N | N | Y | N | Y | Y | Y | Y |
| $d'$ | N | Y | Y | Y | N | N | N | Y | Y | N | Y |
| ROCSS, $A_z$ | N | Y | Y | Y | N | Y | N | Y | Y | N | Y |
| EDS | N | N | N | Y | Y | N | N | N | Y | N | N |
| SEDS | N | Y | Y | N | Y | Y | N | N | Y | N | N |
| SEDI | N | Y | Y | N | Y | Y | N | N | N | Y | Y |
| SEDI$_{ET}$ | Y | Y | Y | N | Y | Y | N | N | Y | N | Y |

expected score awarded by measure $M$ to a random forecasting system is equal to the score that would be awarded to the *expected contingency table* for that forecasting system:

$$\lim_{n \to \infty} E[M(a, b, c, d)] = M(a_r, b_r, c_r, d_r) \quad (3.18)$$

where the expectation is calculated over all possible sample combinations of $a, b, c$ and $d$ associated with a particular population forecast rate, and we use expressions such as Equation 3.13 to calculate the expected values $a_r, b_r, c_r$ and $d_r$. These 'expected random forecasts' lie along the 'no-skill' $H = F$ line in Figure 3.1a and the PSS $= 0$ line in Figure 3.1b. A performance measure that awards zero at these points is said to be *asymptotically equitable* (Hogan et al., 2010) because it is equitable in the limit of large $n$. Figure 3.1 shows that PC is clearly inequitable, although there are several asymptotically equitable measures in Figure 3.2 and Figure 3.3.

In order to be *truly equitable*, a measure must be equitable for all values of $n$. In this case, in calculating the expected score we cannot just consider the scores along the expected no-skill line in Figure 3.1, because a particular sample of random forecasts might appear to have positive or negative skill compared to the average, i.e. lie above or below the no-skill line. Therefore, to calculate the expected random score for a particular base rate $s$ we must sum over all possible contingency tables that have that value of $s$, each multiplied by the probability of it occurring by chance. Note that it has frequently been implicitly assumed in the past that the equality in Equation 3.18 holds for all $n$, and therefore that all asymptotically equitable measures are truly equitable. It was shown by Hogan et al. (2010) that if a score is *linear* (defined in Section 3.4.2) and asymptotically equitable, then it is also truly equitable. Examples of such measures are the Heidke and Peirce Skill Scores. Unfortunately, linearity is incompatible with another desirable property: non-degeneracy for rare events. There are several non-linear, asymptotically equitable measures defined in Table 3.3, whose inequitability can be demonstrated. Consider Finley's set of $n = 2803$ tornado forecasts. It was found by Hogan et al. (2010) that a random forecasting system that predicts tornados

at the same rate as Finley ($r = 0.037$) would have expected values for Yule's Q and the Symmetric Extreme Dependency Score of around $-0.15$, i.e. significantly different from the nominal no-skill value of zero for these two measures. They become much more negative for lower $n$. Hogan et al. (2010) proposed a 'rule of thumb' that such measures can be treated as equitable (specifically that the expected score of a random forecast has an absolute value less than 0.01) if the expected value of $a$ for a random forecasting system (given by Equation 3.13) is at least 10. For Finley's base and forecast rates, this implies that $n > 15\,000$ would be required for these measures to behave equitably.

Hogan et al. (2010) showed that an inequitable measure, $M$, could be transformed to make a truly equitable 'equitably transformed' measure, $M_{ET}$, for example using

$$M_{ET} = \frac{M - E[M \mid s, r]}{\max(M) - E[M \mid s, r]} \quad (3.19)$$

where $E[M \mid s, r]$ is the expected score that would be attained by a random forecasting system with the same base rate and sample forecast rate and usually needs to be calculated numerically. This demonstrates that the properties of true equitability, non-linearity and non-degeneracy for rare events can be combined in a single measure. However, scores for probability forecasts cannot be both *equitable* and *proper* (Jolliffe and Stephenson, 2008), and so proper probability scores (see Chapter 7) are not equitable.

### Difficulty to hedge

A measure should not encourage a forecaster to 'hedge', i.e. to issue a forecast that differs from his or her 'true judgement', in order to improve either the score awarded or its expectation (Jolliffe, 2008). As highlighted in the introduction to this chapter, Gilbert (1884) identified that Finley's *percent correct* is easy to hedge, and the desire to find measures that are not so easy to hedge has motivated much of the subsequent work on binary verification. To define more formally whether a particular measure is hedgable, we need to consider possible hedging strategies. First, consider the case when a deterministic binary forecast is not based on an underlying probability forecast, or if it is then the forecaster who might be tempted to hedge has no knowledge of the underlying probabilities. In this case a hedging strategy could be to randomly change some forecasts of occurrence to non-occurrence, or vice versa, as considered by Stephenson (2000). The effect of this is shown graphically in Figure 3.1b. If the initial set of forecasts lies at point A, corresponding to an overprediction of occurrence by just over a factor of 2, then randomly changing forecasts of occurrence to non-occurrence would move the system (or more precisely, its expectation) leftwards along the dashed line. At point B, enough forecasts would have been changed to remove the initial bias; the forecasts are then said to have been *hedged towards climatology*. In the case of PC, shown by the shaded contours in Figure 3.1b, there is no need to stop there: changing *all* forecasts of occurrence to non-occurrence at the far left of the dashed line yields the maximum score possible by a hedged forecast. PC happens to be a measure that rewards underprediction for $s < 0.5$, but for measures that reward overprediction the best hedging strategy would be to change forecasts of non-occurrence to occurrence and move rightwards along the dashed line from A. Figure 3.3g shows a performance measure that would reward this behaviour. Real forecasting systems can be recalibrated more effectively than this random reassignment strategy would suggest, since they do have an underlying knowledge of the likelihood of the event occurring. Therefore, it could be argued that ability to hedge should be judged by examining how the score awarded changes along the thin solid curve in Figure 3.1b, which is the expected ROC from signal detection theory (see Section 3.3). However, in practice the real ROC of the system can deviate significantly from this line and so we can only be sure that the measure can be hedged if it can be increased by following a dashed line. The third column of properties in Table 3.4 indicates whether or not measures are trivial to hedge in this way.

It is useful to consider the relationship between the concept of hedging in the case of binary forecasts and probability forecasts. A performance measure for probability forecasts that does not encourage hedging is referred to as *proper* (Winkler and Murphy, 1968). If a binary forecast has actually been made by applying a decision thresh-old to an underlying probability forecast, then it is rather more difficult to define the meaning of a 'hedged' forecast, because the forecast of one of the two possible outcomes with apparently 100% certainty clearly differs from the forecaster's true belief (which is probabilistic), and so all such forecasts are hedged in some way (Jolliffe, 2008). Murphy and Daan (1985) recognized this difficulty and defined a weaker property of *consistency*. In this paradigm, a forecaster's judgement is represented by a probability distribution that is converted into a single value by applying a 'directive'. A performance measure consistent with that directive is one that is maximized by applying it. In the case of forecasts of a quasi-continuous variable like temperature, the directive might be 'forecast the mean of the probability distribution'. The performance measure consistent with this directive is mean square error, since this is minimized by forecasting the mean. For binary predictands, the directive might specify a threshold probability, $p^*$, and if the current probability for the event is less than $p^*$ then non-occurrence is forecast, whereas if it is greater, occurrence is forecast (Mason, 1979). It can be shown that a performance measure consistent with this directive is one that is optimal at this threshold probability (see Section 3.5.3). For example, when the directive is 'forecast occurrence when the forecast probability of the event exceeds the base rate', a consistent score is provided by the Peirce Skill Score. It is possible to define an optimum $p^*$ for any given performance measure (Mason, 1979), and vice versa. Therefore, it should be borne in mind that consistency usually offers us little guidance on which measures are better than others, since we can always find an optimum $p^*$ for a given measure (Jolliffe, 2008). The exceptions are some measures that are always improved by hedging, for which $p^*$ takes a trivial value of 0 or 1.

### Base-rate independence

It was shown in Section 3.2.3 that all binary verification measures can be written in terms of hit rate $H$, false alarm rate $F$ and base rate $s$. A measure is *base-rate independent* if it can be written only in terms of $H$ and $F$. This is desirable because it makes the score awarded invariant to natural variations in climate, i.e. to multiplying columns of the contingency table by positive constants (Yule, 1912). A base-rate independent measure forms a unique plot on ROC axes, and indeed the measures shown in Figure 3.2 have this property. By contrast, the measures shown in Figure 3.3 are dependent on $s$, and for different $s$ their appearance in Figure 3.3 would change. Further discussion of base-rate independence is provided by Ferro and Stephenson (2011).

### Non-degeneracy for rare events

It was illustrated by Stephenson *et al.* (2008a) that all performance measures in use at the time tended to a meaningless value (usually zero) for vanishingly rare events. This creates the misleading impression that rare events can never be skilfully forecast, no matter what forecasting system is used. We refer to such measures as *degenerate* for rare events. Stephenson *et al.* (2008a) proposed the Extreme Dependency Score, which provided an informative assessment of skill for rare events for the first time. Improved versions were subsequently proposed by Hogan *et al.* (2009) and Ferro and Stephenson (2011) that enabled more of the other properties discussed in this section to be combined with non-degeneracy (see Table 3.4 and Section 3.5.1). Further information on this property is provided in Chapter 10.

### Boundedness

Most performance measures are *bounded*, having finite upper and lower limits (usually $-1$ and $+1$), but some, such as the odds ratio $\theta = ad/(bc)$ (Stephenson, 2000), become infinite when $b$ or $c$ are zero and very large when they are close to zero. In practical terms this is usually tolerable since almost all real-world forecasting systems have non-zero values for all four contingency-table cell counts. However, it is impossible to determine whether an unbounded measure is truly equitable, since calculation of the expected score by a random forecasting system consists of a probability-weighted average of the score from all possible contingency tables, including the infinite scores awarded to tables containing a zero.

A related but less important property is *definedness*: some measures become undefined when one or more of the contingency-table cell counts is zero, because a zero divided by zero appears somewhere in their definition. For example, Yule's Q and the ROC Skill Score, defined in Table 3.3, are both undefined when the event is never forecast ($a = b = 0$) or when it is always forecast ($c = d = 0$). The practical solution is simply to *define* them to take the no-skill value (usually zero) at these points.

### 3.4.2 Other properties

The properties now described should be considered in the design of new performance measures, but may not be of overriding importance depending on the application, for example because they are incompatible with another property.

### Linearity

A linear performance measure is one that is equally sensitive to a change in skill throughout its range. More precisely, 'near-perfect' and 'near-random' forecasts, with the same bias, will be rewarded by the same change in score if 1 is added to $a$ and $d$ and 1 is subtracted from $b$ and $c$. In a skill-bias diagram, a measure is linear if a vertical line intersects contours at evenly spaced intervals. Hogan *et al.* (2009) found that the strong non-linearity of Yule's Q made it unsuitable for certain types of analysis, such as calculating the rate at which forecast skill decays with increasing lead time, although weakly non-linear measures (such as the Symmetric Extreme Dependency Score) could still be used. Unfortunately, perfect linearity is fundamentally incompatible with the property of non-degeneracy for rare events, so it is likely that we will have to make do with weakly non-linear measures.

### Regularity

The concept of a *regular* ROC was introduced by Birdsall (1966) to mean a graph of $H$ against $F$ that passes through (0,0) and (1,1), is complete (i.e. for each value of $F$ there is just one value of $H$) and convex (i.e. the graph will be on or above straight line segments connecting any two points on it). The term was used in a less restricted sense by Swets (1986a), and here, to mean that the graph passes through (0,0) and (1,1) and is elsewhere interior to the ROC unit square, i.e. it does not touch the axes except at these points. All empirical ROCs derived from probabilistic forecasting systems by varying the decision threshold have so far been found to be regular in Swets' sense. The convexity condition of Birdsall's definition is not always satisfied, although deviations are in general slight and near the extremes of the graph.

Since verification measures for binary forecasts may be plotted on ROC axes, we may define a regular measure as one whose isopleths are regular in Swets' sense. All of the measures plotted in Figure 3.2 are regular, except for the Peirce Skill Score, while none of those in Figure 3.3 are regular (on skill-bias axes, the $H = F = 0$ and $H = F = 1$ points lie at the extreme left and extreme right of the horizontal no-skill line, and it can be seen that although all the contours of the false alarm ratio (FAR; see Section 3.5.1) converge at the extreme left point, none of the measures has all its contours converging at both points). Whether or not regularity is desirable depends on how one wishes to treat a biased forecasting system that is the best it can be, given its bias. For example, suppose a forecaster underpredicts occurrence but all forecasts issued are correct, i.e. there are misses but no false alarms. A regular measure would award this forecaster the same (perfect) score as a perfect unbiased forecaster who issues no misses or false alarms. If it is intended that the performance measure should rank the second forecaster above the first then an irregular measure may be preferred to a regular one (as argued by Woodcock, 1976). Table 3.4 indicates which measures are regular. This is a subset of the measures that can award a perfect score to a biased forecast (also shown as a column in Table 3.4).

Usually the bias of a forecasting system is reported together with its skill, in which case it may be intended that these two pieces of information are orthogonal, i.e. the skill reported by the performance measure should be invariant if the same forecasting system were to be corrected to remove its bias (e.g. by recalibration). In order to do this, we need to have a model for the trajectory of a forecasting system on a ROC diagram when it is recalibrated. Such a model is provided by Signal Detection Theory (see Section 3.3) and predicts the curve through point A in Figure 3.1a. The ROC Skill Score shown

in Figure 3.2c has isopleths that follow these fitted ROC curves.

### Symmetry

Stephenson (2000) discussed two forms of symmetry. A measure is *transpose symmetric* if it is invariant to a relabelling of what is a forecast and what is an observation, equivalent to swapping $b$ and $c$ in its definition; it therefore ensures that 'misses' and 'false alarms' are treated equally. This is useful if a measure is to be used to compare the correspondence between two estimates of the occurrence of a sequence of binary events, but with neither estimate being regarded as any more valid than the other. In the normal situation of verifying a forecast against observations (assumed to be 'truth'), there appear to be no other reasons for requiring measures to be transpose symmetric (Ferro and Stephenson, 2011).

A measure is *complement symmetric* if it is invariant to a relabelling of events and non-events, equivalent to swapping $a$ and $d$ in its definition as well as swapping $b$ and $c$. Complement asymmetric measures have often been developed with a view to verification of rare and extreme events, such as tornados, where it is a much higher priority to get the occurrences right than the non-occurrences. However, these measures can usually be made complement symmetric without losing any of their advantages. An example is the Gilbert Skill Score (GSS), which was proposed by Gilbert (1884) as a more robust performance measure for tornado forecasts than Proportion Correct. This measure can be made complement symmetric by adding extra terms in the numerator and denominator, and the result is the Heidke Skill Score, which has the additional advantage of being truly equitable.

## 3.5 Performance measures

The previous sections have described the desirable properties that performance measures should have and convenient ways to visualize them. Section 3.5.1 makes use of this context to discuss the merits and demerits of various measures that have been proposed in the literature. The measures we discuss are defined in Table 3.3, and their properties summarized in Table 3.4. Figure 3.2 and Figure 3.3 present most of them on ROC or skill-bias diagrams. Section 3.5.2 shows how confidence intervals may be calculated for each measure. Finally, Section 3.5.3 demonstrates how optimal threshold probabilities may be calculated.

### 3.5.1  Overview of performance measures

*Basic inequitable measures*

The quantities hit rate and false alarm rate have been described in Section 3.2.3, and while they are very useful in the analysis of other performance measures, it is clear from Table 3.4 that they are unsuitable as performance measures themselves, being inequitable, trivial to hedge and degenerate for rare events. These same failings are shared by *Proportion Correct,* originally introduced by Finley (1884) and discussed in Section 3.2.4 to illustrate the use of the diagrams in Figure 3.1.

*False Alarm Ratio,* FAR $= b/(a + b)$, not to be confused with the false alarm rate $F$ (Barnes *et al.,* 2009), is a sample estimate of the conditional probability of a false alarm given that occurrence was forecast. It can be seen in Figure 3.3a that its isopleths consist of a family of straight lines that pass through the origin. One of these straight lines lies along the no-skill line, but the expected score awarded to a random forecast is $1 - s$, and therefore base-rate dependent. This means that the measure is inequitable because random forecasts of rare events will score higher than random forecasts of common events. Therefore, as with the other measures discussed in the previous paragraph, it is not recommended.

*The Critical Success Index and the Gilbert Skill Score*

Gilbert (1884) proposed two performance measures, the first of which is now commonly called the *Critical Success Index* and is defined as CSI $= a/(a + b + c)$. Gilbert himself referred to it as the *Ratio of Verification.* Schaefer (1990) noted that it has been rediscovered and renamed at least twice, by Palmer and Allen (1949), who called it the *Threat Score,* and by Donaldson *et al.* (1975), who introduced the term CSI that we shall adopt here. It can be regarded as a sample estimate of the conditional

probability of a hit given that the event of interest was either forecast, or observed, or both. The fact that its definition lacks any dependence on the number of correct rejections, $d$, has led to it being used widely as a performance measure for rare events, based on the view that non-occurrence of rare events is trivially easy to forecast and should not be counted. Certainly it is unsuitable for very *common* events because then the number of correct rejections becomes important. However, CSI is not suitable for rare events either, since it tends to a meaningless limit of zero as $s \to 0$ (Stephenson *et al.,* 2008a); in such situations one of the non-degenerate measures discussed below in this section would be preferable. It is clear from Figure 3.3b that CSI is inequitable since the value of CSI corresponding to no skill can vary between 0 and $s$, depending on the bias of the forecasting system. Therefore, a forecasting system that is awarded a CSI of less than $s$ can be trivially hedged by instead forecasting occurrence all the time (see also Mason, 1989).

Gilbert (1884) recognized the inequitability of CSI (although this property was not referred to as equitability until over a century later) and proposed a modification to allow for the number of hits that would be expected purely by chance. We follow Schaefer (1990) and refer to the resulting measure as the *Gilbert Skill Score* (GSS), defined in Table 3.3. It was discussed in detail by both Schaefer (1990) and Doswell *et al.* (1990). Probably the most widely used name for it is the *Equitable Threat Score* (ETS), first used by Mesinger and Black (1992). However, it was shown by Hogan *et al.* (2010) that it is not quite equitable. Figure 3.3c shows that while GSS $= 0$ along the no-skill line, the isopleths are more tightly packed above this line than below (i.e. GSS is non-linear). This means that in calculating the expected score for a random forecasting system, there is incomplete cancellation between the positive and negative scores above and below the line, and the expected GSS for a random forecasting system is always positive but decreases with increasing $n$. Therefore GSS is only asymptotically equitable, and so we recommend the use of the name GSS rather than ETS. It should be noted that for practical purposes the difference from true equitability is quite small: for $n < 30$, the expected value for a random forecasting system falls below 0.01.

## The Peirce, Heidke, Clayton and Doolittle Skill Scores

We now discuss four linear, truly equitable performance measures. The first, the *Peirce Skill Score* (PSS), was originally proposed by Peirce (1884). It was independently rediscovered for the general case of multi-category contingency tables (rather than just two categories) by Hanssen and Kuipers (1965), and is often referred to as *Kuipers' performance index* or *Hanssen and Kuipers' score*. It has also been rediscovered by Flueck (1987), who called it the *True Skill Statistic* (TSS). In medical statistics, an identical measure of diagnostic discrimination is known as Youden's (1950) index, and it has also been used in psychology (Woodworth, 1938). In the absence of any consensus on terminology, it seems reasonable that it should be named after the earliest known discoverer. It is most simply expressed as $PSS = H - F$, so that isopleths on ROC axes are a family of parallel lines with unit slope, as shown in Figure 3.2a. It is independent of base rate. In Section 9.2.1 of this book, it is shown how PSS can be related to the economic value of binary event forecasts in the cost/loss model.

A related measure, the *Heidke Skill Score* (HSS), was proposed by Heidke (1926) for multi-category verification. A binary-forecast version of HSS had been proposed 41 years earlier by Doolittle (1888), but we will follow most previous authors and call it HSS (we prefer to apply Doolittle's name to another measure described below). HSS has a moderately complicated definition in Table 3.3, but the thinking behind it is simple. We may define a *generalized skill score* in the form of Equation 2.27:

$$S = (x - x_r)/(x_p - x_r) \qquad (3.20)$$

where $x$ is some linear function of the elements of the contingency table, $x_r$ is the expected value of $x$ from a random forecast with the same bias as the actual forecasting system, while $x_p$ is the value of $x$ that would be obtained by a perfect unbiased forecast. This should yield an equitable measure that awards random forecasts 0 and perfect forecasts 1. The Heidke Skill Score is obtained by setting $x = a + d$, or equivalently $x = PC$. Therefore it can be thought of as a version of PC that has been scaled and calibrated to make it equitable. It was also noted

by Schaefer (1990) to be uniquely related to the Gilbert Skill Score via

$$GSS = HSS/(2 - HSS) \qquad (3.21)$$

and therefore can also be regarded as a version of the GSS that has been rescaled to make it linear. The linearity of HSS is evident in Figure 3.3d by the fact that a vertical line intersects contours at evenly spaced intervals, which is not the case for GSS.

Both PSS and HSS are truly equitable, awarding random and constant forecasts an expected score of zero. They are equal for unbiased forecasts, and when $s = 1/2$ they are equal for all forecasts. They therefore differ only in the way they treat biased forecasts for $s \neq 1/2$. Figure 3.3d shows that when $s < 1/2$, isopleths of HSS are further apart than isopleths of PSS for forecasting systems that overpredict occurrence, but closer together for forecasting systems that underpredict. Therefore, for systems with positive skill, PSS will treat overpredicting systems more generously than HSS and underpredicting systems more harshly. The opposite is true when $s > 1/2$. Being truly equitable and difficult to hedge, both measures are more robust indicators of skill than the previous ones discussed in this section. In terms of properties listed in Table 3.4, the only difference is that HSS is transpose symmetric while PSS is base-rate independent.

The two other measures in this category are much less commonly used than PSS and HSS. The *Clayton Skill Score* (CSS) was proposed by Clayton (1934) and can be thought of as the calibration-refinement analogue of PSS. It is defined in terms of the calibration-refinement factors given in Section 3.2.3: $CSS = FOH - DFR$. In the cost/loss model for weather-related decisions, it is equal to the range of cost/loss ratios over which the forecasts have positive economic value (see Section 9.2.1). The *Doolittle Skill Score* is related to a chi-squared measure of correlation, and is given by

$$DSS = \sqrt{\chi^2/n}$$

where $\chi^2$ is defined in this case following Stephenson (2000), and the resulting expression in terms of contingency-table cell counts is given in Table 3.3. As with PSS and HSS, both CSS and DSS are linear and truly equitable, but none of the four is regular, which means that they all tend to penalize forecasts with bias more than some of the regular measures that we discuss next.

## Measures from signal detection theory

A number of measures of discrimination have been developed based on the ROC and signal detection theory, discussed, for example, by Swets (1996). The *discrimination distance*, $d'$, given by Equation 3.17, provides an estimate of the distance between the means of the 'noise' and 'signal plus noise' distributions, normalized by their standard deviations (which are assumed equal). It was first proposed as a signal detection measure by Tanner and Birdsall (1958). The possible range of $d'$ is $(-\infty, +\infty)$, but the range encountered in weather forecasts is generally from zero to 4.

The arguments given in Section 3.4.1 against unbounded measures suggest the desirability of rescaling $d'$ to make it bounded. Such a rescaled measure is $A_z$, the area under the *modelled* ROC on probability axes ($H$ vs $F$). In other words, the straight-line ROC on normal deviate axes (Figure 3.5c) is transferred to probability axes and $A_z$ is the area under this curve (Figure 3.5b). The subscript $z$ serves as a reminder that the measure was taken from a binormal graph (Swets, 1988). The possible range of $A_z$ is [0,1]. Zero skill is indicated by $A_z = 0.5$, when the modelled ROC lies along the positive diagonal, and 1.0 indicates perfect skill, when the ROC rises from (0,0) to (0,1) then to (1,1). A value of $A_z$ less than 0.5 corresponds to a modelled ROC curve below the diagonal, indicating the same level of discrimination capacity as if it were reflected about the diagonal, but with an inverted calibration. When only a single point on the ROC is available, the variances of the underlying distributions must be assumed equal and $A_z$ can be calculated from $d'$ as the area under the normal distribution up to the normal deviate value equal to $d'/\sqrt{2}$, that is

$$A_z = \Phi(d'/\sqrt{2})$$

(Swets and Pickett, 1982). By this definition, $A_z$ is undefined for constant forecasts of a single category, i.e. at either $H = F = 0$ or $H = F = 1$, but it seems reasonable to *define* it to be 0.5 at these points.

A final trivial transformation is to generate an equivalent measure that follows most of the others listed in Table 3.3 and awards zero to expected random forecasts, rather than 0.5. Thus we define the *ROC Skill Score* as ROCSS = $2A_z - 1$ (Wilks, 2006b), which has the range $[-1,1]$. This measure is contoured in Figure 3.2c; note that $d'$ and $A_z$ have exactly the same appearance as ROCSS on these axes, but with different labels for the contours. In terms of properties, these measures are regular, asymptotically equitable and not trivial to hedge. Moreover, if the forecasting system happens to have underlying distributions that are normal and of equal-width (see Section 3.3) then, in principle at least, these measures provide the most accurate indication of the underlying skill. In practice, if we only have one contingency table available (e.g. Finley's) then we do not know the shape of the true ROC and it may deviate from the symmetry assumed in the design of these measures. It should be noted that these measures are degenerate for rare events.

## Yule's Q and the Odds Ratio

*Yule's Q* is a measure of association in $2 \times 2$ contingency tables developed by Yule (1900) and named after the nineteenth century Belgian statistician A. Quetelet. It was advocated as a performance measure for binary forecasts by Stephenson (2000), who referred to it as the *Odds Ratio Skill Score* (ORSS). Q has been discussed in the context of ROC analysis by Swets (1986a). It has a simple definition, given in Table 3.3, and in Figure 3.2b its isopleths can be seen to form a family of hyperbolas all passing through (0,0) and (1,1). ROCs of this form are generated by a signal detection model with equal variance *logistic distributions* and are similar to those produced by equal variance *normal distributions* (Swets, 1986a). This can be seen in comparing Figures 3.2b and 3.2c. For this reason, Q has very similar properties to ROCSS, as shown in Table 3.4. A difference to note is that isopleths of Q approach the (0,0) point in Figure 3.2 further from the axes than isopleths of ROCSS. This means that, if the underlying distributions of a particular forecasting system are closer to normal than logistic in shape, then Q will tend to award biased forecasts a higher score than the equivalent system without a bias.

Q was derived by Stephenson (2000) as a transformation of the *Odds Ratio* $\theta$. If the odds of a hit are denoted by $\omega_H = H/(1 - H)$ (i.e. given that the event occurred this is the ratio of the probability of forecasting it to the probability of not forecasting it) and the odds of a false alarm are denoted by $\omega_F = F/(1 - F)$ then the odds ratio is defined by $\theta = \omega_H/\omega_F = ad/(bc)$. It is related to Q via $Q = (\theta - 1)/(\theta + 1)$. The odds ratio is widely used in medical statistics (Agresti, 2007) and is equal to the square of the $\eta$ measure developed in psychology (Luce, 1963). It has the range $[0, +\infty)$, and $\theta = 1$ corresponds to no-skill.

A related measure is the *Log of Odds Ratio*, given by $\text{LOR} = \ln(\theta)$, which has the range $(-\infty, +\infty)$. Hogan et al. (2009) preferred LOR over Q for measuring the rate of decay of forecast skill with lead time, because Q is much more non-linear, tending to 'saturate' for very skilful forecasts. This is shown in Figure 3.2b by contours lying much further apart in the upper-left than the middle of the diagram. LOR has the convenience of a very simple sampling distribution: asymptotically it is a normal distribution with standard deviation $1/\sqrt{n_h}$ where $n_h$ is the effective number of degrees of freedom, given by $1/n_h = 1/a + 1/b + 1/c + 1/d$ (Stephenson, 2000). This property can be used to estimate confidence intervals on Q and $\theta$.

*Non-degenerate measures*

All of the measures discussed above are degenerate for rare events, in that they tend to a meaningless value (usually zero) as the base rate, $s$, tends to zero. To overcome this, Stephenson et al. (2008a) proposed the *Extreme Dependency Score* (EDS), defined in Table 3.3. For rare events, EDS measures the rate at which $a$ (or equivalently, the hit rate) tends to zero as $s$ tends to zero, assuming the forecast remains unbiased ($r = s$). The expected value of $a$ for a random but unbiased forecasting system is $a_r = ns^2$, while for a perfect forecasting system it is $a_p = ns$. The exponent on $s$ can therefore be regarded as a measure of skill that is not degenerate for rare events. Substituting $a = ks^\delta$ into the definition of EDS given in Table 3.3 (where $k$ is an arbitrary constant of proportionality), and taking the limit as $s \to 0$, we find that $\delta = 2/(\text{EDS} + 1)$. Thus, $\text{EDS} = 0$ leads to an exponent of 2, the value for random forecasts, while $\text{EDS} = 1$ leads to an expo-

nent of 1, the value for perfect forecasts. Stephenson et al. (2008a) demonstrated the non-degeneracy of this measure to evaluate rainfall forecasts, applying a range of threshold rain rates in order to assess skill as a function of rarity. They found that EDS remained approximately constant (within the confidence interval) as the threshold was raised, while other measures such as PSS tended to zero.

In order to use EDS it is necessary to first recalibrate the forecasts to remove bias (Stephenson et al., 2008a). If this is not done then it has some undesirable properties illustrated in Figure 3.3g: the lack of a constant value along the no-skill line means that the measure is inequitable, and the fact that scores increase to the right means that it is easy to hedge by simply predicting occurrence all the time (Primo and Ghelli, 2009). Hence it is unsuitable as a general performance measure. In an attempt to overcome these drawbacks, Hogan et al. (2009) proposed a transpose symmetric version they referred to as the *Symmetric Extreme Dependency Score* (SEDS), defined in Table 3.3. This measure is asymptotically equitable and not trivial to hedge. It is therefore more suitable for general use.

Despite its name, SEDS has a curiously asymmetric appearance in Figure 3.3h; the fact that its isopleths do not converge at the $H = F = 0$ and $H = F = 1$ points indicates that it is irregular. Ferro and Stephenson (2011) proposed the *Symmetric Extremal Dependence Index* (SEDI), which combines non-degeneracy with several other useful properties such as regularity, complement symmetry and base-rate independence. Furthermore, it is not only non-degenerate for rare events, but *non-degenerate for overwhelmingly common events*. It is plotted in Figure 3.2d, and while its isopleths appear similar to Yule's Q and ROCSS (which are also regular, complement symmetric and base-rate independent), the logarithmic dependence on the proximity to both the (0,0) and (1,1) points is what imparts this non-degeneracy property. Further information on non-degeneracy is provided in Chapter 10.

SEDI combines more desirable properties in one measure than any other presented in this chapter, but lacks true equitability. One way to overcome this is to apply the equitable transform given by Equation 3.19. While the 'equitably transformed SEDI' is both truly equitable and non-degenerate, the fact that the transformation depends on both the base rate and the forecast rate means that it is no longer base-rate independent or perfectly regular. This is summarized in Table 3.4. Its minimum value also depends in a complex fashion on both these rates. Further work is therefore needed to combine as many as possible of the desirable properties described in Section 3.4 into a single measure.

### 3.5.2  Sampling uncertainty and confidence intervals for performance measures

A particular verification data set can be regarded as just one of many possible samples from a population with certain fixed characteristics. The various verification measures are therefore only finite-sample estimates of 'true' population values, and as such are subject to sampling uncertainty. It has been unusual in weather-forecast verification studies for any attempt to be made to assess this sampling uncertainty, although without some such attempt it is not possible to be sure that apparent differences in skill (e.g. after changing a parameterization in a numerical forecast model) are real and not just due to random fluctuations. This is a deficiency in verification practice, since as Stephenson (2000) has pointed out, an estimate or measurement without some indication of precision has little meaning.

In estimating sampling uncertainty, we consider the observations fixed. This is appropriate for the common situation of two forecasting systems being compared against the same set of observations. Hence both the sample size $n$ and the base rate $s$ are fixed, and following the discussion in Section 3.2.3, the range of possible samples occupies a two-dimensional space described by the hit rate, $H$, and the false alarm rate, $F$. Both $H$ and $F$ are sample estimates of probabilities, so can be expected to have the sampling distribution of a proportion. It is possible to calculate exact confidence limits for a probability $p$ based on a sample proportion $\hat{p}$ using the binomial distribution. The theory is discussed in most statistical textbooks and is summarized by Seaman et al. (1996). The confidence interval for $p$ obtained by inverting the binomial expression is known as the Clopper–Pearson interval. This interval, while sometimes labelled 'exact' is in fact quite conservative, containing $p$ on more than its nominal (e.g. 95%) proportion of occasions, due to the discreteness of the binomial distribution. Consequently, the 'exact' interval is not regarded as

optimal for statistical practice (Agresti and Coull, 1998).

It is common for large sample size $m$ to use the normal approximation to the binomial distribution, giving the Wald confidence interval

$$\hat{p} \pm z_{\alpha/2}\sqrt{\hat{p}(1-\hat{p})/m} \qquad (3.22)$$

for the true probability of success $p$ estimated by $\hat{p}$, the fraction of 'successes' in $m$ trials. The quantity $z_{\alpha/2}$ is the appropriate quantile of the standard normal distribution, equal to 1.96 for a 95% interval. The Wald interval tends to contain $p$ less than 95% of the time unless the sample is quite large. As discussed by Agresti and Coull (1998), a more satisfactory confidence interval is given by Wilson's (1927) score method,

$$\frac{\hat{p} + z_{\alpha/2}^2/(2m) \pm z_{\alpha/2}\sqrt{\hat{p}(1-\hat{p})/m + z_{\alpha/2}^2/(4m^2)}}{1 + z_{\alpha/2}^2/m}$$
$$(3.23)$$

A simple modification of the Wald interval that gives results nearly as good as the score method is to use the Wald formula but replace the estimate $\hat{p}$ in Equation 3.22 by $\tilde{p} = (y + 2)/(m + 4)$, where $y = m\hat{p}$ is the measured number of successes, i.e. add two successes and two failures (Agresti and Coull, 1998).

In the case of hit rate, the sample size is the number of occurrences $ns = a + c$. Assuming this to be large enough to use Equation 3.22, the estimated standard error of hit rate, $S_H$, satisfies $S_H^2 = H(1 - H)/(a + c)$, while for false alarm rate the equivalent expression is $S_F^2 = F(1 - F)/(b + d)$. $H$ and $F$ have the key property that they are *independent* random variables. Therefore we can take any measure defined in Table 3.3 as a function of $H$, $F$ and $s$, calculate the error contribution due to uncertainties in $H$ and $F$, and sum these errors in quadrature to obtain the sampling error in the measure. Thus for measure $M$ we can write

$$S_M^2 = S_H^2 \left(\frac{\partial M}{\partial H}\right)^2 + S_F^2 \left(\frac{\partial M}{\partial F}\right)^2 \qquad (3.24)$$

This has been done for most of the performance measures listed in Table 3.3 and the resulting expressions are shown in Table 3.5. Note that for

**Table 3.5** The optimal threshold probability and the error variance for various measures. Note that in several cases the error variances are expressed in terms of the error variances for hit rate and false alarm rate, $S_H^2$ and $S_F^2$. The symbols used for each measure are defined in Table 3.3

| Measure ($M$) | Optimal threshold probability ($p^*$) | Error variance ($S_M^2$) |
|---|---|---|
| H | $0$ | $S_H^2 = H(1-H)/(a+c)$ |
| F | $1$ | $S_F^2 = F(1-F)/(b+d)$ |
| FAR | $a/(a+b)$ | $FAR^4[(1-H)/a + (1-F)/b]a^2/b^2$ |
| PC | $0.5$ | $sH(1-H)/n + (1-s)F(1-F)/n$ |
| CSI | $CSI/(1+CSI)$ | $CSI^2\left[\dfrac{1-H}{a} + \dfrac{b(1-F)}{(a+b+c)^2}\right]$ |
| GSS | $\dfrac{1-GSS}{s}\dfrac{GSS}{1+GSS} + \dfrac{GSS}{1+GSS}$ | $\dfrac{4S_{HSS}^2}{(2-HSS)^4}$ |
| HSS | $s + (1-2s)HSS/2$ | $S_F^2 HSS^2\left[\dfrac{1}{H-F} + (1-s)(1-2s)\right]^2 + S_H^2 HSS^2\left[\dfrac{1}{H-F} - s(1-2s)\right]^2$ |
| PSS | $(a+c+1)/(n+2)$ | $S_H^2 + S_F^2$ |
| CSS | $\dfrac{c(a+b)^2 + a(c+d)^2}{(a+b)(c+d)n}$ | $S_H^2\left[\dfrac{b(a+c)}{(a+b)^2} + \dfrac{d(a+c)}{(c+d)^2}\right]^2 + S_F^2\left[\dfrac{a(b+d)}{(a+b)^2} + \dfrac{c(b+d)}{(c+d)^2}\right]^2$ |
| LOR | $\dfrac{sH(1-H)+(1-s)F(1-F)}{sH(1-H)}$ | $\dfrac{1}{a}+\dfrac{1}{b}+\dfrac{1}{c}+\dfrac{1}{d}$ |
| Q/ORSS | $\dfrac{sH(1-H)+(1-s)F(1-F)}{sH(1-H)}$ | $S_{LOR}^2\,\dfrac{4\theta^2}{(\theta+1)^4}$ |
| $d'$ | $[1+\exp(d'^2/2)(1-s)/s]^{-1}$ | $2\pi\{S_H^2\exp[(\Phi^{-1}H)^2] + S_F^2\exp[(\Phi^{-1}F)^2]\}$ |
| ROCSS | $[1+\exp(d'^2/2)(1-s)/s]^{-1}$ | $S_{d'}^2\exp(-d'^2/2)\pi$ |
| EDS | $0$ | $S_H^2\left[\dfrac{2\ln s}{H(\ln Hs)^2}\right]^2$ |
| SEDS | $\dfrac{a}{(a+b)(SEDS+1)}$ | $\dfrac{S_H^2}{X^2} + \dfrac{S_F^2}{H^2}\left(\dfrac{SEDS}{\log Hs} + \dfrac{F}{X}\right)^2$, where $X = [F + Hs/(1-s)]\ln Hs$ |
| SEDI | $\left\{1 + \dfrac{[SEDI(1-2H)+1](1/a+1/c)}{[SEDI(1-2F)-1](1/b+1/d)}\right\}^{-1}$ | $\dfrac{S_H^2\left[\dfrac{SEDI(1-2H)+1}{H(1-H)}\right]^2 + S_F^2\left[\dfrac{SEDI(1-2F)-1}{F(1-F)}\right]^2}{[\ln F + \ln H + \ln(1-F) + \ln(1-H)]^2}$ |

clarity of display, the error expressions for several of the measures have been expressed in terms of the errors for related variables.

The validity of these expressions relies on two assumptions, which should ideally be tested in any data set to which they are applied. These are that the time series of forecast outcomes is stationary, and that successive outcomes are independent. Stationarity means for present purposes that there is no significant change over the period of the data in the skill or decision threshold of the forecasting system or in the climatological probability of the event being forecast. If there are trends or discontinuities in these quantities, the validity of any statistic based on the whole sample is questionable. Methods for dealing with non-stationarity are described in many textbooks (e.g. Jones, 1985); see also Hamill and Juras (2006) and Hogan et al. (2009). The second assumption, independence, implies that in the series of forecast outcomes, the probability of an outcome at any particular step conditional on any other outcomes is equal to the unconditional probability of that outcome. Seaman et al. (1996) discuss this issue and comment that the independence condition seems unlikely to be completely satisfied in practice, but confidence intervals based on the assumption are still useful, providing at least lower bounds on the uncertainty in the estimated proportions. If it is possible to estimate the *equivalent number of independent samples*, then n and the contingency table cell counts can be scaled down appropriately before estimating confidence intervals (see Section 3.2.5). Hamill (1999) illustrated several methods for assessing serial correlation in binary forecast verification. For situations where non-stationarity and serial correlation are present, non-parametric methods such as resampling may be more appropriate for estimating the confidence intervals on verification scores (Wilks, 2006b, sections 5.3.3 and 5.3.4). Despite these caveats, it is our hope that the straightforwardness of Equation 3.24 will encourage those engaged in forecast verification to always accompany reported performance measures with estimated confidence intervals.

The most obvious use of confidence intervals is to help diagnose whether forecasting system A is significantly better than forecasting system B in terms of a particular performance measure $M$; but it is important to avoid a couple of common pitfalls. Firstly, it is not valid simply to check whether the confidence intervals overlap; rather, we need to estimate a confidence interval for the difference in the measure between the two forecasting systems $\Delta M = M_A - M_B$. If the two systems are being evaluated against different observations then we may assume them to be independent, in which case the confidence interval of $\Delta M$ is simple to estimate from the individual confidence intervals for $M_A$ and $M_B$ (see Section 3.2.5). More commonly, forecasting systems are evaluated against the same observational data and independence cannot be assumed. The *bootstrap method* (Wilks, 2006b) is a numerical way to account for this. Suppose we have a time-series of $n$ observations of a binary phenomenon, O, together with the corresponding forecasts from forecasting systems A and B. We create 1000 different *bootstrap samples*, each of length $n$, by repeatedly taking $n$ random triplets of O, A and B from the original data *with replacement*, i.e. allowing individual events to be selected more than once. For each bootstrap sample $i$, the difference in the value of the measure $\Delta M_i$ is calculated. We then calculate the 95% confidence interval of $\Delta M$ as the interval between the 25th smallest and the 25th largest value of $\Delta M_i$. If this interval includes zero then we may say that forecasting systems A and B are not significantly different for measure $M$.

### 3.5.3 Optimal threshold probabilities

In Section 3.4.1 the property of *consistency* was discussed for binary forecasts generated by applying a decision threshold to a probability forecast; the decision threshold is *consistent* with the measure if it maximizes the measure's expected value. Mason (1979) demonstrated the calculation of this *optimal threshold probability* $p = p^*$ for any particular measure. This approach has strong similarities to the approaches used to maximize *forecast value* discussed in Chapter 9.

Here we show how the optimal threshold probability can be derived. The original forecast probability is denoted $\hat{p}$. All performance measures for binary forecasts can be regarded as average payoffs determined by some payoff matrix of the general form shown in Table 3.6. This table represents the change to the overall score that was awarded after

**Table 3.6** Schematic payoff table. $U_i$ is the reward or penalty (the change in the value of a measure) resulting in the next forecast-observation pair falling in category $i$

|         |     | Observed |     |
|---------|-----|----------|-----|
|         |     | Yes      | No  |
| Forecast | Yes | $U_a$    | $U_b$ |
|          | No  | $U_c$    | $U_d$ |

$n$ forecasts have been made, from forecast $n + 1$ resulting in an increment to $a, b, c$ or $d$. Thus if a particular measure is represented by $M$, then

$$U_a = M(a + 1, b, c, d) - M(a, b, c, d) \quad (3.25)$$

and similarly for $U_b$ and the other members of the payoff matrix. The expected payoff (contribution to the score) for forecasting an occurrence ($\hat{x} = 1$) is given by

$$E_1 = \bar{p}U_a + (1 - \bar{p})U_b \quad (3.26)$$

where $\bar{p}$ is the probability of the event occurring given an original forecast probability of $\hat{p}$, i.e. $\bar{p} = E(x = 1 \mid \hat{p})$. For perfectly calibrated probability forecasts, $E(x \mid \hat{p}) = \hat{p}$ (see Chapter 2), and so for such forecasts $\bar{p} = \hat{p}$. The expected payoff for a forecast of non-occurrence is similarly given by

$$E_0 = \bar{p}U_c + (1 - \bar{p})U_d \quad (3.27)$$

Therefore, a higher mean score can be obtained after forecast $n + 1$ by forecasting occurrence when $E_1 \geq E_0$. This can be seen by rearrangement of

Equations 3.26 and 3.27 to correspond to the forecast probability exceeding the optimal threshold probability $p^*$ given by

$$\hat{p} \geq p^* = \left(1 + \frac{U_a - U_c}{U_d - U_b}\right)^{-1} \quad (3.28)$$

In the case of PC, for example, the payoff matrix is given by Table 3.7, which contains the increments in PC from the $n^{\text{th}}$ to the $(n + 1)^{\text{th}}$ forecast for each possible combination of forecast and observation. Making the appropriate substitutions in Equation 3.28 it is straightforward to show that for PC, $E_1 \geq E_0$ when $\hat{p} \geq 0.5$, so the optimal threshold probability for PC is 0.5. Therefore, a forecaster evaluated using PC can maximize their score by only forecasting occurrence when the probability of the event is greater than 0.5. Table 3.5 provides optimal threshold probabilities for most of the scores discussed in this chapter. It can be seen that for measures that are always improved by hedging towards a constant forecast of occurrence or non-occurrence (e.g. $H$, $F$ and EDS), the threshold probability takes a trivial value of 0 or 1. For some of the measures with more complicated expressions, the use of Equation 3.25 can lead to an unwieldy expression for $p^*$. In this case we have made the approximation that

$$U_a \approx \left.\frac{\partial M}{\partial a}\right|_{b,c,d} \quad (3.29)$$

(i.e. where the partial derivative is taken holding $b$, $c$ and $d$ constant), and similarly for the other members of the payoff matrix. This is equivalent to assuming large $n$, and indeed when applied to PSS it predicts

**Table 3.7** Implied payoff table for the measure 'proportion correct' on the $(n + 1)^{\text{th}}$ forecast. PC here is the proportion correct over the previous $n$ forecasts

|         |     | Observed |     |
|---------|-----|----------|-----|
|         |     | Yes | No |
| Forecast | Yes | $(1 - \text{PC})/(n + 1)$ | $-\text{PC}/(n + 1)$ |
|          | No  | $-\text{PC}/(n + 1)$ | $(1 - \text{PC})/(n + 1)$ |

$p^* = (a + c)/n$, which is what the expression in Table 3.5 tends to for large $n$.

## Acknowledgements

We thank many of the participants of the EU-funded Cloudnet project who were involved in producing the cloud observations and model data used to produce Figure 3.4, specifically: Peter Clark, Anthony Illingworth, Martial Haeffelin, Henk Klein Baltinck, Erik van Meijgaard, Ewan O'Connor, Jean-Marcel Piriou, Thorsten Reinhardt, Axel Seifert, Adrian Tompkins, Ulrika Willén, Damian Wilson and Charles Wrench.

# 4

# Deterministic forecasts of multi-category events

**Robert E. Livezey**
*Retired (formerly US National Weather Service)*

## 4.1 Introduction

The previous chapter examined the verification of forecasts of binary events, i.e. forecasts of two categories of discrete events that are collectively exhaustive. Such forecasts reduce to forecasts of either 'yes' or 'no' for some event. Here verification is extended to forecast problems with more than two categories or classes, again mutually exclusive and collectively exhaustive. The focus is strictly on discrete events, but these can be defined in terms of ranges of continuous variables; for example, forecasts of low, near-normal or high temperatures where the categories are defined as temperatures less than or equal to some threshold, between this threshold and a higher one, and greater than or equal to the second threshold respectively.

One approach to verification of forecasts of three or more categories is to apply the results and procedures of signal detection theory described in Section 3.3 separately to all of the embedded two-category forecast sets. For the three-class temperature forecast example above, the embedded two-class forecast sets are for the occurrence of a low temperature or not, a near-normal temperature or not, and

a high temperature or not. Dealing with multiple Relative Operating Characteristic (ROC) diagrams simultaneously can be cumbersome and visualization of their joint implications difficult. Likewise the Odds Ratio score (Yule's Q) for binary forecasts described in Section 3.5.1 cannot be generalized simply to more than two categories. However, log-linear models whose parameters can be related to Q can be constructed for these verification situations. Discussion of these models is beyond the scope of this text, but extensive information about them can be found in Agresti (2007, Chapters 6 and 7).

The difficulties inherent in verifying multi-category forecasts have led to a plethora of proposed scores and misinformation about them in the literature. Fortunately, in the early 1990s, three related sets of equitable skill scores were proposed that embody almost all of the desirable attributes various authors have highlighted. Two of these are specific subsets of the third set elegantly developed by Gandin and Murphy (1992). Their only deficiencies may be that they are not regular (see Section 3.4.2) and are not best at detecting positive trends in already relatively skilful forecasts (Rodwell *et al.*, 2010). Nevertheless the use of the

*Forecast Verification: A Practitioner's Guide in Atmospheric Science*, Second Edition. Edited by Ian T. Jolliffe and David B. Stephenson.
© 2012 John Wiley & Sons, Ltd. Published 2012 by John Wiley & Sons, Ltd.

Gandin and Murphy scores is the main emphasis here, and other scores (with one exception, SEEPS – Rodwell *et al.*, 2010, covered in Section 4.3.5 below), like the higher-dimension extensions of the Heidke (HSS) and Peirce (PSS) Skill Scores, are presented only for purposes of clarification or illustration of points. Aside from the exception, these other scores have considerable deficiencies compared to the Gandin and Murphy scores and are not recommended.

The discussion of skill scores in Section 4.3 makes up the bulk of this chapter. It is preceded in Section 4.2 by material about contingency tables and associated measures of accuracy, and is followed in Section 4.4 by a discussion of the important topic of sampling variability of skill. The latter addresses the question of whether an apparently successful set of categorical forecasts represents real skill or is just the result of luck. Before proceeding to the next section, two critical issues need to be understood to approach the application of the measures and scores presented below.

The first issue was noted in Section 2.2 but needs to be re-emphasized here, namely that a set of forecast event categories can be either ordinal or nominal. These terms refer to whether the order of categories does or does not matter respectively. There are preferred attributes for scores for ordinal forecast categories in addition to those for nominal ones, and the Gandin and Murphy skill scores are constrained in different ways in the two cases. These points are made where appropriate in Section 4.3.

The other important consideration is the estimation of probabilities in determining skill. An example is the application of probabilities estimated from a large sample of previous observation-forecast pairs to a verification of a new set. This requires the assumption that the new sample came from the same population as the previous sample, i.e. that the statistics of the observation-forecast pairs are stationary. In this chapter the exclusive use of sample probabilities (observed frequencies) of categories of the forecast-observation set being verified is recommended, rather than the use of historical data. The only exception to this is for the case where statistics are stationary and very well estimated. This is rarely the case in environmental verification problems, especially those that deal with forecasts on seasonal to interannual or longer timescales. Clear

examples of this are presented in Section 4.2 for seasonal temperature forecasts made in the 1980s by the US National Weather Service (NWS). The use of observed current frequencies for estimating probabilities implies that forecast sample sizes be large enough for reasonably accurate estimates. For three- or more class forecast problems it can be argued that a sample size of the order of $10K^2$ or more 'independent' data points is required to properly estimate the $K^2$ possible forecast-observation outcomes in a $(K \times K)$ contingency table. Smaller sample sizes would be sufficient to estimate the marginal probabilities of the $K$ different categories.

## 4.2 The contingency table: notation, definitions, and measures of accuracy

### 4.2.1 Notation and definitions

The basis for the discussion of verification of categorical forecasts and the complete summary of the joint distribution of forecasts and observations is the contingency table. Let $\hat{x}_i$ and $x_i$ denote a forecast and corresponding observation respectively of category $i$ ($i = 1, \ldots, K$). Then the relative sample frequency (i.e. the cell count $n_{ij}$ divided by the total forecast-observation pair sample size $n$) of forecast category $i$ and observed category $j$ can be written as

$$\hat{p}(\hat{x}_i, x_j) = p_{ij}; \; i, j = 1, \ldots, K \qquad (4.1)$$

The two-dimensional array of all the $n_{ij}$ is referred to as the contingency table. The sample probability distributions of forecasts and observations respectively then become

$$\hat{p}(\hat{x}_i) = \sum_{j=1}^{K} p_{ij} = \hat{p}_i; \; i = 1, \ldots, K$$

$$\hat{p}(x_i) = \sum_{j=1}^{K} p_{ji} = p_i; \; i = 1, \ldots, K \quad (4.2)$$

The sample frequencies (Equation 4.2) are sometimes referred to as the empirical marginal distributions. Although the notation on the left-hand sides

**Table 4.1** Relative frequencies $p_{ij}$ in percent (total sample size $n = 788$) for US mean temperature forecasts for February through April 1983–90

| | Observed | | | |
|---|---|---|---|---|
| Forecast | Below normal | Near normal | Above normal | Forecast distribution |
| Below normal | 7 | 14 | 14 | 35 |
| Near normal | 4 | 9 | 16 | 29 |
| Above normal | 4 | 8 | 24 | 36 |
| Observed distribution | 15 | 31 | 54 | 100 |

of Equations 4.1 and 4.2 is technically desirable, indicating as it does with '^' that the quantities are *estimated* probabilities, it is somewhat cumbersome. Hence we will use the simpler notation on the right-hand sides of the equations for the remainder of the chapter.

Examples of contingency tables showing $p_{ij}$ in percent with accompanying marginal distributions are given for NWS US seasonal mean temperature forecasts in three categories in Tables 4.1 and 4.2. The forecasts were made at almost 100 US cities for the years 1983–90 for February through April (FMA) and June through August (JJA) respectively, so each table is constructed from a sample of almost 800 (not entirely independent) forecasts-observations. The three classes of below-, near-, and above-normal temperatures were defined in terms of class limits separating the coldest and warmest 30% of the 1951–80 record from the middle 40% for each city and season. Thus the three categories are not equally probable, given the observed climate. Based on the distributions of the observations (the bottom rows) in both tables it seems reasonable to conclude

that the seasonal temperature climate of the USA was not stationary between 1951–80 and 1983–90; specifically the climate has warmed between the two periods leading to considerable non-uniformity in the marginal distributions of the observations.

As can be seen in the right-hand columns, curiously the NWS forecasters were cognizant of this climate change for the warm season forecasts (Table 4.2) but not for the cold season forecasts (Table 4.1); the former set (JJA) is clearly skewed towards forecasts of the warm category at the expense of the cold category whereas the latter (FMA) exhibits no clear preference for either extreme category. This may be related to a tendency to issue forecasts of 'least regret' on the part of the forecasters; i.e. the forecasters may have been reluctant to predict relatively warm conditions too frequently in the winter but not in the summer. This kind of forecaster bias is based on the perception that an erroneous prediction of above normal has more serious consequences for most users than an erroneous forecast of below normal in the winter and vice versa in the summer.

**Table 4.2** Relative frequencies $p_{ij}$ in percent (total sample size $n = 788$) for US mean temperature forecasts for June through August 1983–90

| | Observed | | | |
|---|---|---|---|---|
| Forecast | Below normal | Near normal | Above normal | Forecast distribution |
| Below normal | 3 | 8 | 4 | 15 |
| Near normal | 8 | 13 | 18 | 39 |
| Above normal | 7 | 14 | 25 | 46 |
| Observed distribution | 18 | 35 | 47 | 100 |

The ability here to glean useful information from the marginal distributions of forecasts and observations is an example of the power of the distributions-oriented approach to verification first discussed in Section 2.10. Throughout this chapter the use of all of the information in the contingency table will be emphasized.

### 4.2.2 Measures of accuracy

An extensive list of measures of accuracy for binary forecasts based on the entries of the contingency table and the marginal distributions was introduced in Section 3.2 and described in Sections 3.2 and 3.5. Only three of these, the proportion correct (PC) (Section 3.5.1), bias (B) (Section 3.2.1), and probability of detection (POD) or hit rate (Section 3.2.3), are both appropriate and frequently used when $K > 2$. Using the notation in Equations 4.1 and 4.2 they are defined for $K$ categories respectively:

$$PC = \sum_{i=1}^{K} p_{ii} \qquad (4.3)$$

$$B_i = \hat{p}_i / p_i; \; i = 1, \ldots, K \qquad (4.4)$$

$$POD_i = p_{ii} / p_i; \; i = 1, \ldots, K \qquad (4.5)$$

The biases reveal whether some forecast categories are being over- or under-forecast while the probabilities of detection quantify the success rates for detecting different categorical events. These quantities along with PCs are presented in Table 4.3 for the forecast-observation sets from Tables 4.1 and 4.2.

The two seasonal mean temperature forecast-observation sets have similar proportions correct (around 40%) with little bias for near normal forecasts (value close to 1) and only modest probabilities of detection (less than 40%). As discussed earlier, the sets are quite dissimilar for the extreme categories. The winter forecasts have a large cold bias and similar probabilities of detection for above- and below-normal categories. In contrast, for summer the below-normal class is modestly under-forecast and severely under-detected.

It should be noted that, although the next section concentrates on a small number of skill scores with desirable attributes, there are other 'measures of association' for contingency tables that could, in theory, be used as verification scores. Goodman and Kruskal (1979) collect together four papers published by them between 1954 and 1972 to illustrate the range of possibilities. Many of the measures discussed by Goodman and Kruskal are for $(2 \times 2)$ tables (including those covered in Chapter 3) or $(2 \times K)$ tables, but some are for the $(K \times K)$ tables of this chapter.

## 4.3 Skill scores

In this section, desirable aspects of skill scores for multi-categorical forecasts will first be discussed. Following this, the development of the Gandin and Murphy (1992) equitable scores will be outlined along with a frequently encountered special case. The convenient subset of these scores presented by Gerrity (1992) as well as the subset called LEPSCAT (Potts et al., 1996) will be covered next. The section will conclude by contrasting

**Table 4.3** Measures of accuracy for US mean temperature forecasts in three categories (below, near and above normal) for February through April (FMA, Table 4.1) and June through August (JJA, Table 4.2) 1983–90

|  | PC | Bias Below | Bias Near | Bias Above | POD Below | POD Near | POD Above |
|---|---|---|---|---|---|---|---|
| FMA | 0.40 | 2.30 | 0.94 | 0.67 | 0.47 | 0.29 | 0.44 |
| JJA | 0.42 | 0.78 | 1.11 | 0.98 | 0.17 | 0.37 | 0.53 |

PC, proportion correct; POD, probability of detection.

the development of SEEPS (Rodwell *et al.*, 2010), a new score for daily precipitation forecasts motivated by the needs of the forecast modelling community, with that of Gandin and Murphy scores.

## 4.3.1 Desirable attributes

The Heidke and Peirce skill scores, introduced in Section 3.5.1, extend to forecasts for more than two categories, and are useful to illustrate some of the attributes (by their presence or absence) exhibited by the Gandin and Murphy scores. The Heidke and Peirce skill scores are given respectively by

$$\text{HSS} = \left( \sum_{i=1}^{K} p_{ii} - \sum_{i=1}^{K} p_i \hat{p}_i \right) \bigg/ \left( 1 - \sum_{i=1}^{K} p_i \hat{p}_i \right)$$

(4.6)

$$\text{PSS} = \left( \sum_{i=1}^{K} p_{ii} - \sum_{i=1}^{K} p_i \hat{p}_i \right) \bigg/ \left( 1 - \sum_{i=1}^{K} p_i p_i \right)$$

(4.7)

Except for the use of different estimates for the number of correct forecasts expected by chance in their denominators, these skill scores are the same. They are both measures of the proportion of possible improvement in number of correct forecasts over random forecasts.

Note first that both scores are *equitable* (see Sections 2.8 and 3.4.1); HSS and PSS have zero expectation for random forecasts and for constant forecasts of any single category ($\hat{p}_i = 1.0$). These are highly desirable properties. The form of both of these scores explicitly depends on the forecast distribution, a less desirable property, in contrast to the other equitable scores featured in this chapter.

An example of a non-equitable score is the one formerly used by NWS to verify the forecast sets represented by Tables 4.1 and 4.2. Recall that the forecast categories were defined by class limits for three unequally distributed categories with probabilities of 0.30, 0.40 and 0.30 estimated from 1951–80 data. The NWS replaced the estimates of expected hit frequency for random forecasts in both the numerators and denominators of Equations 4.6 and 4.7 with 0.34 ($= 0.30^2 + 0.40^2 + 0.30^2$), the theoretical probability of random hits. With a sta-

**Table 4.4** Skill scores for US mean temperature forecasts in three categories for February through April (FMA, Table 4.1) and June through August (JJA, Table 4.2) 1983–90. The scores are described in the text

|  | Score | | | |
|---|---|---|---|---|
|  | Inequitable | HSS | PSS | GS |
| FMA | 0.09 | 0.09 | 0.10 | 0.17 |
| JJA | 0.12 | 0.05 | 0.05 | 0.08 |

GS, Gerrity Skill Score; HSS, Heidke Skill Score; PSS, Peirce Skill Score.

tionary climate the expected proportion correct for constant forecasts of near-normal is 0.40, leading to both positive expected numerators and hence positive expected skills in Equations 4.6 and 4.7. For the 1983–90 winter forecasts and observations in Table 4.1 there is little difference between this non-equitable score and HSS and PSS. All three are between 0.09 and 0.10 in Table 4.4. However, the non-equitable score is substantially higher than the other two (0.12 vs 0.05; Table 4.4) for the 1983–90 summer forecast-observations in Table 4.2, because the expected proportion of correct forecasts by chance is:

$$0.38 (= 0.18 \times 0.15 + 0.35 \times 0.39 + 0.47 \times 0.46)$$

compared to 0.34 for the non-equitable score. The Gerrity skill scores (GS) in Table 4.4 are defined in Section 4.3.3.

Next observe that both HSS and PSS weight all correct forecasts the same regardless of the relative sample probabilities. This encourages a forecaster to be conservative by not providing a greater reward for successful forecasts of lower probability events. By the same token a skill score should provide greater or lesser penalties for different types of incorrect forecasts if categories have different sample probabilities. For $K > 2$ neither of these scores utilize off-diagonal information, i.e. the distribution of incorrect forecasts, in the contingency table. More specifically, if the predictands are ordinal, greater discrepancy between forecast and observed classes should ideally be penalized more than a lesser difference between classes. For example, an erroneous cold forecast should be more heavily penalized if warm is observed than if near normal is observed.

Lastly, for ordinal predictands, the magnitude of a score should be relatively insensitive to the type or number of categories when applied to forecasts made by assigning categories to objectively produced continuous forecasts. Satisfying this criterion will also ensure that scores computed from smaller dimensional categorizations embedded within the contingency table will be consistent with the score for the full table and each other. Barnston (1992) graphically illustrates how a score similar to the discontinued NWS score applied to equally probable categories does not exhibit this property (Figure 4.1). He synthetically produced categorical forecasts for different numbers of equally probable categories from continuous forecasts with known linear association (correlation – see Sections 2.4 and 5.4.4) with the observations. Note that a forecaster can achieve a higher score with the same underlying measure of linear association simply by reducing the number of categories, or degree of resolution, of the forecasts. Further note the (perhaps counterintuitive) phenomenon that the correspondence

between scores and underlying linear associations (correlations) becomes closer as the number of categories decrease in Barnston's experiment. Ideally this close association should not be degraded with an increase in the number of categories.

The subsets of the complete family of Gandin and Murphy scores described in Sections 4.3.3 and 4.3.4 are constructed in ways that ensure their consistency from categorization to categorization and with underlying linear correlations. These scores likewise are equitable, do not depend on the forecast distribution, do not reward conservatism, utilize off-diagonal information in the contingency table, and penalize larger errors more when predictands are ordinal. An important distinguishing factor between the two sets is the convenience of their use, with the practical advantage going to the Gerrity (1992) scores rather than LEPSCAT. Also, for at least the case of $K = 3$ equiprobable categories, three other ways Gerrity scores have the advantage (see Section 4.3.4) can be inferred from the analyses of Rodwell *et al.* (2010).

A desirable score attribute not mentioned above is sensitivity to skill improvements for forecasts that already exhibit good skill. As described in Section 4.3.5, this attribute is enhanced for SEEPS scores through tradeoffs with other desirable attributes that are not deemed as important by the scores' authors. Most notably, SEEPS (like HSS and PSS) weights all correct forecasts the same regardless of the relative sample probabilities.

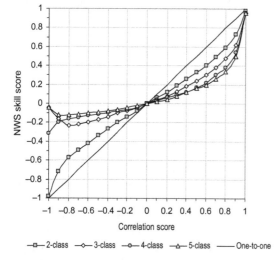

**Figure 4.1** A previously used National Weather Service (NWS) score as a function of correlation score for two (shaded squares), three (open diamonds), four (shaded circles), and five (open triangles) equally likely categories. Note that for a given underlying linear association the score can be increased by reducing the number of forecast categories. After Barnston (1992)

### 4.3.2 Gandin and Murphy equitable scores

In this section, the major steps and assumptions used in the axiomatic approach of Gandin and Murphy to develop their family of equitable scores will be briefly outlined. A scoring matrix $S$, with elements $s_{ij}$, is used to define a general form of a skill score using the contingency table:

$$\text{GMSS} = \sum_{i=1}^{K} \sum_{j=1}^{K} p_{ij} s_{ij} \tag{4.8}$$

The scoring matrix is a tabulation of the reward or penalty every forecast-observation outcome represented by the contingency table is accorded. It is similar in spirit to the expense matrix of the simple cost-loss model used to assess forecast value (see Section 9.2). The remaining problem is to determine what the $K^2$ elements of the scoring matrix should be to ensure that GMSS is equitable. If random forecasts are arbitrarily required to have an expected score of 0 and the score for perfect forecasts is set to 1, then $K + 1$ relationships constraining the $s_{ij}$ are:

$$\sum_{j=1}^{K} p_j s_{ij} = 0, \ i = 1, \ldots, K$$

$$\sum_{i=1}^{K} p_i s_{ii} = 1 \qquad (4.9)$$

The first $K$ relationships in Equation 4.9 simply state that constant forecasts of any category $i$ must have a score of 0 while the other relationship constrains a perfect score to be 1. The Equations 4.9 are insufficient to determine the full scoring matrix: If $K = 3$ they provide only four out of nine necessary relationships and, more generally, only $K + 1$ of $K^2$.

The number of remaining relationships needed can be reduced by $K(K - 1)/2$ by the assumption of symmetry for $S$:

$$s_{ji} = s_{ij} \qquad (4.10)$$

This condition states that it is no more or less serious to forecast class $i$ and observe $j$ than vice versa and is reasonable in the majority of environmental prediction problems. There are user-related situations where the condition would not be appropriate and the development of equitable scores would proceed differently from what is presented here. The description of SEEPS in Section 4.3.5 is an especially illustrative example of such a different developmental path.

Another reasonable constraint on $S$ is to require that the reward for an incorrect forecast be less than a correct one:

$$s_{ij} \leq s_{ii}$$

$$s_{ij} \leq s_{jj} \qquad (4.11)$$

If the predictand categories are ordinal a final constraint requires that the reward for an incorrect forecast be less than or equal to an incorrect forecast that misses the observed category by fewer classes, i.e. for three-category predictands a two-class error is rewarded less than or equal to a one-class error:

$$s_{i'j} \leq s_{ij}, \ |i' - j| > |i - j|$$

$$s_{ij'} \leq s_{ij}, \ |i - j'| > |i - j| \qquad (4.12)$$

Equations 4.9 to 4.12 completely describe the Gandin and Murphy family of equitable scoring matrices. Note that both subsets of this family presented in Sections 4.3.3 and 4.3.4 respectively and SEEPS in Section 4.3.5 are for ordinal predictand categories.

For three-category predictands, Equation 4.10 reduces the number of missing relationships to determine $S$ to two. Thus two elements of $S$ have to be specified, and Gandin and Murphy set $s_{12} = k_1$ and $s_{23} = k_2$. The elements of the scoring matrix then become:

$$s_{11} = [p_3 + p_1(p_3 - p_2)k_1 \\ + \ p_3(p_2 + p_3)k_2] / [p_1(p_1 + p_3)]$$

$$s_{13} = - [1 + (p_1 + p_2)k_1 + (p_2 + p_3)k_2]/(p_1 + p_3)$$

$$s_{22} = -(p_1 k_1 + p_3 k_2)/p_2$$

$$s_{33} = [p_1 + p_1(p_1 + p_2)k_1 \\ + \ p_3(p_1 - p_2)k_2] / [p_3(p_1 + p_3)] \qquad (4.13)$$

To complete the construction of an equitable score for three-category predictands, choices have to be made for $k_1$ and $k_2$. These choices are limited by the constraints (Equation 4.11) and, if the predictands are ordinal, Equation 4.12 as well. Application of the constraints leads to the permissible values defined by the quadrilaterals in Figure 4.2; nominal predictand categories by the larger quadrilateral and ordinal ones by the square.

Gandin and Murphy provide a numerical example in which $p_1 = 0.5$, $p_2 = 0.3$ and $p_3 = 0.2$, where the three categories are far from being equally probable and the categorization is highly asymmetrical. Choosing $k_1 = -0.5$ and $k_2 = -0.25$, the midpoint of the bottom of the square in Figure 4.2, leads

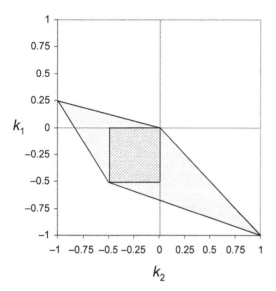

**Figure 4.2** Acceptable domains for numerical values of the specified elements $k_1 (= s_{12})$ and $k_2 (= s_{23})$. After Gandin and Murphy (1992)

to the scoring matrix in the upper left section of Table 4.5. Also shown for contrast are the case for $p_1 = 0.2$, $p_2 = 0.5$ and $p_3 = 0.3$, in which the categorization is more symmetrical, and the scoring matrices in both cases for the scores developed in the next section (where they will be discussed). The most important thing to note about the scoring ta-

ble for the asymmetrical case is that the reward for a correct forecast of event three is more than double that for event two and over three times that for event one. Event three is not only the least probable but also represents one of the two tails of the ordinal continuum encompassed by the three events, so the reward for correctly forecasting it should be substantially higher than forecasting event two (with only a slightly larger probability). Another interesting feature of this scoring matrix is the fact that the penalty for incorrectly predicting event three but observing event one and vice versa is greater than the reward for correctly forecasting the highly probable event one. This is not the case for the more symmetrical categorization (upper right of Table 4.5) because events one and three have more comparable probabilities. However, the rewards for correctly forecasting these two events are quite disparate, suggesting that $k_1 = -0.5$ and $k_2 = -0.25$ are not the best choices for this verification situation. The related scores developed in the next section lead to a more reasonable reward/penalty matrix, so trial and error is not recommended. Before these scores are presented the scoring matrices for a frequently encountered three-event special case are examined to conclude this section.

The special case is that for symmetric event probabilities, i.e. when $p_1 = p_3$. With this condition only one of the elements of the scoring matrix needs

**Table 4.5** Equitable scoring matrices for three-category forecasts with two asymmetric sets of event probabilities

|  | Event probabilities $(p_1,p_2,p_3)$ | |
|---|---|---|
|  | (0.5,0.3,0.2) | (0.2,0.5,0.3) |
| Gandin and Murphy (1992) | $k_1 = -0.5, k_2 = -0.25$ | $k_1 = -0.5, k_2 = -0.25$ |
|  | $\dfrac{1}{28}\begin{bmatrix} 16 & -14 & -19 \\ -14 & -28 & -7 \\ -19 & -7 & 58 \end{bmatrix}$ | $\dfrac{1}{60}\begin{bmatrix} 156 & -30 & -54 \\ -30 & 21 & -15 \\ -54 & -15 & 61 \end{bmatrix}$ |
| Gerrity (1992) | $k_1 = -0.375, k_2 = 0.0$ | $k_1 = -0.286, k_2 = -0.375$ |
|  | $\dfrac{1}{8}\begin{bmatrix} 5 & -3 & -8 \\ -3 & 5 & 0 \\ -8 & 0 & 20 \end{bmatrix}$ | $\dfrac{1}{168}\begin{bmatrix} 372 & -48 & -168 \\ -48 & 57 & -63 \\ -168 & -63 & 217 \end{bmatrix}$ |

to be specified. Gandin and Murphy set

$$s_{12} = s_{23} = k, \text{ leading to:}$$
$$s_{11} = s_{33} = (1 + 2kp_1)/2p_1$$
$$s_{13} = -[1 + 2k(1 - p_1)]/2p_1$$
$$s_{22} = -2kp_1/(1 - 2p_1) \qquad (4.14)$$

The range of permissible values of $k$ is 0 to $-1$ inclusive, but if the events are ordinal and the categories are defined so that both events one and three can be considered equidistant from event two, then $-0.5 \le k \le 0$. The equidistant assumption may not always be reasonable, especially in those cases where the distribution of the underlying variable is highly skewed. An example is precipitation, where the range of values for equiprobable tails will be very dissimilar, with a much narrower range for the below-normal precipitation class compared to above-normal.

When $k = -0.5$, all of the off-diagonal elements of the scoring matrix are equal to $k$, appropriate for nominal events where no distinction is made between incorrect forecasts. The values of all elements of $S$ are presented in Figure 4.3 for ranges of $p_1$ and $k$. Except for very large and very small values of $p_1$, $S$ is not very sensitive to $k$. Scoring matrices are shown in Table 4.6 for both $p_1 = 0.33$ and $p_1 = 0.3$ with a choice of $k = -0.25$, midway in its permissible range for the ordinal case. Also included in the table for the same examples are scores developed in the next two sections that are part of the Gandin and Murphy family of equitable scores. Note that for all $S$ rewards for correct predictions of events one and three are considerably greater than for event two, penalties for two-class errors are considerably greater than for a miss by just one event, and that both of these large rewards and penalties increase with a decrease in $p_1$.

### 4.3.3   Gerrity equitable scores

Gerrity (1992) discovered the construction of a subset of the Gandin and Murphy scores for ordinal categorical event forecasts that ensure their equitability (see Sections 2.8 and 3.4.1), provide convenient and alternative means for their computation, and seem to lead to reasonable choices for various $k$. These scores will be denoted by GS. The construction of

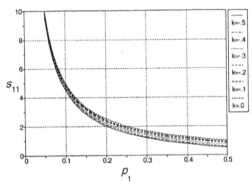
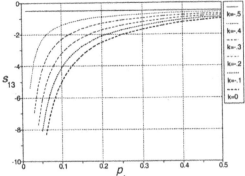
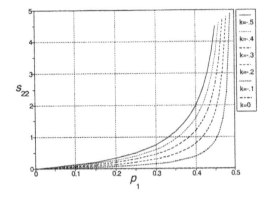

**Figure 4.3**   Numerical values of the elements in the scoring matrix in the special $(3 \times 3)$ situation as a function of the climatological probability $p_1$ for selected values of the specified element $k \, (= s_{12})$: (top) $s_{11}$, (middle) $s_{13}$, (bottom) $s_{22}$. From Gandin and Murphy (1992)

their scoring matrices starts with the following definition:

$$a_i = \frac{1 - \sum\limits_{r=1}^{i} p_r}{\sum\limits_{r=1}^{i} p_r} \qquad (4.15)$$

**Table 4.6**  Equitable scoring matrices for three-category forecasts with symmetric sets of event probabilities

| | Event probabilities $(p_1, p_2, p_3)$ | |
| | 0.33,0.33,0.33) | (0.3,0.4,0.3) |
| --- | --- | --- |
| Gandin and Murphy (1992) | $k = -0.25$ | $k = -0.25$ |
| | $\dfrac{1}{24}\begin{bmatrix} 30 & -6 & -24 \\ -6 & 12 & -6 \\ -24 & -6 & 30 \end{bmatrix}$ | $\dfrac{1}{24}\begin{bmatrix} 34 & -6 & -26 \\ -6 & 9 & -6 \\ -26 & -6 & 34 \end{bmatrix}$ |
| Gerrity (1992) | $k = -0.25$ | $k = -0.286$ |
| | $\dfrac{1}{24}\begin{bmatrix} 30 & -6 & -24 \\ -6 & 12 & -6 \\ -24 & -6 & 30 \end{bmatrix}$ | $\dfrac{1}{21}\begin{bmatrix} 29 & -6 & -21 \\ -6 & 9 & -6 \\ -21 & -6 & 29 \end{bmatrix}$ |
| Potts *et al.* (1996) | $k = -0.167$ | $k = -0.18$ |
| | $\dfrac{1}{36}\begin{bmatrix} 48 & -6 & -42 \\ -6 & 12 & -6 \\ -42 & -6 & 48 \end{bmatrix}$ | $\dfrac{1}{33}\begin{bmatrix} 49 & -6 & -41 \\ -6 & 9 & -6 \\ -41 & -6 & 49 \end{bmatrix}$ |

The elements of S are then given by

$$s_{ii} = b\left(\sum_{r=1}^{i-1} a_r^{-1} + \sum_{r=i}^{K-1} a_r\right)$$

$$s_{ij} = b\left(\sum_{r=1}^{i-1} a_r^{-1} - (j-i) + \sum_{r=j}^{K-1} a_r\right);$$

$$1 \leq i < j \leq K$$

$$s_{ji} = s_{ij}$$

$$b = \frac{1}{K-1} \qquad (4.16)$$

Recall that all Gandin and Murphy scoring matrices (including these) are symmetrical. Note also that the summations are zero for those cases in Equation 4.16 when the upper index is less than the lower. Finally observe in Equation 4.16 that $s_{1K} = -1$ always.

Three-category Gerrity scoring matrices are included in Tables 4.5 and 4.6 for comparison with the Gandin and Murphy versions developed with arbitrarily selected constants. There are no substantial differences between the matrix elements for the symmetric categorizations included in Table 4.6. In fact the tables (and implied $k$s) are identical for the case of three equally probable categories. In contrast, large differences show up for the cases represented in Table 4.5. Both scoring matrices for the highly asymmetrical categorization (0.5,0.3,0.2) seem reasonable, but the Gerrity matrix seems a more logical choice for the more symmetrical case with a highly probable event two (0.2,0.5,0.3). The Gerrity $k$s lead to less disparate rewards for correct forecasts of events one and three, with the former exceeding the latter by much less than a factor of two rather than by much greater.

The fact that the Gerrity scoring matrices all seem to have reasonable rewards/penalties may be related to a remarkable property of GS. The Gerrity Score can be alternatively computed by the numerical average of $K - 1$ two-category scores (which are identical to PSS; when Equations 4.15 and 4.16 are evaluated for $K = 2$ and substituted into Equation 4.8 we get Equation 4.7). These are the $K - 1$ two-category contingency tables formed by combining categories on either side of the partitions between consecutive categories. For three-category temperature forecasts two-category scores would be computed with Equation 4.7 for $K = 2$ for (i) above-normal and a combined near- and below-normal category, and (ii) below-normal and a combined near- and above-normal category, and then simply averaged. This convenient property of GS also guarantees a considerable amount of consistency between the various scores computed from different partitions of the contingency table.

The Gerrity score has been computed for each of the data sets in Tables 4.1 and 4.2. The results are included in Table 4.4 to contrast with the non-equitable score, with HSS, and with PSS. In both cases GS is greater than both HSS and PSS, considerably so for the cold season temperature forecasts. This is mainly because the rewards ($s_{11} = 3.42$ for Table 4.1) for correct below-normal forecasts (7% of the total) by themselves outweigh the penalties for all of the two-class errors (18% of the total; well below the number expected by chance). The reward for a correct forecast of above-normal temperature ($s_{33} = 0.51$) is only about one-seventh of that for a below-normal forecast. However, there are so many of the former (exceeding the expected number) that they also contribute meaningfully to GS. The other scores clearly are deficient in rewarding the NWS forecasters for successfully predicting the less likely below-normal category and for making relatively few large forecast errors. Because of its convenience and built-in consistency, the family of GS are recommended here as equitable scores for most forecasts of ordinal categorical events.

### 4.3.4 LEPSCAT

Another subset of the equitable Gandin and Murphy scores for ordinal predictands that are consistent by construction are those introduced by Ward and Folland (1991) and later refined by Potts *et al.* (1996), called LEPSCAT. LEPS is the acronym for 'linear error in probability space' and CAT stands for 'categorical', because LEPS has been developed for continuous predictands as well. In fact, the latter is given by

$$L = 3 \left( 1 - |F_{\hat{x}} - F_x| + F_{\hat{x}}^2 - F_{\hat{x}} + F_x^2 - F_x \right) - 1 \tag{4.17}$$

where the $F$s are the cumulative distribution functions respectively of the ordinal forecasts and obser-

vations. The family of LEPS scores is based on the linear distance between the forecast and the observation in their sample probability spaces, the second term in $L$. The other terms ensure that $L$ is equitable and does not exhibit certain pathological behaviour at its extremes. Normalization is provided by the factor 3. To develop the scoring matrix S for categorical forecasts, Equation 4.17 is averaged over all possible values of forecasts and observations for each outcome of the contingency table.

The results for equally probable three-category events and the symmetrical categorization $(0.3, 0.4, 0.3)$ are shown for comparison in Table 4.6. The LEPSCAT matrices are renormalized to facilitate comparisons with corresponding ones in Table 4.6. It is clear that the contrasts between rewards/penalties for forecast-observation pairs involving event two and other pairs involving only events one and three are much greater in the LEPSCAT matrices than in either the corresponding Gandin and Murphy or Gerrity scores, perhaps too much so. In particular, LEPSCAT (i) rewards correct forecasts of event two much less and those of events one and three somewhat more, and (ii) penalizes one-category misses (e.g. event one forecast/event two observed) much less and two-category misses (e.g. event one forecast/event three observed) much more.

The much larger internal contrasts in the LEPSCAT scoring matrix than in the Gerrity matrix for three-category, equiprobable forecasts (left side of Table 4.6) have implications for the relative performance of the two scores in the context of four attributes inferred from score assessments in Rodwell *et al.* (2010). These attributes are (i) sensitivity to skill trends, (ii) sensitivity to differences in skill between two sets of forecasts for the same cases, (iii) ability to differentiate between good and bad forecasts, and (iv) susceptibility to enhancement of skill by hedging (see Sections 2.8 and 3.4.1). For the first three of these attributes, the differences between Gerrity and LEPSCAT score performances should not be large. Specifically, the Gerrity score is expected to be more sensitive than LEPSCAT to skill trends and differences in skill between two sets of forecasts for the same cases, but less able to differentiate between good and bad forecasts. Most importantly, the LEPSCAT score in this case is much more vulnerable to hedging than the Gerrity score.

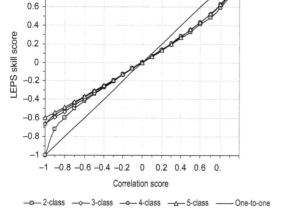

**Figure 4.4** LEPSCAT as a function of correlation score for two (shaded squares), three (open diamonds), four (shaded circles), and five (open triangles) equally likely categories. In contrast to Figure 4.1 there is little difference between the four curves and only moderate differences from the underlying linear correlations over most of their range

These results reinforce the preference here for the family of GS over LEPSCAT scores.

LEPSCAT scores are part of the Gandin and Murphy family of scores because they satisfy Equations 4.9 to 4.12. Substitution of the LEPSCAT $k$s respectively into Equation 4.14 results in exactly the LEPSCAT scoring matrices in Table 4.6. This set of scores provides the opportunity to illustrate the consistency of the two constructed Gandin and Murphy scores and their correspondence with a linear correlation measure of association. J.M. Potts (personal communication) performed the same experiment for LEPSCAT scores for equally probable categorical events as that by Barnston described in Section 4.3.1 for the non-equitable NWS score. The result was practically no difference between scores for two through five categories for underlying linear correlations greater than −0.6, with both sets of scores only moderately different from the correlation in this range (Figure 4.4).

### 4.3.5   SEEPS

The Stable Equitable Error in Probability Space (SEEPS) score was designed (Rodwell *et al.*, 2010)

to meet the needs of numerical weather prediction model developers to monitor the performance of precipitation forecasts to facilitate the identification of model errors and guide further model development. Desired attributes included equitability (Equation 4.9), the ability to accommodate a dry event category along with light and heavy precipitation categories, and sensitivity to forecast improvements. Of little interest was close correspondence to a correlation score and score consistency for different $K$ (constraints on SEEPS limit the number of categories to only three anyway).

A natural way to accommodate the awkward probability distribution of short-period (e.g. daily) precipitation amounts with its spike for dry events and highly skewed shape for non-zero events is the notion of 'linear error in probability space' that is the basis for LEPSCAT. The way to ensure maximum sensitivity to forecast improvements is to modify the second equation of 4.9 to a 'strong' perfect forecast constraint

$$s_{ii} = 1, \; i = 1, \ldots, K \qquad (4.18)$$

The modification to Equation 4.9 is where the development of SEEPS departs from Gandin and Murphy scores. The symmetry condition (Equation 4.10) is no longer appropriate or necessary, but the conditions (Equations 4.11 and 4.12) both still apply in a 'hard' form, i.e. without the equalities, because of the ordinality imposed by LEPS. With Equation 4.18, 'hard' Equations 4.11 and 4.12, and other constraints because errors are formulated in cumulative probability space, the three-category scoring matrix for SEEPS skill with event probabilities $(p_1, p_2, p_3)$ can be written as

$$\begin{bmatrix} 1 & 1-c & 1-c-a \\ 1-d & 1 & 1-a \\ 1-d-b & 1-b & 1 \end{bmatrix} \qquad (4.19)$$

Unlike Gandin and Murphy scoring matrices, Equation 4.19 is asymmetric. The idea of asymmetric equitable scores is not new. Jolliffe and Foord (1975) suggested such a score matrix with $s_{11} = s_{22} = s_{33} = 1$ and $s_{13} = s_{31} = -1$, but it has poor properties if $p_1$ and $p_3$ differ too much.

Note that Rodwell *et al.* (2010) formulated SEEPS as an error score (1 − matrix 4.19), rather than as the skill score here. The next step in the formulation is to apply the constant forecast constraints of Equation 4.9 to Equation 4.19 to write $b$, $c$ and $d$ in terms of $a$:

$$b = \frac{p_3 a}{p_1 + p_2}, c = \frac{1 - p_3 a}{p_2 + p_3}, d = \frac{1 - p_3 a}{p_1}$$

(4.20)

Because $a$ through $d$ are all positive, $0 < a \leq 1/p_3$. Rodwell *et al.* (2010) show that choosing $a = 2/p_3$ produces a score that most rewards forecast sets predicting a full range of possible outcomes. Using this choice and Equation 4.20 in Equation 4.19 gives the SEEPS skill matrix

$$\begin{bmatrix} 1 & 1 - \frac{1}{2(p_2 + p_3)} & 1 - \frac{1}{2p_3} - \frac{1}{2(p_2 + p_3)} \\ 1 - \frac{1}{2p_1} & 1 & 1 - \frac{1}{2p_3} \\ 1 - \frac{1}{2p_1} - \frac{1}{2(p_1 + p_2)} & 1 - \frac{1}{2(p_1 + p_2)} & 1 \end{bmatrix}$$

(4.21)

The version of Equation 4.21 for equiprobable categories is shown in Table 4.7 scaled for easy comparison to its counterparts in Table 4.6. Besides the uniform rewards for hits (smaller for categories 1 and 3 than for the Gerrity skill), note that one-category forecast errors with an observed middle category (light precipitation) are slightly rewarded (rather than slightly penalized in the Gerrity score), but that forecast errors for observed extreme categories (no precipitation and heavy precipitation) are penalized more (than for the Gerrity score). The comparative analyses by Rodwell *et al.* (2010) of skill score attributes for equiprobable categories

**Table 4.7** SEEPS skill scoring matrix for equiprobable event probabilities

| | Event probabilities ($p_1,p_2,p_3$) | | |
|---|---|---|---|
| | (0.33,0.33,0.33) | | |
| Rodwell *et al.* (2010) | $\frac{1}{12}$ | $\begin{bmatrix} 12 & 3 & -15 \\ -6 & 12 & -6 \\ -15 & 3 & 12 \end{bmatrix}$ | |

(referred to in the previous section) show GS is no more or less vulnerable to hedging (see Sections 2.8 and 3.4.1) than SEEPS skill, is just as sensitive to differences in skill for different forecast sets of the same cases, is slightly less adept at differentiating between good and bad forecasts, but is substantially inferior in detecting skill trends.

Based on sampling and operational considerations, Rodwell *et al.* (2010) recommend in Equation 4.21 that (i) dry event sample probabilities be constrained by $0.10 \leq p_1 \leq 0.85$, and (ii) for the wet categories, $p_2 = 2p_3$.

### 4.3.6 Summary remarks on scores

Based on all factors, the Gerrity scores (Section 4.3.3) are an appropriate choice for most ordinal categorical event forecast verification problems. They have almost all of the desirable properties outlined in Section 4.3.1, their scoring matrices or the scores themselves are easy to calculate, and in every instance examined here produced reasonable reward/penalty matrices. SEEPS scores are clearly preferred if $K = 3$ and skill trend detection is especially important. Other major sources for information on verification of forecasts of more than two categorical events are Murphy and Daan (1985), Stanski *et al.* (1989) and Wilks (2006b, section 7.2.6). Each of these sources is rich in information, but the first two are considerably out of date. Wilks devotes several pages to Gerrity scores, including an extensive and informative example from NWS cool season forecasts of precipitation types.

## 4.4 Sampling variability of the contingency table and skill scores

Two related questions about Tables 4.1 and 4.2 and the associated measures of accuracy and skill in Tables 4.3 and 4.4 respectively are (i) whether there is sufficient confidence that they reflect non-random forecasts, and (ii) given sufficient confidence, are the computed estimates of accuracy and skill measures good enough? With only one cautious exception, classical statistics offers little assistance in answering these questions. Nevertheless, with reasonable insight into the spatial-temporal characteristics of the forecast-observation set, various resampling strategies are available that can lead to satisfactory answers.

The standard way to test the null hypothesis that an unevenly populated (or 'extreme') contingency table was the result of independent, randomly matched forecast-observation pairs for categorical events is to perform a 'chi-squared' ($\chi^2$) test. In the notation of Section 4.2 (with $n$ denoting the independent sample size) the test statistic takes the form:

$$X^2 = n \sum_{i,j=1}^{K} (p_{ij} - \hat{p}_i p_j)^2 / \hat{p}_i p_j \qquad (4.22)$$

This expression, called the Pearson chi-squared statistic, is equivalent to the sum over every cell of the $(K \times K)$ contingency table of the squared difference between actual occurrences and the number expected by chance for the cell, divided by the number expected by chance. An alternative to Equation 4.18 with the same asymptotic null distribution is the 'likelihood ratio' chi-squared:

$$G^2 = 2n \sum_{i,j=1}^{K} p_{ij} \log \left( p_{ij} / \hat{p}_i p_j \right) \qquad (4.23)$$

In a sense $G^2$ is more fundamental than $X^2$ as it is a likelihood ratio statistic. However, the two are asymptotically identical and an advantage of the Pearson chi-squared (Equation 4.22) over the likelihood ratio (Equation 4.23) is that the sampling distribution of the former tends to converge more rapidly to the asymptotic distribution as $n$ increases for fixed $K$ (Agresti, 2007, section 2.4.7; see also Stephenson, 2000).

A crucial thing to note about Equations 4.22 and 4.23 is that they are directly proportional to the sample size; two tables of identical relative frequencies will have different $X^2$ and $G^2$ if they are constructed from different size samples. The asymptotic distribution of $X^2$ or $G^2$ for different degrees of freedom $m$ is well known and is tabulated in a large number of sources. The degrees of freedom are the number of cells in the contingency table ($K^2$) minus the number of restrictions on the counts expected by

**Table 4.8**  Values for $X^2$ and $G^2$ for US mean temperature forecasts in three categories for February through April (FMA, Table 4.1) and June through August (JJA< Table 4.2) 1983–90 for two different independent sample sizes, 100 and 788

|  |  | $n = 100$ | $n = 788$ |
|---|---|---|---|
| FMA | $X^2$ | 5.07 | 39.98 |
|  | $G^2$ | 5.12 | 40.41 |
| JJA | $X^2$ | 3.92 | 30.92 |
|  | $G^2$ | 3.97 | 31.25 |

The 10%, 1% and 0.1% critical values of the relevant $\chi^2$ distribution are 7.78, 13.28 and 18.47 respectively.

chance in the cells. For the tables discussed in this chapter each row and column is constrained to sum to its respective observed marginal total. But one of these $2K$ totals is dependent on the rest because it can be determined easily from them. Thus there are $2K - 1$ restrictions and $m = (K - 1)^2$. For Tables 4.1 and 4.2, $m = 4$. For this case, the values that $X^2$ or $G^2$ must exceed for their chance probabilities to be less than respectively 0.1, 0.01 and 0.001, are listed in Table 4.8. These are accompanied by the computed values of $X^2$ and $G^2$ for Tables 4.1 and 4.2 for $n = 788$, the actual number of forecasts used to construct both tables, and for $n = 100$.

The results given by Tables 4.1 and 4.2 are statistically significant at least at the 0.1% level if it can be assumed that all 788 forecasts are independent of each other. In fact, for both the seasonal mean temperature observations and forecasts it is well known that this assumption is not warranted. Each seasonal forecast was made at almost 100 locations over the contiguous USA. The cross-correlations among these sites for seasonal mean temperatures are very strong, so that the 100 stations behave statistically like only about 10 independent locations. This implies that there are perhaps only the equivalent of about 80 independent data points contributing to Tables 4.1 and 4.2. Hence the values corresponding to $n = 100$ in Table 4.8 indicate that the results in Tables 4.1 and 4.2 are probably not even statistically significant at the 10% level.

There is also some serial correlation between one forecast year and the next, but this is weak in this example compared to the uncertainties introduced by the spatial correlations. In some applications temporal correlations in either the forecasts or observations may be strong and introduce additional ambiguities into the application of a chi-squared test.

For these and other situations where spatial (cross-) and/or temporal (serial) correlations cannot be ignored, strategies based on resampling and randomization techniques can often be employed to obtain satisfactory answers to the two questions posed at the beginning of this section. Livezey (1999), von Storch and Zwiers (1999, section 6.8) and Wilks (2006b, sections 5.3 and 5.4) all discuss these situations (with examples) and outline methods for dealing with them. So-called 'bootstrap' techniques (Efron and Tibshirani, 1993) will be outlined below for the US seasonal mean temperature examples used throughout this chapter to illustrate the application of these methods. Although there is no universal reason to prefer bootstrap over other resampling approaches, it is an effective and versatile choice under a broad range of situations.

Let the vectors containing mean temperature observations and forecasts respectively for US locations for particular years, February through April, be denoted by $X^i$, $\hat{X}^i$ ($i = 1983, \ldots, 1990$). The first objective is to develop a null distribution (from random forecasts) for either $X^2$, a measure of accuracy, or a skill score to compare against an actual computed value. This null distribution must preserve the non-trivial spatial and temporal correlations in the observation and forecast samples. In this example only the spatial cross-correlations are important so the sample is built up by randomly mismatching maps (the vectors) of the observations with maps of the forecasts. This is done by randomly selecting a forecast map from the eight possibilities (1983 to 1990) to match against a randomly selected observed map from the eight possibilities (1983 to 1990), replacing the selected maps in their respective pools, and repeating the process eight

times. Symbolically, a randomly selected $\hat{X}^j$ ($j = 1983, \ldots, 1990$) is paired with a randomly selected $X^i$ ($i = 1983, \ldots, 1990$) to produce one year of the 8-year bootstrapped sample. This random selection *with replacement* is what distinguishes the bootstrap technique from other randomization techniques, like permutation methods. The various statistics, measures or scores of interest are computed from this sample and saved. The whole process is then repeated enough times (10 000 should be sufficient) to produce smooth probability distributions for testing actual computed values.

The test is conducted by determining the proportion of the bootstrap sample that is (in the case of $X^2$ or a skill score) larger than the value being tested. Based on the size of this proportion a decision can be made whether or not to reject the null hypothesis.

If a null hypothesis can be reasonably rejected for whatever statistic or measure is being tested, for example the Gerrity score, then another bootstrap procedure can be used in this example to determine confidence limits for the corresponding population score. To do this a large number of bootstrap samples of eight correctly paired forecast and observation maps ($X^i$, $\hat{X}^i$) are constructed, each by eight random draws from the full pool of eight February through April paired maps. For each of these samples the quantity of interest is computed, resulting in a large enough sample for a smooth distribution from which confidence intervals can be determined.

If serial correlation is important these approaches can be appropriately modified to produce applicable distributions as long as the original sample is not too severely constrained in time. Whatever the case, unless a very large number of independent forecast-observation pairs are available or computed accuracies and skills are very large the analyst should view his/her verification results critically and not make unwarranted conclusions about forecast performance.

# 5

# Deterministic forecasts of continuous variables

**Michel Déqué**

*Météo-France CNRM, CNRS/GAME*

## 5.1 Introduction

This chapter will present methods for the verification of real continuous scalar quantities such as temperature, pressure, etc. In practice, not all values are possible for physical variables (e.g. negative precipitation, or temperatures below absolute zero), but it is simpler to consider that all real values in the range $[-\infty, \infty]$ can be reached with the restriction that the probability density is zero outside the physically achievable range. Continuous real variables are commonly produced (e.g. by partial differential equations), and categorical forecasts, which were considered in the previous two chapters, are often obtained by applying thresholds to continuous variables. Computers always represent real numbers at finite precision, and therefore produce, from a mathematical point of view, discrete representations of continuous real variables. However, for high enough machine precision (e.g. 32 bits), the continuous assumption is reasonable unless the data have been discretized by rounding to nearest integers, etc. (e.g. rainfall observations). One can also argue that user-related decision problems are often categorical; for example, a user who wants to prevent frost damage, does not care whether the predicted temperature is exactly 10°C or 11°C. This example highlights the difference between forecast verification, which does not have to impose thresholds, and is thus more general, and the assessment of forecast value, which requires a more categorical decision-based approach.

This chapter is structured as follows. Section 5.2 presents the two forecasting examples used in the rest of the chapter. Verification criteria based on first-order and second-order moments are described in Sections 5.3 and 5.4, respectively. Section 5.5 introduces scores based on the cumulative frequency distribution. Concluding remarks are given in Section 5.6.

## 5.2 Forecast examples

In this chapter, two examples are taken from the European Union project, PROVOST, that involved many seasonal forecast experiments (Palmer *et al.*, 2000). The PROVOST project used four climate models to produce ensembles of 4-month mean forecasts for the winters 1979–1993. Each

*Forecast Verification: A Practitioner's Guide in Atmospheric Science*, Second Edition. Edited by Ian T. Jolliffe and David B. Stephenson.
© 2012 John Wiley & Sons, Ltd. Published 2012 by John Wiley & Sons, Ltd.

ensemble consists of nine model forecasts starting at the best estimate of observed situations, each lagged by 24 hours. The verification of ensemble forecasts is discussed in Chapter 7. Here, we will consider the mean of all the forecasts as a single deterministic forecast. The sea-surface temperatures and initial atmospheric conditions were provided by the European Centre for Medium-Range Weather Forecast (ECMWF) reanalyses (Gibson *et al.*, 1997). The 15-year period 1979–1993 was the longest available at the time of the PROVOST project, having homogeneous three-dimensional atmospheric fields over the globe. Therefore, there are ensembles of $m = 36$ multi-model forecasts for $n = 15$ winters in our examples. The aim of the PROVOST project was to test the feasibility of seasonal forecasting using efficient ocean data assimilation and simulation. The use of *a posteriori* observed instead of *ex ante* forecast sea-surface temperature implies that the predictive skill obtained in these forecasts is a potential maximum skill rather than an achievable skill. Using four different models helps take account of structural model uncertainties in the individual models. In most cases, the ensemble mean forecast performs at least as well if not better than the best of the four individual models. During the decade that followed the PROVOST project, international reforecasting exercises have been undertaken (e.g. Weisheimer *et al.*, 2009), which attempted to predict the sea-surface temperature with coupled ocean-atmosphere models. The predictive skill is of course less than in PROVOST, and therefore very low. Given the illustrative purpose of this chapter, data from PROVOST will be used, except in Section 5.4.8, where we need a wide multi-model framework.

The first example consists of winter mean (January–March, forecasts with lead times of 2–4 months) 850 hPa temperature averaged over France. Forecast and observed values are shown in Figure 5.1. ECMWF reanalyses are used as best estimates of the observed truth. The reason for ignoring the first month of the forecast (December in this case) comes from the need to evaluate the potential of the general circulation model (GCM) to respond to the sea-surface temperature forcing rather than to initial conditions. Including the first month artificially inflates the scores, simply because

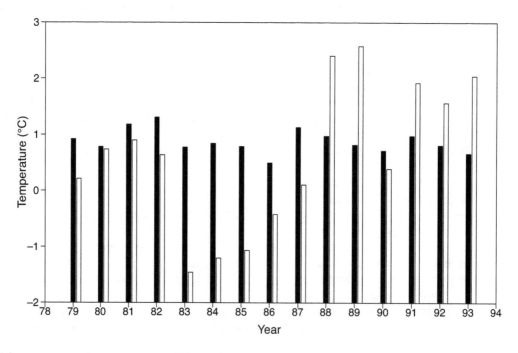

**Figure 5.1**  Mean winter temperature forecasts (filled bars) and mean observations (unfilled bars) over France at 850 hPa (close to 1500 m altitude) from 1979/80 to 1993/94

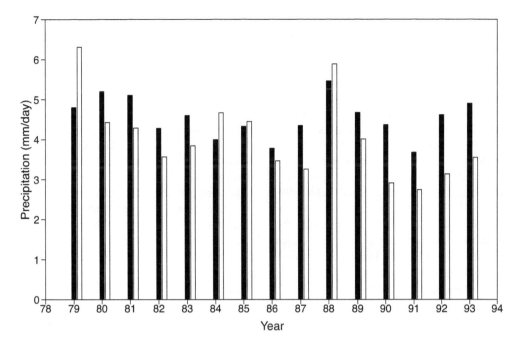

**Figure 5.2** Mean summer precipitation forecasts (filled bars) and mean observations (unfilled bars) over the tropical Atlantic (French West Indies) from 1979 to 1993

the first 10 days can be predicted with good skill (persistence).

The second example concerns forecasts of summer (July–September mean) precipitation over the tropical Atlantic region (French West Indies). Forecast and observed values are shown in Figure 5.2. Observed values were based on merged gauge and satellite precipitation estimates (Xie and Arkin, 1996). In PROVOST, only three models each with ensembles of nine forecasts ($m = 27$) were used for these forecasts.

As can be seen from the figures, the forecast values do not match closely the observed values. This is quite normal, given the long lead time of these seasonal forecasts. The fact that the skill is low makes the use of verification criteria very important. Such criteria help answer questions about whether the scores are superior to those of trivial forecast methods, whether any differences in skill are statistically significant, and whether it might be possible to improve the forecasts by post-processing. If the forecasts matched closely the observations (e.g. as is the case for 12-hour-ahead short-range weather forecasts), quantitative verification criteria would

be of more use to forecasters (e.g. for inter-model 'beauty contests' and evaluation of improvements), but of less use to the forecast users who have to make decisions based on the forecasts.

## 5.3 First-order moments

### 5.3.1 Bias

The forecasts and the observations in Figures 5.1 and 5.2 do not have the same mean levels. In the case of longer-lead prediction with numerical general circulation models, contrary to unbiased statistical predictions, there is little chance that a model mean climatology exactly equals that of the observed climatology. When averaging several models, there is some cancellation of the *mean bias* $E(\hat{X} - X)$. However, as shown by multi-model comparison experiments, models often contain related systematic errors and so multi-model averages can never completely remove all the bias in the mean.

To avoid causing problems for forecast users (who often take the forecasts at face value), it is important to correct the bias before delivering a forecast. This can be done simply by removing the mean bias estimate over a set of previous forecasts. Since the mean bias is only an estimate based on a finite sample of past forecasts, it contains sampling noise, which may degrade the forecast. If a model has a small bias and only a few past forecasts are available, it is better not to correct the forecast (Déqué, 1991). Instead of this *a posteriori* bias correction, one can also add an incremental empirical correction inside the forecasting equations (*a priori* correction: Johansson and Saha, 1989; Kaas *et al.*, 1999). This can be physically more satisfactory since the predicted evolution then remains close to the evolution of the actual climate over a wide range of forecast lead times. However, this is more difficult to set up, and statistically less efficient and transparent than post-processing the forecasts.

In the case of short-range forecasts, bias correction is rarely applied, since biases tend to be small, and bias correction would necessitate that any change in the forecasting system would require the 'reforecasting' of a set of past situations. In the case of longer-lead forecasts, bias correction is essential for correcting model drift in the forecasts.

Another way to consider the mean systematic error is to deal with *anomalies* instead of raw fields. An anomaly, $x' = x - \bar{x}$, is a deviation from the normal value represented by the long-term climatological mean (i.e. a variation centred about the mean). The climatological mean is estimated from past data and should exclude information from the forecast that is being bias corrected. When a hindcast project like PROVOST is performed, the climatological mean is calculated from the 15-year data set, excluding the target year so that only 14 years are averaged for each year (a cross-validation approach – see Efron and Tibshirani, 1993). Forecast anomalies are calculated by subtracting the mean of the other forecast values, and the observation anomaly is calculated by subtracting the means of the observations for the same periods.

The mean systematic error or bias, $E(\hat{X} - X)$, can be estimated from the sample statistic

$$b = \frac{1}{n} \sum_{i=1}^{n} (\hat{x}_i - x_i) \qquad (5.1)$$

where $\hat{x}_i$ is the forecast value and $x_i$ is the observed value at time $i$ (sometimes referred to colloquially as the *verification*). For example, the bias is $0.26°C$ for our temperature forecasts – in other words, the forecasts are too warm on average. Note that calculating the bias for each year with $n - 1$ years, i.e. excluding the year in question, and then averaging the $n$ resulting biases is equivalent to calculating the bias over all $n$ years. Our temperature bias is small compared to the interannual standard deviation in observed temperatures of $1.35°C$ – the bias only represents about 20% of the standard deviation in temperature. For the precipitation forecasts, the bias is $0.51\,mm/day$ and so the forecasts are generally too wet on average. Compared to the mean precipitation amount of $4.03\,mm/day$, the bias in the forecasts of $0.51\,mm/day$ is small relative to the mean climatology. However, the bias is equal to 50% of the observed interannual standard deviation of $1.06\,mm/day$ and so is a substantial bias in the forecasts, whose main objective is to explain the temporal variations not the long-term mean. When the bias is small, its statistical significance can be tested with a two-sample paired $t$-test (Wilks, 2006b).

### 5.3.2   Mean Absolute Error

The mean systematic error is an inadequate measure of skill since negative errors can compensate for positive errors. The simple bias correction of using anomalies described above cancels out all the mean error. However, the corrected forecast can still be far from perfect. A simple way to avoid the compensation of positive and negative forecast errors is to consider the *Mean Absolute Error* (MAE), defined as the mean of the absolute values of the individual forecast errors, $E(|\hat{X} - X|)$. This can be estimated using the sample statistic

$$\text{MAE} = \frac{1}{n} \sum_{i=1}^{n} |\hat{x}_i - x_i| \qquad (5.2)$$

Although MAE is computationally less expensive to compute (no multiplication) and more resistant to outlier errors, the mean squared error (MSE, see Section 5.4.1) is more often used in practice. For our forecasting examples, the MAE is 1.09°C for temperature and 0.87 mm/day for precipitation. This error includes systematic terms that are due to the model being too warm or too moist.

### 5.3.3  Bias correction and artificial skill

Subtracting the estimated model bias is a simple method of forecast recalibration that can be used *a posteriori* to reduce forecast errors. When bias correcting past forecasts in this way, care should be taken to exclude each target year when estimating the mean biases otherwise misleading overestimates of skill can easily be obtained (*artificial skill*). For example, when the overall mean bias of 0.26°C bias is removed from all the temperature forecasts, a new smaller MAE of 1.04°C is obtained, whereas when the bias is calculated and subtracted separately for each year a larger (yet more realistic) MAE of 1.11°C is obtained. This example also shows the detrimental effect of bias correction for small samples of forecasts – the cross-validation bias correction slightly increases the MAE compared to that of the original forecasts. For the example of precipitation forecasts, the MAE is 0.71 mm/day after cross-validation bias correction, and so the correction helps improve the mean score. Note, however, that whereas MSE is minimized by subtracting the *mean error* from the forecasts, the MAE is minimized by removing the *median error*, i.e. the median of the $\hat{x}_i - x_i$ errors. Therefore, when we correct the systematic error by removing the median systematic error instead of the mean systematic error, smaller values for MAE are obtained: 1.05°C for temperature and 0.62 mm/day for precipitation. As precipitation is less well approximated by a normal (Gaussian) distribution than temperature, using the median rather than the mean for the bias correction is more effective. When the forecast events are not independent (e.g. consecutive daily values), a cross-validation estimation of bias can be obtained by excluding a sliding mean centred on the target day with a sufficient width that the values at the edges of the windows are not strongly dependent on the ones at its centre (a typical width is 1 week in meteorological applications).

For the verification of seasonal climate forecasts over a long period (e.g. 50 years), the independence

assumption may also be violated due to long-term trends caused by climate change (e.g. global mean temperature at the end of the twentieth century is generally warmer than at the beginning). In this case, one should remove the long-term trend from both the observations and the forecasts before calculating any score. This procedure will avoid artificial skill caused by any concurrent long-term trends in the forecasts and the observations. Skill due to long-term trends is not useful forecast skill for a user who is interested in knowing in advance year-to-year fluctuations.

### 5.3.4  Mean absolute error and skill

An important thing to ascertain is whether bias-corrected MAE values of 1.05°C and 0.62 mm/day actually signify any real skill. To do this, these scores must be compared with scores of an alternative low-skill forecasting method. The two most commonly used reference methods are *persistence forecasts* and *mean climatology forecasts*. Both these forecasts require no model development (but do require a database of past observations). Persistence can be biased by persistence due to the annual cycle, and so bias correction is mandatory. In our example, we will take the monthly mean for the month just before issuing the forecast, i.e. November for temperature, and May for precipitation. The MAE of the persistence forecasts is larger than the model MAE: 1.50°C for temperature and 1.12 mm/day for precipitation. However, the comparison is not really fair, since the persistence forecast has a larger interannual variability than the multi-model ensemble mean forecast since it is based on a single monthly mean rather than the mean of 36 months. If we apply a linear regression of the persistence on the verification to take account of the extra variance in the persistence forecast, the MAE then becomes 0.98°C for temperature and 0.85 mm/day for precipitation – hence, the multi-model ensemble mean forecast for temperature has a larger MAE than the persistence forecasts.

In long-range forecasting, climatological mean forecasts provide more skilful reference forecasts. The climatological mean forecast is obtained by constantly forecasting the same sample mean of the past observations for every event. For our examples, the climatological mean forecasts give a MAE of 1.17°C for temperature and 0.91 mm/day for precipitation. Hence, the multi-model ensemble forecasts have a smaller MAE than climatological mean forecasts for both temperature and precipitation. The question whether the multi-model performs significantly better or worse must involve a confidence interval about the empirical scores. This point will be discussed in Section 5.4.2.

## 5.4 Second- and higher-order moments

### 5.4.1 Mean Squared Error

Mean Squared Error, $E[(\hat{X} - X)^2]$, is one of the most widely used forecast scores. It is estimated by the sample statistic

$$\text{MSE} = \frac{1}{n} \sum_{i=1}^{n} (\hat{x}_i - x_i)^2 \qquad (5.3)$$

The square root of this quantity, the RMSE, has the same units as the forecast variable. Because of the square in Equation 5.3, the MSE is more sensitive to large forecast errors than is MAE. For example, a single error of 2°C contributes exactly the same amount to MAE as is contributed by two errors of 1°C, whereas for MSE the larger error would contribute four times more to the score. From a user's point of view, it may seem natural to penalize larger errors and to be more indulgent towards small errors. However, the MSE is unduly sensitive to outlier errors in the sample (caused by data corruption, atypical events, etc.) rather than being representative of the forecasts as a whole. MAE is more *resistant* to outliers than is MSE. The scores, MAE and MSE, are specific $p = 1$ and $p = 2$ cases of the more general $L_p$ Minkowski norm $(E[(\hat{X} - X)^p])^{1/p}$ that can be estimated from:

$$\left( \frac{1}{n} \sum_{i=1}^{n} (\hat{x}_i - x_i)^p \right)^{1/p} \qquad (5.4)$$

As $p$ tends to infinity, this norm tends to the maximum error in the sample. The $L_\infty$ norm provides an

upper error bound for a user, but this will increase indefinitely as the sample size of past forecasts increases with time. Unfortunately this score depends on only one value in the sample (the one with largest error), and so for a system with (say) a thousand perfect forecasts and only one erroneous forecast, the score will be entirely determined by the error of the single erroneous forecast.

For our forecasts, the temperature RMSE is 1.27°C and the precipitation RMSE is 0.97 mm/day. As mentioned in Section 5.3.3, the MSE is minimized when the mean error is subtracted from the forecast, or, equivalently, the forecast and observation data are centred about their respective sample means. This simple algebraic result assumes that the mean error is calculated with all cases, including the forecast that is evaluated. When the mean error is calculated without the forecast case (cross-validation method) there is no guarantee to improve the MSE, in particular for small samples (Déqué, 1991). Indeed, with our examples, the RMSE after *a posteriori* correction is 1.33°C for temperature and 0.88 mm/day for precipitation – a similar behaviour to that obtained with the MAE. Therefore, in this case, it is wise to bias correct precipitation, but bias correcting temperature requires more than 15 years of previous forecasts and observations to be effective.

In order to assess predictability, one must compare the RMSE with that obtained using low-skill alternative forecasting methods. The persistence of the anomaly of the latest monthly mean available at the time of the forecast yields an RMSE of 1.86°C for temperature and 1.32 mm/day for precipitation. The climatological mean forecast (the forecast that minimizes the MSE among all the forecasts that are statistically independent of the observation) gives a RMSE of 1.35°C for temperature and 1.06 mm/day for rainfall. As was also the case for MAE, the model outperforms both the climatological and persistence forecasts.

### 5.4.2 MSE skill score

To judge the skill of an RMSE score one must compare it to the RMSE of a low-skill forecast, for example climatological RMSE. Murphy (1988) proposed the *MSE skill score* (MSESS) defined by

$$MSESS = 1 - \frac{MSE}{MSE_{clim}} \quad (5.5)$$

where $MSE_{clim} = E[(E(X) - X)^2] = Var(X)$ is the MSE for climatological mean forecasts $\hat{X} = E(X)$. This skill score is considered as the main deterministic verification score for long-range forecasts in World Meteorological Organization (2007). The basic idea of skill scores is discussed in Section 2.7 of Chapter 2. The maximum value for MSESS is 1 and indicates a perfect forecast; a value of 0 indicates a model forecast skill equal to that of a climatological forecast. A negative value implies that the model forecast skill is worse than climatology, in terms of MSE. When the MSESS is aggregated over time or space, it is wiser to accumulate separately the numerator and the denominator, otherwise a single verification very close to the climatology will contribute a very large value to the final average of Equation 5.5 rather than average the local values of MSESS. Unlike MAE and MSE, MSESS is dimensionless, and increases with forecast skill. In our two examples, the MSESS is 0.12 for temperature and 0.16 for precipitation (before bias correction). After bias correction, the precipitation forecasts have an increased MSESS of 0.31, whereas temperature MSESS decreases to 0.03.

Statistical hypothesis testing can be used to test whether one forecasting system has significantly different true skill than that of another forecasting system. Parametric tests can be used but they are not always very appropriate because of underlying restrictive assumptions. Non-parametric tests offer more flexibility for checking whether the difference between scores for two forecasting systems is statistically significant. A simple way to do this for independent forecasts is by Monte Carlo permutation techniques: the 15 forecasts are reassigned at random among the 15 years. For 15 years, there are about $10^{12}$ permutations, but 1000 random drawings are generally sufficient to estimate a 95% prediction interval for a score based on a no-skill null hypothesis. We use the word prediction interval rather than confidence interval since sample scores are random sample statistics not probability model parameters.

With the above permutation procedure, the 95% prediction interval for the RMSE of the temperature forecasts was found to be [1.25,1.48], which yields a 95% prediction interval on the MSESS of [−0.20,0.14]. The skill score prediction interval includes MSESS = 0.12, and so at the 5% level of significance the null hypothesis that the forecasts have zero skill cannot be rejected. For the precipitation forecasts, the skill score is significantly different from zero at the 5% level of significance, since the MSESS interval is [−0.77,0.28] and so does not include the MSESS = 0.31 obtained for the bias-corrected precipitation forecasts. The interval for the MSESS is not necessarily symmetrical for a random forecast, since MSESS can only take values from minus infinity to 1.

### 5.4.3 MSE of scaled forecasts

Removing the mean bias is one possible way of improving the MSE. Rescaling the forecast anomaly is another simple calibration technique for improving forecasts. In general, it is possible to correct biases in the mean *and* the variance by performing a linear regression

$$\hat{X}' = E(X|\hat{X}) = \beta_0 + \beta_1 \hat{X} \quad (5.6)$$

of the observations on the forecasts (see Section 2.9). A simple geometric interpretation of this operation will be given in Section 5.4.4. For hindcasts, the regression coefficients should be estimated using data that exclude the target year. For our examples, the MSESS of the forecasts recalibrated in this manner are −0.04 for temperature and 0.22 for precipitation. These MSESS values are less than those of the original uncalibrated forecasts, which indicates that in this case the sample size of 15 years is too short to obtain reliable estimates of the regression coefficients. A similar phenomenon was noted in Section 5.4.1 when correcting the mean bias in the temperature forecasts. For the persistence forecasts, the MSESS or the regression-recalibrated forecasts is −0.22 and −0.20 for temperature and precipitation, respectively. If we apply the permutation procedure to the regression recalibrated forecasts, the 95% prediction intervals for the MSESS become [−0.33,0.27] for temperature and [−0.39,0.24] for precipitation. Hence, recalibration in this particular case has unfortunately not produced forecasts with any significant skill at the 5% level. One should therefore be very careful when attempting statistical post-processing of forecasts based on small samples of past forecasts. Post-processing is currently used with success in short- and medium-range forecasting. However, in seasonal forecasting, small sample sizes present a strong limiting factor, and even the bias correction of the mean must be applied with caution.

### 5.4.4   Correlation

Because of its invariance properties, the correlation coefficient is a widely used measure in weather and climate forecasting. The product moment correlation coefficient is defined as

$$\rho = \operatorname{cor}(X, \hat{X}) = \frac{\operatorname{cov}(X, \hat{X})}{\sqrt{\operatorname{var}(X), \operatorname{var}(\hat{X})}} \qquad (5.7)$$

where $\operatorname{cov}(X, \hat{X})$ is the covariance between the observations and forecasts and $\operatorname{var}(X) = \operatorname{cov}(X, X)$ and $\operatorname{var}(\hat{X}) = \operatorname{cov}(\hat{X}, \hat{X})$ are the variances of the observations and forecasts, respectively. Strictly speaking, correlation is not a score measure, as it cannot be written as a sum of contributions from individual forecast times. The covariance can be estimated from the sample of past forecasts and observations; for example, $\operatorname{cov}(X, \hat{X})$ is estimated by the sample statistic

$$\frac{1}{n} \sum_{i=1}^{n} (x_i - \bar{x}_i)(\hat{x}_i - \bar{\hat{x}}_i) \qquad (5.8)$$

Correlation is a dimensionless and positively oriented verification measure. Correlation is invariant to shifts in the mean and multiplicative rescaling of either the forecasts or observations, hence mean bias correction and linear regression post-processing of the forecasts do not change the correlation of the forecasts with the observations. A practical benefit is that there is no need to first remove the bias or correct the amplitude of the forecast anomaly. In fact, one can calculate anomalies with respect to *any* mean climatology. Because correlation is invariant to changes in scale, one can correlate observations measured in Celsius directly with forecasts in Fahrenheit (i.e. conversion of units will not change the result). A correlation of $+1$ or $-1$ implies a perfect linear association between the forecast and observations, whereas a correlation of zero signifies that there is no linear association between the variables. It should be noted that two *uncorrelated* variables are not necessarily *independent* since they can be non-linearly rather than linearly related to one another.

Correlation and MSESS are linked by the relation (Murphy, 1988):

$$\text{MSESS} = 2\rho\sqrt{\frac{\operatorname{var}(\hat{X})}{\operatorname{var}(X)}} - \frac{\operatorname{var}(\hat{X})}{\operatorname{var}(X)} - \frac{(E[\hat{X} - X])^2}{\operatorname{var}(X)}$$

$$(5.9)$$

This equation shows that MSESS increases with increasing correlation, $\rho$, and decreases with increasing mean bias. In particular, when the mean bias is small, and when the variances of forecasts and verifications are similar, a correlation coefficient greater than about 0.5 is necessary to ensure a positive MSESS.

Rescaling the forecasts by Equation 5.6 is equivalent to subtracting the mean bias and multiplying by the correlation coefficient. The maximum MSESS is simply the squared correlation coefficient. When $n$ is small, as in our examples, the cross-validated maximum sample MSESS is smaller than the true value in Equation 5.9 and may even be negative.

From a geometrical point of view, the correlation coefficient can be seen as the cosine of an angle in sample space. Figure 5.3 shows three points, representing the origin (O), which corresponds to a climatological mean forecast with coordinates $\bar{x}$ (repeated $n$ times), the bias corrected forecast (F) with coordinates $\hat{x}_i - \bar{\hat{x}} + \bar{x}$, and the verification observation (V) with coordinates $x_i$, in a projection of the $n$-dimensional sample space. The correlation is the cosine of the angle (FOV) denoted $\theta$ in the figure, the RMSE is the distance VF, and the climatological RMSE is the distance OV. The minimization of RMSE by scaling the forecast anomaly consists of replacing F by F'. Figure 5.3 shows that the squared correlation and the maximum MSESS are identical. For this reason, one can say that correlation measures the potential skill of unbiased forecasts rather than the actual MSE skill that includes

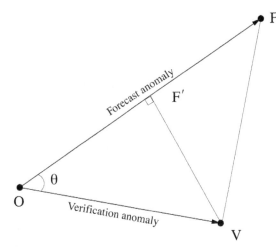

**Figure 5.3** Scheme representing the relative position of origin (0), forecast anomaly (F) and verification anomaly (V). F′ is the best rescaled forecast

**Table 5.1** Summary of main scores for seasonal forecasts of temperature and precipitation. Values in parentheses correspond to bias-corrected forecasts

|  | Temperature forecasts | Precipitation forecasts |
|---|---|---|
| Bias | 0.26°C | 0.51 mm/day |
| MAE | 1.09 (1.05)°C | 0.87 (0.62) mm/day |
| RMSE | 1.27 (1.33)°C | 0.97 (0.88) mm/day |
| MSESS | 0.12 (0.03) | 0.16 (0.31) |
| Correlation | 0.16 | 0.55 |

MAE, mean absolute error; RMSE, root mean squared error; MSESS, mean squared error skill score.

contributions from bias in the mean (Murphy, 1988). We shall come back to this notion of potential skill in Section 5.4.8.

For our examples, the correlation is 0.16 for temperature and 0.55 for precipitation in the case of the model forecasts, and 0.16 and 0.46 respectively for persistence forecasts. Because correlation is restricted to be less than 1, improvements in large correlations are generally smaller than those in correlations closer to zero. For this reason, and another one discussed below, it is better to consider a nonlinear transformation of correlation known as the Fisher $z$-transform:

$$z = \frac{1}{2} \ln \left( \frac{1+r}{1-r} \right) \qquad (5.10)$$

This $z$-score can take any value in the range $-\infty$ to $\infty$, and is better approximated by a normal distribution than is correlation. The statistical significance of the difference between two correlations can be most easily tested using a two-sample $Z$-test with $Z = (z_1 - z_2)/s$ and the asymptotic (large sample) standard error

$$s = \frac{\sqrt{2}}{\sqrt{n-3}}$$

One can also test an empirical $z$-score, $\hat{z}$, against a theoretical one, $z_0$. For the model forecasts, the $z$-scores are 0.16 and 0.62 for temperature and precipitation, respectively. Under the null hypothesis that $z_0 = 0$, the $z$-score divided by $1/\sqrt{15-3} = 0.29$ is asymptotically Gaussian. Hence only the precipitation forecasts have correlations (marginally) significantly different from zero at the 5% level of significance (i.e. a $z$-score more than 1.96 standard errors away from zero).

Table 5.1 summarizes the different scores obtained for temperature and precipitation for the main criteria. Like MSESS, but unlike MAE or RMSE, the correlation coefficient is an increasing function of the skill. The correlation coefficient of the climatological forecast is not uniquely defined, since it involves a ratio of zero by zero. However, by symmetry, one can consider it to be zero, and so a positive correlation is then interpreted as a forecast better than the climatology, provided that its sign is statistically significant. The crucial question is what is the threshold above which a correlation can be considered as statistically significant, since a sample estimate of a correlation coefficient is never exactly zero. In addition, a small statistically significant correlation can lead to no skill at all if the regression of Equation 5.6 is not robust.

The real skill of forecasts can be assessed by statistically testing whether or not the correlation is significantly different from zero. Both parametric and non-parametric approaches can be used. A simple parametric approach is to try to reject the no-skill null hypothesis that the forecasts and observations are independent and normally distributed. Under this null hypothesis, the sample statistic

$$t = \frac{r\sqrt{n-2}}{\sqrt{1-r^2}} \qquad (5.11)$$

is distributed as a Student $t$-distribution with $n-2$ degrees of freedom. Inverting this expression for $n = 15$ forecasts gives a 95% prediction interval for a no-skill correlation of $[-0.57, 0.57]$. Both our temperature and precipitation forecasts have correlations that lie within this interval and so at the 5% level of significance are *not inconsistent with being no-skill forecasts*. Note that we can only say that the previous data are *not inconsistent* with the null hypothesis not that the null hypothesis *is* true (i.e. *the forecasts have no-skill*) – more data in the future may allow us to reject the null hypothesis and then say that the forecasts have skill. One can see that a no-skill hypothesis for precipitation is rejected with the $Z$-test and not rejected with the $t$-test. To avoid the need for the normality assumption, we can use a non-parametric (distribution-free) permutation procedure similar to that described earlier for the MAE and MSE scores. For our forecasts, the permutation method gives a slightly narrower 95% prediction interval for correlation of $[-0.51, 0.51]$. Although it may appear sensible to exclude permutations in which one or more years is unchanged, such censoring is a bad idea since it biases the interval estimate towards negative values. For example, censoring of permutations gives intervals of $[-0.53, 0.50]$ and $[-0.53, 0.47]$ for the temperature and precipitation forecasts, respectively. Based on the permutation interval, it can be concluded that our precipitation forecasts (but not the temperature forecasts) have marginally significant skill at the 5% level of significance, in agreement with the results obtained for the MAE and the RMSE scores.

Another important issue for correlation is how best to aggregate several correlations obtained, for example, by pooling skill over a geographical region. Because of its additive properties, it is well known that one should average the MSE score for all the cases before taking the square root, rather than averaging all the RMSE scores. Nevertheless, it is not unusual in meteorological studies to calculate time averages of spatial correlations or spatial averages of time correlations. Alternatively, some studies average the Fisher $z$-scores, which then give more weight to correlations further from zero (e.g. Miller and Roads, 1990). A better, less biased approach is to calculate the components of the $2 \times 2$ (co-)variance matrix between forecasts and observations by averaging over all cases, and then calculate the overall correlation coefficient using these four values (Déqué and Royer, 1992; Déqué, 1997). This aggregated covariance approach has several advantages. Firstly, the aggregation of the covariances over space and time can be done in either order and so it treats spatio-temporal aggregation in a symmetrical manner, which is not the case when taking the spatial average of correlations over time or time averages of correlations over space (see discussion of anomaly correlation coefficient (ACC) in Section 6.3). Secondly, the resulting correlation corresponds to a weighted average of the individual correlation coefficients, with larger weights being given to forecast and observation anomalies with larger magnitudes. This makes sense from the user's perspective since large magnitude anomalies are the ones that have the most impact. Thirdly, the geometrical interpretation remains applicable and the squared overall correlation is equal to the MSESS for unbiased forecasts.

### 5.4.5    An example: testing the 'limit of predictability'

To illustrate how correlation can be used to test the skill of forecasts, we will demonstrate its use in this section on daily temperature forecasts from the PROVOST experiment. Figure 5.4 shows the correlation between daily temperature forecasts and observations as a function of the lead time – days 1–30 correspond to the days in December, the first month of forecasts. With only a small sample of 15 winters, the correlation curve does not exhibit perfectly smooth monotonic decay due to sampling fluctuations. The correlations lie within the 95% no-skill prediction interval of $[-0.57, 0.57]$ (shaded area) for lead times longer than 6 days. The limit of predictability for daily temperature forecasts (crudely) estimated from these 15 years is therefore 6 days. Beyond this timescale, there is no significant skill for daily temperature values but there can be skill

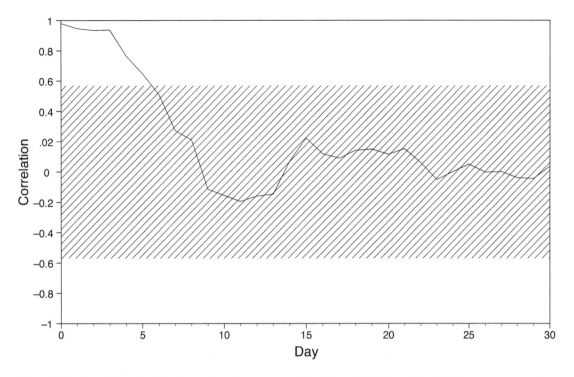

**Figure 5.4**   Correlation of daily temperature forecasts as a function of lead time; the shaded area corresponds to the 95% no-skill prediction interval

in forecasting statistical quantities such as monthly and seasonal mean temperatures.

### 5.4.6   Rank correlations

The product moment correlation discussed above measures the strength of the *linear association* between forecasts and observations. Linear and non-linear *monotonic association* between forecasts and observations can be measured using statistics based on the ranks of the data (i.e. the position of the values when arranged in increasing order). For example, the Spearman rank correlation coefficient is simply the product moment correlation coefficient of the ranks of the data (see Wilks, 2006b).

A rank correlation of 1 means that the forecast values are an increasing function of the observations – i.e. there is a perfect monotonic association between the forecasts and the observations. The rank correlation coefficient is also more resistant (less sensitive) to large outlier values than is the product moment correlation coefficient. Since

the sample mean of the ranks is $(n + 1)/2$ and the sample variance of the ranks is $(n^2 - 1)/12$, the Spearman rank correlation coefficient can be expressed as

$$r_s = \frac{12}{n(n^2 - 1)} \sum_{i=1}^{n} \hat{R}_i R_i - \frac{3(n + 1)}{n - 1} \quad (5.12)$$

where $\hat{R}_i$ and $R_i$ are the ranks of the $i$th forecast and observation, respectively. In the asymptotic large sample limit for normally distributed data, the product moment correlation coefficient and the rank correlation coefficient can be shown (Saporta, 1990) to be related by

$$r \approx 2\sin\left(\frac{r_s \pi}{6}\right) \quad (5.13)$$

For our examples, the rank correlation is 0.20 for temperature and 0.46 for precipitation. The fact that these values are close to the product moment correlation coefficients (0.16 and 0.55) suggests that the association between forecasts and observations has a dominant linear component. Tests based on the rank correlation are distribution-free (robust) since they make no assumption about the marginal distributions of the data. For large samples of independent forecasts and observations (the asymptotic no-skill null hypothesis), the rank correlation coefficient is normally distributed with a mean of zero and a variance of $1/(n-1)$. For a sample of $n = 15$ forecasts this result gives a 95% prediction interval of $[-0.53, 0.53]$ for the rank correlation coefficient (the exact calculation yields $[-0.52, 0.52]$). Both temperature and precipitation forecasts have a rank correlation inside this interval and so do not have significant skill at 95% confidence.

Dependency can also be tested non-parametrically using Kendall's tau correlation statistic. For each pair of times $i$ and $j$, a new sign variable $s_{ij}$ is created that takes the value 1 when $(x_i - x_j)(\hat{x}_i - \hat{x}_j)$ is positive, i.e. when the forecast and observation evolve in the same direction, and $-1$ otherwise. The average of the sign variable can then be used to construct the correlation

$$\tau = 1 - \frac{4}{n(n-1)} \sum_{i=1}^{n} \sum_{j=1}^{i-1} s_{ij} \qquad (5.14)$$

When the forecasts and observations are independent, the distribution of $\tau$ can be tabulated without any assumptions about the distribution of the forecast and observation. It can be demonstrated that $\tau$ is asymptotically normally distributed with a mean of zero and a variance of $2(2n+5)/(9n(n-1))$. One advantage compared to Spearman's rank correlation is that the convergence to the asymptotic normal distribution is faster for Kendall's tau correlation, which is approximately normally distributed for sample sizes as small as eight. For our examples, Kendall's correlation is 0.18 and 0.40 for the temperature and precipitation forecasts, respectively. This is a little less than Spearman's correlation. The 95% no-skill asymptotic prediction interval is $[-0.38, 0.38]$, which is narrower than the interval for Spearman's correlation. Therefore, once again the precipitation forecasts over the tropical Atlantic have significant skill at the 5% level of significance whereas the temperature forecasts over Europe are not significantly skilful.

As explained in Saporta (1990), the three correlation coefficients (Pearson, Spearman and Kendall) are in fact particular cases of the so-called Daniels' correlation (Daniels, 1944):

$$r_d = \frac{\sum_{i=1}^{n} \sum_{j=1}^{n} d_{ij} \hat{d}_{ij}}{\sqrt{\sum_{i=1}^{n} \sum_{j=1}^{n} d_{ij}^2 \sum_{i=1}^{n} \sum_{j=1}^{n} \hat{d}_{ij}^2}}. \qquad (5.15)$$

where $d_{ij}$ is a distance between the objects $i$ and $j$ defined for the three correlations as follows:

- Pearson $d_{ij} = x_i - x_j$
- Spearman $d_{ij} = R_i - R_j$ where $R_i$ is the rank of $x_i$
- Kendall $d_{ij} = \text{sgn}(x_i - x_j)$

From Equation 5.15 it is obvious that Daniels' correlations are invariant under any data affine linear transformation (i.e. bias correction or rescaling) that just scale $d_{ij}$ (Pearson) or that do not modify it (Spearman or Kendall). It is interesting to note that the persistence forecast for precipitation that has a rather high, yet non-significant Pearson correlation for precipitation (0.46), has almost zero values of Spearman's (0.08) and Kendall's (0.04) correlations. This feature is explained by the scatter plot of these forecasts and observations shown in Figure 5.5. The first year with a persistence forecast of 7.7 mm/day and an observed value of 6.3 mm/day is a clear outlier from the cloud of other points, and thereby contributes excessively to the Pearson correlation but not to the more resistant Spearman and Kendall correlations. This example emphasizes the need to use resistant scores when judging small samples of forecasts.

### 5.4.7 Comparison of moments of the marginal distributions

In forecast evaluation, it is desirable that the overall marginal distribution of forecasts is similar to that of the observations, irrespective of their pairwise relationship with the observations. A high correlation can misleadingly be obtained with forecasts that have a very different statistical distribution from

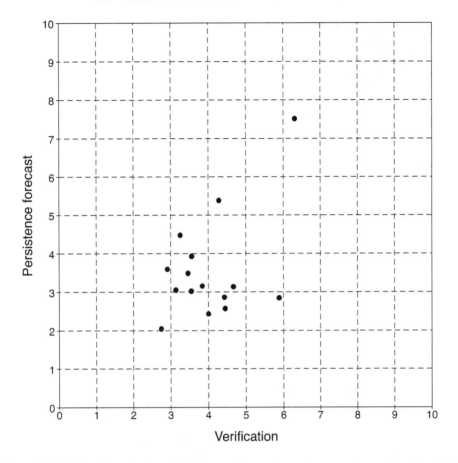

**Figure 5.5**   Scatter diagram of precipitation persistence forecasts versus observations (mm/day)

the observations. If forecasts are used at face value as input variables for an impacts model, for example of crop production, or are used as a basis for a search for analogues, the results can be misleading. Therefore, in addition to estimating association, one must also compare properties of the marginal distributions of the forecasts and the observations.

The first-order moments of the marginal distributions are the means of the forecasts and observations, and these can be compared by taking the difference to obtain the mean bias, discussed in Section 5.3.1. The centred second-order moments of the marginal distributions are the variances of the forecasts and the observations. These can be tested for equality by performing an $F$-test on the ratio of variances. However, this test is not widely used since the assumption of equal variances is hard to reject for small samples: with sample sizes of 15, the ratio

of variances must be greater than 2.5 in order to reject the null hypothesis at the 5% level of significance (the 97.5% quantile of an $F$-distribution with 15/15 degrees of freedom is 2.4). The variance of the forecasts and observations is more easily interpreted when expressed in terms of standard deviations that have the same units as the predictand. For ensemble mean forecasts, the variance of the mean forecast will invariably be less than that of the observations due to the averaging that has taken place to create the mean of the ensemble of forecasts. For example, the standard deviation of our temperature forecasts is 0.22°C compared to 1.35°C for the observations. For the precipitation forecasts, the standard deviation of the forecasts is 0.53 mm/day compared to 1.06 mm/day for the observations. Before the forecasts are delivered to unsuspecting users, it is important to rescale (inflate) them to account for this loss of variance. The scaling factor that yields the correct interannual variance (6.1 for temperature and 2.0 for precipitation) is invariably not the best choice for minimizing the MSE ($\beta_1$ in Equation 5.6, which is 0.2 for temperature and 0.4 for precipitation). However, rescaling does not change measures of association such as the correlation between the forecasts and observations.

Comparisons of higher order moments of the marginal distributions such as skewness and kurtosis can also be revealing for forecasts and observations that are not normally distributed. The asymmetry of the distributions can be estimated using the third-order moment about the mean, known as the moment measure of skewness:

$$b_1 = E[(X - E(X))^3]/(var(X))^{\frac{3}{2}} \qquad (5.16)$$

Skewness is a dimensionless measure of the asymmetry of the distribution. Positive skewness indicates that the right tail of the distribution is fatter than the left tail and that more values lie below the mean than above. Symmetric distributions such as the normal distribution have zero skewness, but zero skewness does not imply that the distribution has to be symmetric. Recalibration procedures such as removing the mean bias or rescaling do not change the skewness. More resistant measures of skewness also exist and are more reliable than the moment measure of skewness when dealing with small samples (see Wilks, 2006b). The fourth-order moment about the mean is known as kurtosis and is defined by the dimensionless non-negative quantity

$$b_2 = E[(X - E(X))^4]/(var(X))^2 \qquad (5.17)$$

Kurtosis does not measure the spread of the variable, since it is insensitive to scaling or bias correction. It measures the frequency of rare events, with respect to the standard deviation. If the tails of the probability density decrease rapidly, kurtosis is small, whereas if the tails decrease slowly (e.g. as a negative power law) then kurtosis is large. The normal distribution has a kurtosis of 3 and exhibits a rapid decrease in the tails. Skewness and kurtosis can be used to test whether a variable is normally distributed (normality test). However, both these higher moments of the distribution are very sensitive to large outlier values, and more resistant

approaches based on the cumulative distribution are preferable.

### 5.4.8  Graphical summaries

In some cases the number of forecast methods or models to be compared is such that a visual approach is more helpful than a large table of numerical scores. Taylor diagrams (Taylor, 2001) have not been introduced for forecast verification, but for climate model intercomparison. The Atmospheric Model Inter-comparison Project (Gates, 1992) produced a database of simulations from more than 20 climate models. Plotting 20 maps of mean wintertime temperature over the globe on a single picture was too congested. Displaying a table with 20 RMSE versus the observed climatology was insufficient. The aim of summary diagrams is to use a 2D diagram in which each model is represented by a point and the distance to a reference point is proportional to a metric in the phase space. In its original version, the metric was the spatial mean square difference between two maps. However, the property of this diagram (a consequence of basic Euclidean geometry) is also valid if we use a metric based on time mean square differences.

Let us introduce the unbiased mean square error

$$var(\hat{X}) + var(X) - 2\rho\sqrt{var(\hat{X})var(X)}$$

which can be estimated as a modification of Equation 5.3 by:

$$MSE' = \frac{1}{n}\sum_{i=1}^{n}(\hat{x}_i - \bar{\hat{x}}_i - x_i + \bar{x}_i)^2 \qquad (5.18)$$

The correlation $\rho$ is defined by Equation 5.7. The diagram is constructed as follows.

From the origin O, along the $x$-axis, a point V with abscissa $OV = \sqrt{var(X)}$ represents the verifications. From O, along a line having an angle $\theta$ so that $\rho = \cos(\theta)$, a point F so that $OF = \sqrt{var(\hat{X})}$ represents the forecasts. If F is close to the $x$-axis, the correlation is high. The larger the (OV,OF) angle, the smaller the correlation. Consider the circle with centre O and radius OV. If F is inside the circle, the forecast system underestimates the variability.

If F is outside, the forecast system overestimates the variability.

$$VF^2 = OV^2 + OF^2 - 2\cos\theta\, OV, OF \qquad (5.19)$$

Hence VF corresponds to RMSE'. This representation is similar to Figure 5.3, except that we are in a 2D space (not in a particular projection from a 15D space), and therefore we can introduce as many points F as we want.

Let us illustrate this with results from a recent multi-model intercomparison hindcast experiment named ENSEMBLES (Weisheimer *et al.*, 2009). Here we have five coupled ocean-atmosphere models, and we are interested in the best predicted feature in the seasonal range, the December-January-February (DJF) surface temperature in the equatorial Pacific Ocean. Figure 5.6 shows that the five models (F1, F2, F3, F4, F5) have a similar correlation (close to 0.9), but F2 underpredicts the year-to-year variability of the predictand, whereas the other four overpredict it. In terms of MSE', F2 and F3 are equivalent (i.e. they are equidistant to V). If we consider F1, F4 and F5, they have a similar ratio of variance, but F5 has the worst correlation. In the diagram, the two dotted quarter-circles correspond to ratios of 0.5 and 1.5, and the dotted diagonal corresponds to a correlation of 0.7 ($\theta = 45°$); these dotted lines are just aids for numerical evaluation.

The Taylor diagrams do not display the model bias. Another summary diagram, the target diagram, has been introduced by Jolliff *et al.* (2009) to remedy this. Here the origin and the verification are the same point O. A forecast system F is represented by a point with RMSE' as the abscissa and the bias as the ordinate. To complete the information, the sign of the abscissa is changed when the variance of the forecasts is less than the variance of the verification.

Forecast systems in the upper part (lower) of the diagram have a positive (negative) bias. Forecast systems in the right-hand (left-hand) part of the diagram overpredict (underpredict) the variance. The distance OF is, thanks to Pythagoras, the RMSE. One can notice that here the correlation has no geometrical equivalent, so this diagram is more adapted to model evaluation than to forecast evaluation.

## 5.5 Scores based on cumulative frequency

### 5.5.1 Linear Error in Probability Space (LEPS)

The *Linear Error in Probability Space* (LEPS) score is defined as the mean absolute difference between the cumulative frequency of the forecast and the cumulative frequency of the observation:

$$LEPS_0 = \frac{1}{n}\sum_{i=1}^{n}|F_X(\hat{x}_i) - F_X(x_i)| \qquad (5.20)$$

where $F_X$ is the (empirical) cumulative distribution function of the observations. The basic idea is that an error of 1°C in temperature in the tail of the distribution is less important than the same error nearer the mean where it corresponds to a greater discrepancy in cumulative probability. For uniformly distributed forecasts and observations in the range [0,1], the $LEPS_0$ score becomes equivalent to the MAE score. $LEPS_0$ is the probability of obtaining a value between the forecast and the observation. $LEPS_0$ is a negatively oriented score

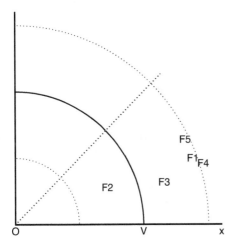

**Figure 5.6** Taylor diagram for five models' (F1, F2, F3, F4 and F5) forecasts of December-January-February (DJF) surface temperature in the equatorial Pacific Ocean, from the ENSEMBLES project. The distance to V from F1 indicates the unbiased root mean squared error (RMSE), and the angle to the x-axis indicates the correlation (the closer to this axis, the better). The solid quarter-circle indicates a right ratio of variances for V and F1

with values in the range [0,1]. It is only zero when the forecasts are perfect. If we assume the forecasts are completely independent of the observations, and replace in Equation 5.20 the average by an expectation with respect to $X$ and $\hat{X}$, simple algebra yields the no-skill expected $\text{LEPS}_0$

$$\text{LEPS}_0 = \frac{1}{2} - E_{\hat{X}}[F_X(\hat{X})(1 - F_X(\hat{X}))] \quad (5.21)$$

which is minimized by repeatedly forecasting the median of the observations $\hat{x} = F_X^{-1}(0.5)$ as was the case for the MAE score. $\text{LEPS}_0$ can be used for continuous as well as discrete variables (see Section 4.3.3). Potts et al. (1996) introduced a positively oriented score based on $\text{LEPS}_0$:

$$\text{LEPS} = 2 - 3(\text{LEPS}_0 + \overline{F_X(\hat{x})(1 - F_X(\hat{x}))} \\ + \overline{F_X(x)(1 - F_X(x))}) \quad (5.22)$$

where the overbar indicates the mean over a previous sample of forecasts and matched observations.

Unlike $\text{LEPS}_0$, this score is equitable (Gandin and Murphy, 1992) – when the forecast is independent of the verification (random or constant forecast), the expected score is zero. Correct forecasts of extreme events contribute more to the score than the other correct forecasts. For our example forecasts, due to the small sample size, the mean LEPS score of climatological mean forecasts is 0.04 and is 1.07 for perfect forecasts. $\text{LEPS}_0$ is 0.24 and 0.28 for the temperature and precipitation forecasts, respectively, while LEPS is 0.12 and 0.17. The significance of the LEPS scores can be tested by a permutation procedure and all the above values are inside the 95% no-skill prediction interval. As was noted for the MAE score, the LEPS score is sensitive to the model bias and is generally the best when the median of the forecasts is close to the median of the observations. With such a bias correction, LEPS decreases for temperature (0.01) and increases for precipitation (0.21) in agreement with similar behaviour noted for the RMSE score.

### 5.5.2   Quantile-quantile plots

The *quantile-quantile plot* (q-q plot, or qqplot) is a visual way for comparing the marginal cumula-

tive probability distribution of a sample with either that of a theoretical distribution or with that of another sample of data (see Wilks, 2006b). For comparison with a theoretical distribution (e.g. the normal distribution), the method consists of plotting the empirical quantiles of a sample against the corresponding quantiles of the theoretical distribution – i.e. the points $(x_{[i]}, F^{-1}(i/(n+1)))$ where $x_{[i]}$ is the $i'$th value of the sequence of $x$ values sorted into ascending order. If the points lie along the line $x_{[i]} = F^{-1}(i/(n+1))$ then the cumulative probability distribution of $x$ is equal to the theoretical distribution $F(x)$. The panels in Figure 5.7 show plots of normal quantiles versus the quantiles for standardized temperature and precipitation forecasts (filled circles) and observations (unfilled circles). The temperature and precipitation forecast and observed data have been standardized to have zero mean and unit variance, so that they can be directly compared with standard normal variables. All the points lie close to the 45° line, which indicates that the forecasts and observations are approximately normally distributed (thereby justifying some of our earlier normality assumptions in this chapter). However, the points do deviate below the line for large anomalies in the data, which indicates some presence of positive skewness especially in precipitation.

### 5.5.3   Conditional quantile plots

The conditional quantile plot is an extension of the above method to forecast verification. The empirical conditional distribution of the observations for given forecast values is estimated and the 25%, 50% and 75% percentiles are then plotted against the forecast values. Mathematically, the conditional quantile plot is a scatter plot of the conditional observation quantiles $F_{X|\hat{X}=\hat{x}}^{-1}(q)$ versus the conditioning forecast values $\hat{x}$ for a chosen probability, for example $q = 0.25$. If the sample of observations is large enough, percentiles in the tails of the distribution such as 10% and 90% can then be reliably estimated and plotted (Wilks, 2006b). A very good forecast shows the points lying close to the 45° diagonal. An easy to improve forecast shows the percentiles close to each other, but far from the 45°

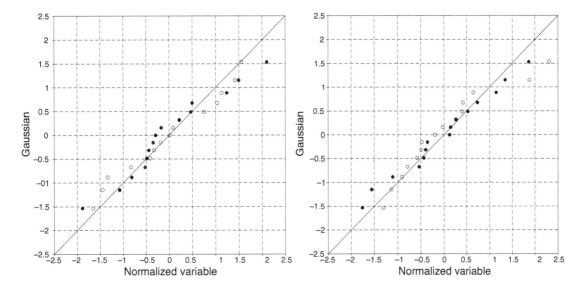

**Figure 5.7** Quantile-quantile plots of standardized temperature (left) and precipitation (right) forecasts. The filled circles indicate the 15 ranked forecasts and the unfilled circles the 15 ranked observations

diagonal. In this situation, a good recalibrated forecast can be obtained by using the 50% percentile of the observation conditioned on the forecast instead of the original forecast. The worst case is when the 25% and 75% percentiles are far from each other, even if the 50% percentile is along the 45° line. However, in this case, there may be some forecasts for which the range is narrower, so that there is some skill in some instances.

Constructing conditional distributions is in fact the best way (in a least squares variation) to produce a probabilistic forecast from a deterministic forecast (Déqué *et al.*, 1994). However, a practical difficulty arises from the finite size of the sample of past forecasts and observations. If the forecast is only crudely binned into 10 distinct categories, one needs at least 100 pairs of previous forecasts and observations to guarantee an average of at least 10 forecast-observation pairs in each bin. With our example based on only 15 forecasts, it is not possible to consider more than a maximum of three bins (e.g. below-normal, normal, above-normal) in order to be able to have at least the very minimum of five observations in each bin needed to estimate quantiles. The conditional quantile plots illustrated in Figure 5.8 were constructed by using a larger data sample obtained by regional pooling of data from all model grid points located in wider

boxes: Europe (35N–75N, 10W–40E) for temperature, and the western tropical Atlantic (10S–30N, 100W–50W) for precipitation. Each local grid point variable is standardized to have zero mean and unit variance, in order to avoid artificial effects due to geographical contrasts (e.g. a good correspondence between forecast and observed cold areas).

Figure 5.8 shows for 15 ranked subsets of the forecasts (the number 15 is arbitrarily chosen to be equal to the number of years, but a different number could easily have been taken), the conditional median (50% percentile) and the conditional 50% interval about the median (the 25% and 75% percentiles) of the corresponding observations. The geographical domain has 352 grid points, so that 5280 (15 × 352) forecast values have been sorted, and the quantiles have been estimated with subsamples of 352 observations. The conditional quantiles in Figure 5.8 do not lie on perfectly smooth curves due to the presence of sampling variations. The total variation of the 50% quantile with forecast value is comparable to the width of the conditional interquartile range, which confirms the lack of strong dependency between the observation and these forecasts previously noted in this chapter. For a climatological mean forecast, the observation interquartile range would be the 25% and 75% percentiles of the observed distribution [−0.7,0.7] no matter

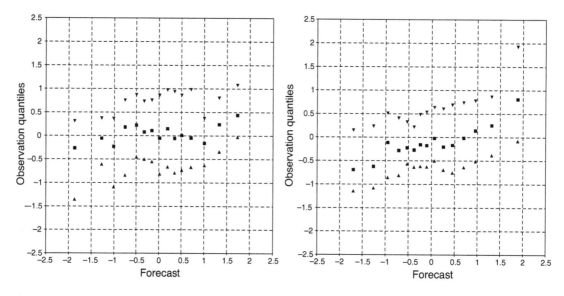

**Figure 5.8** Conditional quantile plots for standardized forecasts of temperature (left) and precipitation (right). The downward pointing triangles indicate the 75% percentiles; the squares indicate the 50% percentiles, and the upward pointing triangles indicate the 25% percentiles

what value the forecast took. The fact that some slight trends can be seen in the conditional quantiles in Figure 5.8 suggests some potential for making (low-skill but not no-skill) probability forecasts on seasonal timescales.

## 5.6 Summary and concluding remarks

We have seen in this chapter that there are many ways of measuring the skill of a forecast of a continuous variable. Although the product moment correlation coefficient may appear to be a good choice for measuring the association between forecasts and observations, this score is sensitive to outliers, is not adapted to finding non-linear relationships between predictor and predictand, and is difficult to test analytically for variables that are not normally distributed. For these reasons, many verification measures have been developed and the most widespread have been presented in this chapter: bias, mean absolute error, mean squared error, product moment

correlation, rank correlation, tau correlation, Daniels' correlation, and linear error in probability space.

We have considered here predictands consisting of a single continuous scalar variable. However, meteorological as well as climate forecasts often consist of fields of numbers representing spatial maps. However, from a specific user's point of view, the predictand of interest is often local, and the verification of a forecast at a specified location is generally of utmost importance. In fact, forecast evaluation over a region can be considered the aggregation of local scores. When the number of local forecasts is small, spatial aggregation can sometimes help to increase the sample size and thereby reduce the sampling uncertainties in the scores (provided that the area is large enough and the spatial dependency weak enough). In this situation, one can then test whether forecasts over a whole region are skilful without being able to determine whether local forecasts have any real skill. This rather paradoxical situation is often encountered in seasonal forecasts, which frequently have very small sample sizes (e.g. 15 winter forecasts).

# 6

# Forecasts of spatial fields

**Barbara G. Brown[1], Eric Gilleland[1], and Elizabeth E. Ebert[2]**

[1]*National Center for Atmospheric Research, Boulder, Colorado, USA*
[2]*Centre for Australian Weather and Climate Research, Bureau of Meteorology, Melbourne, Australia*

## 6.1 Introduction

Forecasts often are created and presented in a form that emphasizes their spatial nature. For example, forecasts from numerical weather prediction (NWP) models are typically computed on a grid and thus can naturally be presented, interpreted and applied using a grid or contours. While some forecasts are only available at specific points, spatial forecasts are actually becoming much more commonly applied in meteorological practice. An example of a spatial forecast is shown in Figure 6.1. In this figure, the variations of the forecast values across the grid are evident, as are certain distinguishable features (e.g. particularly hot regions in the southwestern USA).

While spatial forecasts can simply be treated as a set of forecasts at points, and evaluated using the methods described in the previous chapters, some attributes of these forecasts make it beneficial, at least in some cases, to treat them somewhat differently. Historically, many spatial forecasts have been evaluated using a few specific summary measures, and some of these measures are still in use, particularly in operational meteorological settings. In recent years new approaches have been developed that specifically consider the spatially coher-

ent nature of the forecast fields, and these methods are now starting to be applied both in research and in operational settings. This chapter describes both traditional and new approaches for spatial forecast verification and discusses the advantages of the new spatial verification approaches.

The next section discusses matching forecast values with observation values. Much of the rest of the chapter considers the case of gridded sets of forecast and observed fields. Section 6.3 discusses the traditional approaches to spatial forecast verification, including methods that investigate distributional properties of forecast performance. The background and motivation for alternative approaches are given in Section 6.4, and the subsequent Sections 6.5 through 6.8 describe several spatial verification methods belonging to four categories: neighbourhood, scale separation, features-based and field deformation, respectively. Although the bulk of the newer spatial methods have been demonstrated on precipitation data, largely because of the availability of high-resolution radar rainfall estimates for verification, most of the methods can be applied to any gridded scalar field. Section 6.9 offers a brief comparison of these newer methods along with information about further reading while Section 6.10

*Forecast Verification: A Practitioner's Guide in Atmospheric Science*, Second Edition. Edited by Ian T. Jolliffe and David B. Stephenson.
© 2012 John Wiley & Sons, Ltd. Published 2012 by John Wiley & Sons, Ltd.

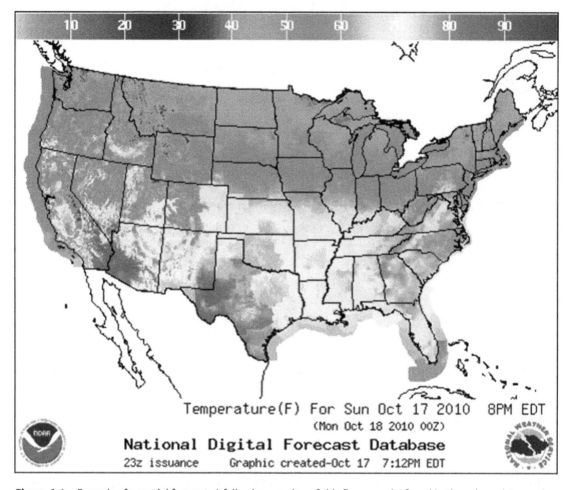

**Figure 6.1**   Example of a spatial forecast. A full colour version of this figure can be found in the colour plate section

provides information about possible future directions and ongoing work. The main points of the chapter are summarized in the final section.

## 6.2   Matching methods

Spatial forecasts are normally given on a spatial grid (in two or three dimensions, though typically only two-dimensional forecast grids are evaluated). The observations that are available to evaluate the forecasts can be on a grid (the same or different from the forecast grid) or they can be located at points that may or may not be coincident with the forecast grid. Ideally, the forecast and observed points should be coincident. However, this situation is frequently not the case, and thus it is necessary to alter

one or more of the grids (in the case of gridded forecasts and observations) or select a scheme to match the observation points to the forecast grid. Two possible combinations of grids and points are most common: (i) Forecast Grid matched to Observation Grid (FGtoOG); and (ii) Forecast Grid matched to Observation Point (FGtoOP).

In the case of FGtoOG it is usually adequate to apply an appropriate interpolation scheme to place the forecasts and observations on the same grid. Of course, the choice of which grid resolution should be used is not always straightforward, and often depends on the goal of the forecast evaluation. For example, if the native grid for the forecasts is coarser than the grid for the observations, it would be possible to (i) re-grid the forecasts to the higher-resolution observation grid; (ii) re-grid (i.e. smooth) the observations to the coarser forecast grid; or (iii) select some other grid resolution that is in between the forecast and observation grid scales, and re-grid both fields. The outcome of the verification analysis can be highly dependent on which method is selected, and the choice truly depends on the questions to be answered through the verification analysis. One other factor that needs to be considered, however, is making consistent estimates of the values assigned to the grid points, for example to ensure that the total volume of precipitation is conserved across the grid.

In the case of FGtoOP, it is necessary to select the method of matching the forecast to the observed point. Two choices are 'upscaling' and 'downscaling'. Both of these options are illustrated in Figure 6.2. In this figure a set of four gridpoints is shown surrounding an observed point. The observed point is upscaled in the top example – i.e. it is simply assigned to the closest forecast gridpoint. In the bottom example in Figure 6.2, the forecast is downscaled to the observed point, by averaging the values of the forecasts at the four gridpoints (note that this is a very simple form of downscaling, and

a variety of others could be used, including spatial interpolation, using the median value, etc.). The important implication of this figure is that the results – matched pairs of forecast and observed values – are very different (forecast of 0 vs forecast of 15) depending on which method is used.

It is possible to create an analysis grid from irregularly spaced observation points. Many caveats must be considered when embarking on such an effort or using such grids generated elsewhere. First, the feasibility of creating meaningful or correct grids is highly dependent on the structure of the variable being gridded. Variables that vary relatively smoothly in space, such as temperature, are amenable to being gridded. In contrast, variables like precipitation or clouds that are discontinuous and more highly variable in space are much more difficult to place on a grid, and the results of gridding may not be representative of the true variations in the variable. Extreme caution is warranted when considering the possibility of creating grids from point observations of convective precipitation or precipitation in complex terrain.

Finally, another approach that is frequently adopted, especially in modelling centres, is to use the model analysis field as the 'observation'. While this approach has many advantages because both fields are automatically on the same grid, the interpretation of the verification results is much different than when the observation grid is independent of the model. In fact, it has been shown that model forecasts that are verified using the analysis field from a different model may have much less skill than when the model forecasts are verified using the model's own analysis field (e.g. Park *et al.*, 2008). Thus, the results of a comparison of a model forecast to its own analysis (i.e. 0-h forecast) field must be considered only an estimate of potential skill.

## 6.3  Traditional verification methods

### 6.3.1  Standard continuous and categorical approaches

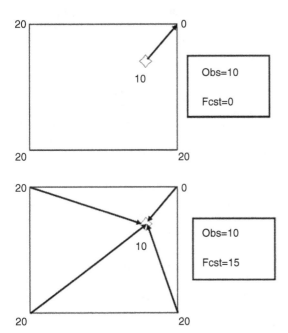

**Figure 6.2**  Example of impacts of choices for matching forecast grids to observed points. Obs, observed; Fcst, forecast

Forecasts on a spatial grid may be evaluated in the same manner as forecasts at individual points, using the appropriate measures defined in the previous chapters to evaluate continuous, categorical and probabilistic forecasts. Common measures that are used include correlation, bias, mean absolute error (MAE) and root mean squared error (RMSE) to evaluate continuous variables like temperature, pressure and wind speed; and Critical Success Index (CSI) or Gilbert's skill score and frequency bias to evaluate binary variables. Computation of verification measures across a grid essentially requires overlaying an observation grid on a forecast grid, and comparing the forecast and observed values at each point.

Because a single gridded forecast may be associated with a grid of observations, or with many observations at point locations within the domain, it is possible to evaluate a *single* spatial forecast (i.e. a forecast valid at a single time), which is much less meaningful when considering a forecast at an individual point. For example, the RMSE for a single forecast at time $t$ would be computed as

$$\text{RMSE}_t = \sqrt{\frac{1}{M} \sum_{i=1}^{M} \left( \hat{x}_{i,t} - x_{i,t} \right)^2} \qquad (6.1)$$

where $M$ is the number of points in the grid (and $x_{i,t}$ is the observed value at time $t$ and location $i$, and $\hat{x}_{i,t}$ is the corresponding forecast value). When such an evaluation is made it is important to correctly interpret the verification result as an *evaluation across a spatial grid for a single time*. In contrast, verification at a point typically represents an *evaluation of forecasts across time at a single location*.

Verification results for spatial forecasts (based on statistics like those mentioned in the previous paragraph) may be aggregated across time in a variety of ways: (i) as verification statistics for a set of individual points across time (i.e. considering only single points in the grid); (ii) as an overall or accumulative 'score' across all points in the grid and across all times; or (iii) as an average (or other statistic) applied to the statistics across time. For example, for (i) one might compute the statistics at each point across time and plot the values on a map to demonstrate variations in performance across a spatial domain. While it may seem that approaches (ii) and (iii) would provide the same answers, it is important to note that this is not necessarily the case.

For example, for RMSE the results associated with the two approaches are

$$(ii) \qquad \text{RMSE} = \sqrt{\frac{1}{MN} \sum_{t=1}^{N} \sum_{i-1}^{M} \left( \hat{x}_{i,t} - x_{i,t} \right)^2} \qquad (6.2)$$

$$(iii) \qquad \text{RMSE} = \frac{1}{N} \sum_{t=1}^{N} \text{RMSE}_t \qquad (6.3)$$

Clearly, since RMSE is a non-linear function, the two approaches could potentially provide quite different estimates of the quality of the forecasts. It is most important to recognize that they answer two different questions about forecast performance. In particular, (ii) answers the question 'What is the overall RMSE for data pooled across all gridpoints and times?'; (iii) answers the question 'What is the mean grid-based RMSE, averaged across all times?'

### 6.3.2 S1 and anomaly correlation

In addition to standard methods for continuous, categorical and probabilistic forecasts, two particular verification measures require special attention in the context of spatial forecast verification. These two methods were specifically developed for the evaluation of spatial forecasts, and especially for gridded forecasts of upper-air variables. These two measures are the *S1 score* and the *anomaly correlation coefficient (ACC)*. While these measures are used relatively infrequently in most research-related verification activities, they are still commonly used for evaluation of operational forecasts by many major national weather services. Moreover, they are of interest from a historical perspective, for tracking the improvement in forecasting capabilities over time; in many cases, they are the only measures available of the quality of early NWP forecasts.

The S1 score was introduced by Teweles and Wobus (1954) for the 'evaluation of prognostic charts'. Rather than considering the actual forecast values at points on a grid, this score focuses on the gradients of values across the grid. In particular, it compares the differences in forecast and observed gradients, $\Delta \hat{x}_i$ and $\Delta x_i$, relative to the maximum of either the forecast or observed gradient. S1 is given by

$$S1_t = 100 \frac{\sum\limits_{i}^{M} |\Delta \hat{x}_i - \Delta x_i|}{\sum\limits_{i}^{M} \max\{|\Delta \hat{x}_i|, |\Delta x_i|\}}, \quad (6.4)$$

where $M$ here is the number of pairs of adjacent points that are included in the evaluation (note that $i$ also could represent adjacent stations, as was the case in early applications of this measure). The numerator in Equation 6.4 represents the MAE of the gradients, and the denominator is a scaling factor that considers the maximum gradient that was forecast or observed for a given pair of gridpoints. As noted earlier, the S1 score is still used to measure trends in forecast performance, but it is not widely used in most operational or research-related forecast verification analyses. This limitation in its use is perhaps partly due to some of the less desirable attributes of the score. For example, Thompson and Carter (1972) showed that the measure could be optimized by forecasters choosing a somewhat larger forecast gradient than they expect will actually occur. Moreover, the measure is very sensitive to seasonal variations in gradients. It also is very dependent on the size of the domain and the number of gridpoints, and may have unstable results over smaller regions.

The ACC is simply the correlation between forecast and observed anomalies, based on a grid of climatological values. The correlation can be either centred or uncentred (e.g. Miyakoda *et al.* 1972), and the verification results can differ depending on which approach is applied, as discussed in Wilks (2006b, section 7.6.4). While the uncentred version of the measure has been used in some applications, it appears to be less commonly applied operationally and in research studies. Here we introduce the centred version of the ACC:

$$ACC_t = \frac{\sum\limits_{i=1}^{M} (\hat{x}_i' - \bar{\hat{x}}')(x_i' - \bar{x}')}{M s_{\hat{x}'} s_{x'}}. \quad (6.5)$$

In Equation 6.5, $x_i'$ and $\hat{x}_i'$ are the observed and forecast anomalies, respectively. That is,

$$x_i' = x_i - x_{ci}$$

and

$$\hat{x}_i' = \hat{x}_i - x_{ci},$$

where $x_{ci}$ is the climatological value at location $i$. Then $s_{x'}$ and $s_{\hat{x}'}$ are the standard deviations of the anomaly values across the spatial grid. This formulation is equivalent to the Pearson product-moment correlation coefficient of the anomalies. The anomaly correlation measures how well a forecast captures the magnitude of anomalies from climatology, and is thus highly dependent on the gridded estimate of the climatology. As a correlation, it ignores biases in the forecast anomalies and thus it is more appropriately considered a measure of potential performance (Murphy and Epstein, 1989).

### 6.3.3 Distributional methods

The realism of a forecast can be measured by comparing its textural characteristics with those of the observations. This approach does not measure whether a forecast is accurate or not, merely whether it looks realistic and has plausible structures. A real world example might be assessing whether a rainfall forecast from a high-resolution numerical model could be mistaken for a radar rainfall field. In this case scale-dependent spatial statistics for forecast and observation fields are computed separately and compared against each other to check for similar behaviour.

A variety of approaches can be used to obtain scale-dependent distribution information about the forecast and observed fields. This section highlights some of them, including the Fourier spectrum, generalized structure function (of which the variogram is a special case), principal component analysis, and a composite approach introduced specifically for forecast verification purposes. This last approach differs from the others, which all involve utilization of basis functions either on the original fields or on their covariance (or correlation).

The Fourier spectrum and the generalized structure function are two widely used multi-scale statistical methods for evaluating the scale-dependent similarity of two fields. The Fourier spectrum is computed using a two-dimensional fast Fourier transform and then azimuthally averaged about wavenumber 0 to give the power spectral density as a function of scale (e.g. Schowengerdt, 1997). The power spectrum indicates how much energy is associated with each spatial scale, i.e. how strongly that scale is represented in the original data. For weather grids the greatest power is usually found in the larger scales, falling off with decreasing spatial scale.

The generalized structure function measures the smoothness and degree of organization of the field. For a two-dimensional field $x$ the structure function can be expressed as

$$
\begin{aligned}
S_q(&\Delta s_1, \Delta s_2) \\
&= E\left(|x(s_1 + \Delta s_1, s_2 + \Delta s_2) - x(s_1, s_2)|^q\right)
\end{aligned}
$$
(6.6)

where $\Delta s_1$ and $\Delta s_2$ are the lags in the $s_1$ and $s_2$ directions, respectively, and the expectation, $E$, denotes averaging. Note that to avoid confusion with $\hat{x}$ and $x$, which in this book are used to denote the prediction and observation, respectively, we adopt the notation $s = (s_1, s_2)$, often used in the spatial statistics literature, to refer to spatial coordinates, rather than $(x, y)$. The exponent $q$ in Equation 6.6 controls the degree to which extreme differences are emphasized. Harris *et al.* (2001) use a value of $q = 1$ to facilitate moment-scale analysis, while Marzban and Sandgathe (2009) use $q = 2$ to produce the well-known variogram. $S_q$ is typically azimuthally averaged around $\Delta s_1 = \Delta s_2 = 0$, which converts it to a one-dimensional quantity, and plotted as a function of scale (e.g. Cressie, 1993). The usual scale-dependent behaviour for meteorological fields is for $S_q$ to increase steadily with scale, then level off when values are no longer correlated.

The structure function is frequently used in a meteorological context to assess precipitation fields, hence the restriction to non-zero grid boxes. However, Marzban and Sandgathe (2009) showed that when this restriction is lifted and non-precipitating grid boxes are included, the variograms are sensitive to size and location of the precipitation structures. Depending on the application, this may or may not be desirable.

Principal Component Analysis (PCA) is a method for analysing the variability of a data set, and is often referred to as empirical orthogonal function (EOF) analysis in meteorology (e.g. Wilks, 2006b, chap-

ter 11), except that EOFs find both time series and spatial patterns. Possibly correlated random variables are transformed by combining them linearly into new, uncorrelated variables, showing the modes of variability. PCA provides the optimal linear transformation to yield the subspace with maximum variance. Detailed information at local spatial points, however, is lost. Independent Component Analysis (ICA) is a refinement of PCA whereby the possibly correlated random variables are rotated into independent components, rather than ones with only zero correlation (Stone, 2004). Such techniques allow verification of whether or not the dominant spatial patterns from model simulations and observations have similar features. They can also be used to determine whether the relationship between the leading modes and other variables has similar features in the model output and observations.

Nachamkin (2004) introduced a composite approach that locates and composites well-defined spatial entities such as heavy rain areas or strong wind events, then measures (i) properties of the forecast field given that an observed event occurred, and (ii) properties of the observed field given that a forecast event was predicted. By overlaying contours of the composite forecasts with those of the observations, systematic phase and bias errors in the forecast are easily diagnosed (see also Nachamkin *et al.*, 2005).

## 6.4  Motivation for alternative approaches

Traditional verification scores such as RMSE, anomaly correlation, critical success index, etc. have been used for many years to evaluate spatial forecasts. The implicit assumption is that perfect forecasts should be possible, and that departures from a perfect forecast indicate deficiencies in the model or forecasting system. This is not too unreasonable when forecasts are made for predictable scales (e.g. synoptic scale flow during the next day or two). However, in recent years models with increasingly fine spatial resolution routinely produce forecasts for scales that are theoretically unpredictable (e.g. Hohenegger and Schär, 2007). It is unreasonable to expect high-resolution forecasts to be perfect. At the same time, they are undeniably

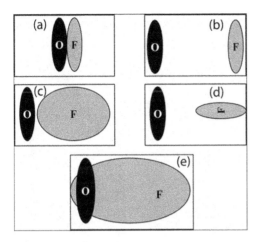

**Figure 6.3** Schematic example of various forecast (F) and observation (O) combinations. (a–d) These all yield a critical success index (CSI) = 0, whereas (e) has a positive CSI, but would probably not be evaluated as the best subjectively. From "Intercomparison of spatial forecast verification methods" by Gilleland, Ahijevych, Brown, Casati, and Ebert, Weather and Forecasting, 24, 1416–1430, 2009. © American Meteorological Society. Reprinted with permission

useful, providing information on weather extremes and variability even when the details are not precisely known.

It is difficult to quantify the improvement of more detailed forecasts over coarser ones using traditional verification approaches. A common problem encountered in verifying spatial forecasts occurs when there is a slight offset in the predicted position of a weather event relative to where it actually occurred. An example is shown in the first panel of Figure 6.3. Such a forecast incurs a 'double penalty', where it is penalized for predicting an event where it did not occur, and again for failing to predict the event where it occurred. This situation arises more frequently in higher-resolution forecasts than in lower-resolution ones, resulting in poorer scores (Rossa *et al.*, 2008). Moreover, traditional grid-scale scores do not distinguish between a 'near miss' and much poorer forecasts, as illustrated by panels (b) to (d) in Figure 6.3. According to the critical success index the forecast in panel (e) is the best performer, though subjectively it does not resemble the observations.

The need to reconcile subjective evaluations of forecast quality with objective measures of forecast accuracy has motivated a great deal of recent research and development of diagnostic methods for spatial verification that more adequately reflect their worth. Many of these methods have their origins in image processing applications, where the need to condense a large amount of spatial information into a more manageable size or produce special effects for the film industry has led to efficient algorithms for processing two-dimensional fields. Other spatial verification approaches try objectively to mimic how a human might interpret and evaluate spatial forecasts. An added benefit of the new spatial methods, in general, is that they provide more diagnostic information regarding the causes of poor or good performance than can be obtained from traditional approaches. For example, some of the methods would be able to distinguish the kinds of errors that are illustrated in Figure 6.3.

To come to grips with the proliferation of new spatial verification methods, and to provide some guidance to potential users of the new methods, a Spatial Verification Method Intercomparison Project was initiated in 2007. This project has resulted in a better understanding of these spatial methods, including their strengths and weaknesses, and the types of forecast applications that they are best suited to evaluate. Gilleland *et al.* (2009, 2010) and Ahijevych *et al.* (2009) provide an overview of the project including a review of the methods that were tested. They classified the majority of spatial verification methods into two general types: *filtering methods*, which apply a scale-dependent filter to the forecast and (usually) the observations before calculating verification statistics on the filtered data; and *displacement methods*, which measure the spatial displacement and/or distortion that would need to be applied to the forecast field to make it match the observed field. These two types can be further subdivided into *neighbourhood* and *scale separation* (filtering methods), and *features-based* and *field deformation* (displacement methods) (Figure 6.4).

The filtering approach includes neighbourhood methods and scale separation methods, both of which measure forecast quality as a function of scale. Neighbourhood methods relax the requirement for an exact match by allowing forecasts located within spatial neighbourhoods of the observation to be counted as (at least partly) correct. Statistics are computed for a sequence of neighbourhood sizes starting from a single grid box

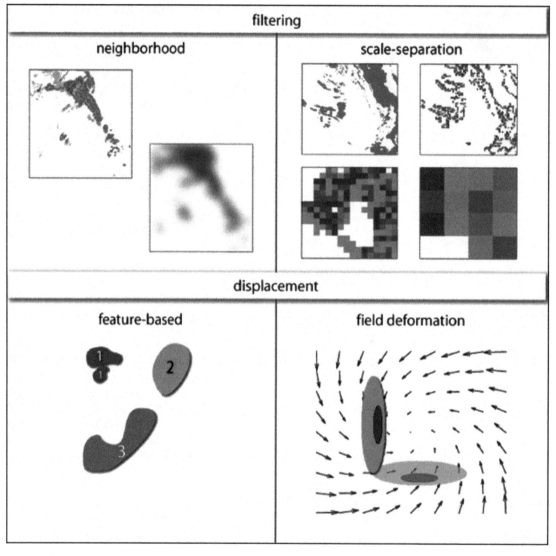

**Figure 6.4**  Four types of spatial verification methods. From "Wavelets and field forecast verification" by Briggs and Levine, Monthly Weather Review, 125, 1329–1341, 1997. © American Meteorological Society. Reprinted with permission. A full colour version of this figure can be found in the colour plate section

(equivalent to traditional verification) and increasing to some much larger, but still meaningful, scale. An advantage of this approach is that it can provide forecast users with practical information on the scale at which an acceptable level of skill is attained.

Scale separation methods, on the other hand, diagnose the errors at distinct scales. This allows, for example, the performance at small, medium and large scales to be examined separately. As different physical processes are associated with different spatial scales, this information is potentially extremely useful for developers.

Forecast displacement approaches include feature-based methods that identify and evaluate objects of interest in the spatial fields; and field deformation methods that alter the forecast to better match the observations, with less deformation indicating a better forecast. Feature-based methods focus on the attributes of well-defined spatial objects such as rain areas or wind maxima. When matching is possible (i.e. when the forecast sufficiently resembles the observations), errors in the location, size, shape, etc. of the forecast features give useful information to users and developers alike. Even when matching is not possible, distributions of forecast and observed feature attributes can be compared to assess forecast realism.

Field deformation methods use morphing techniques to stretch and squeeze the forecast to resemble the observations. In contrast to feature-based methods, which measure location errors for particular objects, field deformation methods yield an array of vector location errors for the field. Penalty functions or search restrictions are imposed to prevent unphysical displacements from being made. Once the forecast has been aligned with the observations, errors in amplitude can be measured. These methods are ideally suited for diagnosing phase errors.

It is worth mentioning that not all spatial verification methods that have been proposed fall neatly into one of these four categories. For example, a method that mainly fits into a particular category may also have attributes of another approach. This situation will be discussed further in Section 6.10.

Although more complex than traditional methods, the newer spatial verification methods provide objective statistics that generally correspond better with subjective assessments (Ahijevych *et al.*, 2009) when applied to high-resolution forecasts. The next four sections delve into the four categories of diagnostic spatial verification, providing detail on some of the methods that are now emerging as particularly useful (or perhaps accessible). Where necessary, relevant tools and subroutines can be found in commonly used software packages like R, IDL and MATLAB, or in *Numerical Recipes* (Press *et al.*, 2007). Software for many of the methods is also publicly available (see Appendix).

## 6.5  Neighbourhood methods

Neighbourhood methods are particularly well suited to verifying high-resolution forecasts, where the realistic spatial structures in the forecast fields provide useful information for users, but getting the forecast exactly right at fine scales is difficult or even impossible due to the chaotic nature of the atmosphere. High-resolution forecasts are usually interpreted with a certain degree of uncertainty, i.e. 'about this time, around this place, about this magnitude', rather than taken at exactly face value. Neighbourhood verification methods account for this uncertainty by evaluating the forecasts within a spatial (and sometimes temporal) window surrounding each observation, thereby enabling an assessment of how 'close' the forecasts are, and giving partial credit to near misses. These methods usually assess forecast performance for several neighbourhood sizes ranging from grid scale to many times the grid scale. Models run at different resolutions can be compared in a fair way using neighbourhood verification.

The forecast performance generally improves as the neighbourhood size is increased and the smaller scales are effectively filtered out. The scale at which an acceptable level of performance is achieved can be thought of as a 'skilful scale' (Mittermaier and Roberts, 2010); this might suggest an appropriate scale at which to present forecasts to users.

While traditional verification matches the observed grid value with the corresponding forecast grid value, neighbourhood verification assumes that forecast values in nearby grid boxes are equally likely and allows these 'close' forecasts to be considered as well. Some quantitative information is extracted from the neighbourhood, such as mean value, median, number of grid values exceeding a critical threshold, etc., for comparison with the observation. In evaluating model performance it is common to compare like with like, and so many neighbourhood verification methods compare the neighbourhood of forecasts with the corresponding neighbourhood of observations. The neighbourhoods may be circular or square; in practice the results are not very sensitive to this choice and square neighbourhoods are much easier to implement.

To apply neighbourhood verification, a window of a given size is centred on each grid box and a measure comparing the forecast and observed values is computed, which might be the mean difference, binary category (hit, miss, false alarm, correct rejection) or some other simple statistic. After moving the window over all grid boxes in the domain, an aggregate score for that window size (scale) is computed. This process is repeated across a range of scales, allowing the scale-dependence of forecast performance to be assessed.

Several neighbourhood verification methods have been developed independently over the last decade or so; for a review see Ebert (2008). Depending on the method, a variety of continuous, categorical and probabilistic scores are used to measure the correspondence between forecasts and observations. Each addresses a particular aspect of forecast quality, such as 'How accurate is the forecast when averaged to successively larger scales?', or 'How likely is the forecast to predict an event of a certain magnitude within a certain distance of an observed event?' This section briefly describes a few of the most popular and useful neighbourhood verification methods. For convenience we describe their use for evaluating rain forecasts, but in principle they can be applied to any scalar field.

### 6.5.1 Comparing neighbourhoods of forecasts and observations

Methods that compare forecast and observed neighbourhoods apply the same processing to both neighbourhoods, then measure the similarity of the processed fields. A commonly used approach is to average, or upscale, the forecast and observed fields onto successively coarser scales and to verify using traditional approaches (see Section 6.3) for each scale (e.g. Zepeda-Arce *et al.*, 2000; Yates *et al.*, 2006). In some cases the mean values are of less interest than the extremes, in which case other types of neighbourhood processing might be more appropriate. Marsigli *et al.* (2008) proposed a distributional approach that compares various moments of the forecast and observed distributions within neighbourhoods: mean, maximum, and 75th and 95th percentiles.

A probabilistic approach that has recently been adopted by many modelling centres to evaluate high-resolution precipitation forecasts against radar observations is the Fractions Skill Score (Roberts and Lean, 2008). Using a threshold to define 'raining' grid boxes, this method compares the forecast and observed rain frequencies using a variant of the Brier score (see Chapter 7). The Fractions Brier Score (FBS) is defined as

$$\text{FBS} = \frac{1}{N} \sum_N (P_{fcst} - P_{obs})^2 \qquad (6.7)$$

where $P_{fcst}$ and $P_{obs}$ are the fractional forecast and observed rain areas in each neighbourhood (analogous to the probability that a grid box in the neighbourhood contains rain), and $N$ is the number of neighbourhoods in the domain. Computing a skill score with respect to the FBS for the perfectly mismatched case results in the positively oriented Fractions Skill Score (FSS):

$$\text{FSS} = 1 - \frac{\text{FBS}}{\frac{1}{N}\left[\sum_N P_{fcst}^2 + \sum_N P_{obs}^2\right]} \qquad (6.8)$$

The FSS ranges from 0 for a complete mismatch to 1 for a perfect match.

The FSS is normally computed for several rain thresholds and a range of spatial scales to show scale- and intensity-dependent performance. In the example in Figure 6.5 the forecast skill for rain exceeding lower rain thresholds exceeds that for high-intensity rain. As one would expect, skill also improves with increasing spatial scale. For a given rain threshold the 'skilful scale' can be defined as the scale at which FSS exceeds $0.5 + f_{obs}/2$, which is the grid-scale fractions skill score for a uniform forecast of the observed rain fraction, $f_{obs}$, everywhere in the domain. Code for this method is freely available (see Appendix).

### 6.5.2 Comparing spatial forecasts with point observations

For some applications it is necessary to evaluate the performance of high-resolution forecasts for a point location or single observation grid box. A case in point is assessment of forecasts and warnings for cities, airports or other locations where the impact of an event may be particularly high (see also Chapter 10). One neighbourhood approach that is particularly well suited to this application is the spatial multi-event contingency table method of Atger (2001). This binary approach evaluates the 'closeness' of the forecasts to the observations according to multiple criteria.

It is easiest to illustrate this method by way of an example. Suppose we are interested in how well a model performs in predicting heavy rain at a particular location. A traditional approach is to construct

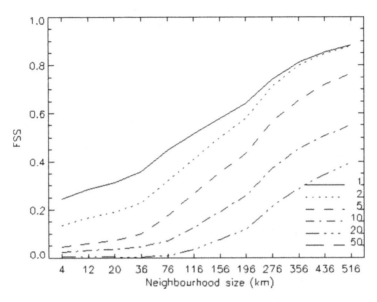

**Figure 6.5** Example of an application of the Fractions Skill Score (FSS) to the verification of a 24-h forecast of hourly precipitation over the continental USA from a high-resolution model. The curves show performance as a function of neighbourhood size for rain exceeding various thresholds (mm)

a time series of matched forecast and observed rain values for that location, then compute one or more binary scores (see Chapter 3) for rain exceeding a critical threshold, based on the accumulated hits, misses, false alarms and correct rejections. This approach does not account for small differences in rain intensity, timing or location, which can lead to poor scores for forecasts that are otherwise very useful. Atger (2001) proposed expanding the contingency table to additional dimensions to also allow for closeness in intensity, distance and time, and testing a range of thresholds for each dimension. For each choice of multi-event thresholds (e.g. rain intensity of at least 10 mm/h, within 30 km and ±1 hour of the observation) a separate $2 \times 2$ contingency table is constructed. A hit is counted when the observation meets the intensity threshold, and at least one forecast grid box in the space-time neighbourhood meets the multi-event criteria. A miss is counted when the observation meets the intensity threshold but no forecast within the neighbourhood meets all of the criteria. Similarly, a false alarm means that the forecast meets all the criteria but no event was observed.

The contingency tables are simple to interpret and contain a wealth of information. However, the large number of tables is somewhat unwieldy. To summarize this information the hit rate and false alarm rate can be computed from each of the contingency tables and plotted as points on a relative operating characteristic (ROC) diagram (see Chapters 3 and 7). The 'cloud' of points describes the relative operating characteristic for the forecasts. Those points with the greatest hit rates and lowest false alarm rates represent the optimum intensity, space and timescales for the forecasts. As a bonus, the ROC diagram facilitates direct comparison with ensemble forecasts, providing useful information for modellers who must decide whether to invest computing resources in higher-resolution deterministic forecasts or in ensemble prediction systems, or some combination of the two.

## 6.6 Scale separation methods

Scale separation methods can be used to assess the components of error associated with different spatial scales. These methods apply a single-band spatial filter such as Fourier or wavelet decomposition to the field to identify structures (or texture) at particular scales. These structures are often associated with physical processes that have different degrees of predictability or model skill. For example, an atmospheric model might simulate cloud-scale convective processes at the finest scales and local circulations at moderate scales, while the largest scales might represent the synoptic scale flow. Verification using scale separation methods can provide information to guide model developers as to which processes are most in need of improvement.

Scale separation methods differ from neighbourhood methods in an important aspect, in that they *isolate* the scale-dependent behaviour at each wavelength. Neighbourhood methods, on the other hand, *smooth* the fields, filtering out the smaller scales while retaining the larger scale information.

Scale separation methods generally follow one of two approaches. The first measures the similarity of forecast and observed structures without actually matching the fields, and is therefore more of a validation of whether the forecast produces realistic looking fields. The Fourier spectrum and generalized structure function discussed in Section 6.3.3 are examples of diagnostics that measure field similarity. The second approach matches the forecast and observed fields and provides error information for each scale. Two methods that use wavelet decomposition to enable the assessment of scale-dependent error are described below.

Wavelet decomposition allows an image to be separated into orthogonal component images with different levels of detail (scale) in a fashion similar to Fourier decomposition, but where the wavelet basis functions are non-zero only around a localized area. In this way, wavelet decomposition, unlike Fourier decomposition, maintains information about spatial locations allowing for the potential to inform about displacement errors at (possibly physically meaningful) independent scales (e.g. Figure 6.6). The first dyadic scale (scale 1) in Figure 6.6 is fairly coarse, and represents the overall structure of the two fields. Scale 2 represents only those features that are present in each half of the fields. Subsequent scales represent further divisions of the space in two (because they used a dyadic decomposition). Caution must be taken to check that the scales are physically meaningful. For fields that are not smooth, special care is required to ensure that the detail fields do not represent artefacts of the wavelet basis functions themselves, rather than something

real. For example, precipitation fields are often characterized by having numerous zero-valued grid points. It is possible to obtain a more realistic wavelet decomposition if coefficients are set to zero wherever the original values were zero. Wavelet software is widely available, making the methods described below relatively easy to implement.

The use of two-dimensional wavelet decomposition to measure scale-dependent forecast error was pioneered by Briggs and Levine (1997). They applied a 2D discrete wavelet transformation to forecast and observed fields of 500 hPa height and isolated the scale-dependent structures in each. For each scale the forecast component fields were then verified against the corresponding observed component fields using standard scores such as anomaly correlation and mean squared error. A very nice feature of this approach is the fact that the sum of the scale-dependent MSE values is equal to the total MSE of the original field because of the orthogonality of the wavelet-decomposed structures. This means that the fractional contribution of error at each scale is easy to compute and interpret.

The intensity-scale method of Casati *et al.* (2004), with improvements described by Casati (2010), augments the wavelet-based verification with a dimension for intensity (rain rate, wind strength, etc.). Instead of operating on fields in physical space, the forecast and observed fields are first converted to binary fields (1s and 0s) by applying an intensity threshold. Two-dimensional wavelet decomposition using a Haar (square) wavelet is then performed to obtain the scale-dependent structures. The scale-dependent MSE of the component binary fields can be computed as above. For the $l^{th}$ scale (equal to a square of size $2^l$ grid lengths) the MSE-based skill with respect to a random forecast can be computed as

$$SS_l = 1 - \frac{\text{MSE}_l}{[B\varepsilon(1 - f_{obs}) + f_{obs}(1 - Bf_{obs})]/L} \tag{6.9}$$

where $f_{obs}$ is the fractional coverage of 1s in the observed field, $B$ is the bias (ratio of forecast to observed fractional coverage) and $L$ is the dyadic dimension of the image (a square array of dimension $2^L$). Casati *et al.* (2004) showed that this expression is equal to the Heidke skill score (see Chapter 3).

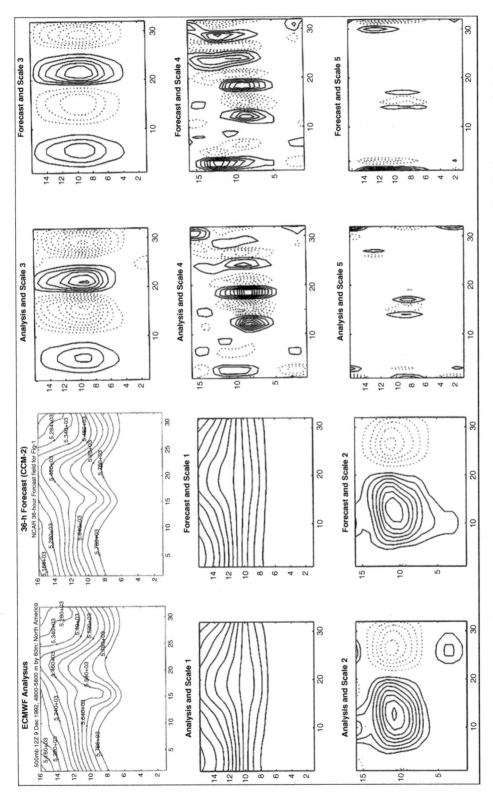

**Figure 6.6**  Wavelet decomposition of analysed and forecast fields of 500hPa height over North America on 9 Dec 1992. The first two panels in the top row are the original analysis and forecast fields. Subsequent panels show their *detail* fields (reconstructions using only coefficients for a particular scale) for progressively finer scales. From "Wavelets and field forecast verification" by Briggs and Levine, Monthly Weather Review, 125, 1329–1341, 1997. © American Meteorological Society. Reprinted with permission

It is also useful to assess whether the forecast has an appropriate amount of energy at each scale. The energy, $En_l$, of the $l^{th}$ binary scale component is given by its mean squared value over the grid. A practical measure of forecast bias is the energy relative difference,

$$\text{En rel diff}_l = \frac{En_l(F) - En_l(O)}{En_l(F) + En_l(O)}. \qquad (6.10)$$

Positive values indicate over-forecasting and negative values indicate under-forecasting of precipitation over threshold at that scale.

Repeating this procedure using a range of thresholds to define the binary fields allows the skill and bias to be assessed as a function of both intensity and scale. Code for the intensity-scale method is freely available (see Appendix).

## 6.7 Feature-based methods

Perhaps the most intuitive of the spatial verification approaches are the feature-based methods. These approaches try to replicate the human's ability to identify and characterize coherent spatial objects (or features; these terms are used interchangeably in this section) in forecast and observed fields. Many of these methods also attempt to match features in the two fields. Once a match has been made, one can diagnose the errors in various attributes of the forecast feature – its location, size, mean and maximum intensity, orientation, etc. Candidate meteorological features for this type of verification include rain systems, cloud systems, low pressure centres and wind speed maxima.

Feature-based methods require that forecast and observed fields be available on a common grid. Although research is still active in this area, a few feature-based methods are now considered mainstream and are described here.

### 6.7.1 Feature-matching techniques

The earliest feature-based approaches were developed to describe errors in distinct forecast features that could be matched to observed features. The Contiguous Rain Area (CRA) approach of Ebert and McBride (2000) identifies features as contiguous areas enclosed within a specified threshold and meeting a minimum size criterion. Forecast and observed features are associated if they overlap. The location error of the forecast is determined by translating the forecast field within a constrained search radius until an optimum match with the observations is achieved, either through minimizing the MSE or maximizing the spatial correlation. Once the features are spatially aligned, their attributes (size, intensity, etc.) are directly compared. The total error can be decomposed into contributions from location error, intensity error, pattern error and (optionally) orientation error (Moise and Delage, 2010). Forecast performance for *features* can also be summarized categorically in terms of hits, missed events, etc. When aggregated over many cases, this feature-based error information can inform users on systematic forecast errors (e.g. Tartaglioni *et al.*, 2008), and guide modellers as to which model processes need to be improved.

The Method for Object-based Diagnostic Evaluation (MODE) is a more sophisticated feature-based method developed by Davis *et al.* (2006a, 2009). Features of interest are identified separately in forecast and observed fields using convolution (smoothing) followed by thresholding. Nearby objects may be merged into clusters and treated as composite objects. In order to match objects in the forecasts and observations, MODE applies a fuzzy logic approach to derive a total interest value that combines information on the similarity of forecast and observed objects in terms of their centroid distance separation, minimum separation distance of object boundaries, orientation angle difference, area ratio and intersection area (and possibly other attributes that can be user-defined). Different weights can be given to these quantities to emphasize those that the user considers to be most important (e.g. over a watershed the centroid distance separation might receive high weight). Each forecast/observed object pair scores a total interest value between 0 and 1, where values closer to 1 signify a greater likelihood that the objects represent the same phenomenon.

Figure 6.7 shows an example of how MODE handles features. In this example, three object clusters were matched between the forecast and observed fields. Several other relatively small areas were not matched between the forecast and observed fields and can be considered false alarms (in the case of

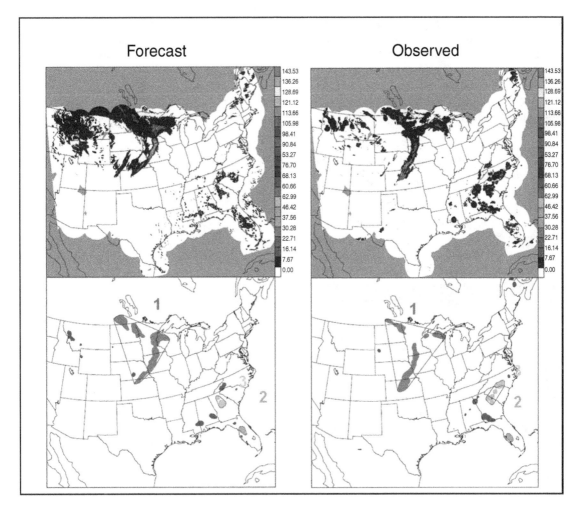

**Figure 6.7** Example of an application of the Method for Object-based Diagnostic Evaluation (MODE). See text for details. A full colour version of this figure can be found in the colour plate section

the forecasts) and misses (in the case of the observations). Comparison of the object features for Object 1 reveals that the forecasted area was somewhat too large and was displaced by about 190 km to the west. The distribution of forecast precipitation intensities for this object cluster was approximately correct. Similar results could be obtained for the other matched objects. In contrast, traditional verification results for this case indicate that probability of detection (POD) = 0.22, false alarm ratio (FAR) = 0.86, and critical success index (CSI) = 0.09, which suggest it was a rather poor forecast.

In order to compare forecast and observed object attributes, Davis *et al.* (2009) adopt a total interest value of 0.7 as a minimum criterion for a match.

The overall goodness of a field is measured by the strength of the interest values for the collection of objects. For each forecast (observed) object the maximum of the interest values achieved by pairing with all observed (forecast) objects is retained. The median of the distribution of forecast and observed maximum interest values, referred to as the median of maximum interest (MMI), is used as a measure to rate the forecast quality.

As discussed in Davis *et al.* (2009) the user must make some subjective choices when applying these methods. This flexibility allows the verification to be tailored to particular types of forecasts or applications (e.g. mesoscale model predictions of heavy rain events). To assist in the choice of the convolution radius and threshold, the MMI can be computed for different combinations of those two quantities. A radius and threshold combination that produces high MMI values may be considered more useful.

The CRA and MODE object-matching methods are both available online (see Appendix).

### 6.7.2 Structure-Amplitude-Location (SAL) technique

In some cases the attributes of forecast features are important but they cannot be sensibly matched to observed features. An example might be forecasts of scattered convection where small, highly variable precipitation features are predicted and observed but no reasonable one-to-one matching is possible. One option would be to cluster the objects into composite objects, as is done in MODE (Davis *et al.*, 2009). Another possibility is to characterize and compare the objects without trying to match them. Wernli *et al.* (2008) took this approach when developing the Structure-Amplitude-Location (SAL) method to verify high-resolution precipitation forecasts over a predefined area such as a large river catchment. The SAL method is now used to evaluate precipitation forecasts in several European meteorological centres.

Given a fixed domain with forecast and observed precipitation fields $\hat{x}$ and $x$, respectively, the SAL method starts by identifying precipitation objects using a dynamic threshold of $R^* = 1/15 \times R^{95}$, where $R^{95}$ is the 95th percentile value of all rainfall exceeding 0.1 mm in the domain (Wernli *et al.*, 2009). Each object $n$, $n = 1,\ldots,M$, is defined by its spatial centre of mass, $s_n$, within the domain, and its integrated precipitation, $R_n$, which is the sum of its grid box values.

Three non-dimensional error measures are then computed to characterize the forecast quality. The first is the amplitude error, $A$, which is simply the normalized difference in mean rainfall,

$$A = \frac{D(\hat{x}) - D(x)}{0.5[D(\hat{x}) + D(x)]} \quad (6.11)$$

where $D(R)$ denotes the domain mean value. $A$ can vary between $-2$ and $+2$ with 0 representing an unbiased forecast.

The location error, $L$, has two components. The first term represents the error in the overall centre of mass, while the second represents the spacing between objects:

$$L = \frac{|s(\hat{x}) - s(x)|}{d} + 2\frac{|r(\hat{x}) - r(x)|}{d}. \quad (6.12)$$

In this expression $s(.)$ is the centre of mass of the field, $r(.)$ is the mass-weighted mean distance from each object's centre of mass to the overall centre of mass,

$$r(R) = \frac{\sum_{n=1}^{M} R_n |s - s_n|}{\sum_{n=1}^{M} R_n} \quad (6.13)$$

and $d$ is the maximum distance across the domain. Note the second term in $L$ is 0 when there is only one object in the domain. $L$ varies from 0 to $+2$, with 0 representing a perfectly located forecast.

The third non-dimensional error is the structure error, $S$, which verifies the peakiness or smoothness of the objects. For each object a scaled volume can be defined as $V_n = R_n / R_n^{max}$ where $R_n^{max}$ is the maximum rain value within the object. The structure error is defined similarly to the amplitude error as

$$S = \frac{V(\hat{x}) - V(x)}{0.5[V(\hat{x}) + V(x)]} \quad (6.14)$$

where $V(R)$ is the mass-weighted scaled volume,

$$V(R) = \frac{\sum_{n=1}^{M} R_n V_n}{\sum_{n=1}^{M} R_n}. \quad (6.15)$$

$S$ ranges from $-2$ to 2, where a value of 0 indicates perfect structure, negative values mean the forecast is too peaked, and positive values indicate that the forecast is too smooth.

Taken together, $S$, $A$ and $L$ provide a reasonably inclusive description of rainfall forecast quality that is relevant to forecast users. The SAL method has demonstrated the value of high-resolution modelling in predicting realistic rain structures (e.g. Wernli *et al.*, 2008, 2009).

## 6.8   Field deformation methods

Field deformation methods consider the forecast to be an estimate of the observed field, with errors in the intensities, spatial extent and patterns. Location errors can be caused by temporal or spatial displacements, and most methods will not distinguish these cases. Small-scale scatter, structure errors, as well as misses and false alarms complicate the determination of location errors.

Gilleland *et al.* (2009) classified field deformation approaches as methods that attempt to determine location errors for an entire field at once, as opposed to location errors for individual entities. Here they are further subdivided into metrics and true field deformation methods. Metrics are mathematical summaries of the overall similarity in shape, spatial extent and/or location of events in the verification set. They can be considered to be a summary of distributional characteristics related to a forecast's overall performance in terms of spatial placement. True field deformation approaches involve directly deforming the forecast field until it is optimally close in space to the observed field.

### 6.8.1   Location metrics

Metrics are mathematically defined measurements for comparing two processes. While metrics exist for real-valued spatial processes, we illustrate only those for binary fields as they are simpler to discuss, and more widely studied in the literature.

A metric, $M$, for a pair of binary images (or verification set) is a function that operates on subsets (say $A$, $B$ and $C$) of the binary images in such a way as to satisfy the properties of (i) symmetry $M(A,B) = M(B,A)$, (ii) the triangle inequality $M(A,B) + M(B,C) \geq M(A,C)$ and (iii) positivity $M(A,B) \geq 0$, where $M(A,B) = 0$ only if $A = B$. Many metrics have been proposed for comparing binary images (see, e.g., Baddeley, 1992; Gesù and Starovoitov, 1999), but here the focus is on those utilized specifically for spatial forecast verification.

The Hausdorff distance between forecast (binary) event areas, $A$, and observed binary event areas, $B$, contained in the verification domain is defined to be the maximum absolute difference of the shortest distances between a two-dimensional grid-point $s$ and the two sets $A$ and $B$; i.e.,

$$H(A, B) = \max |d(s, A) - d(s, B)|, \quad (6.16)$$

where the maximum is over all points in the domain, and $d(s, \cdot)$ is the shortest distance between $s$ and the forecast (observed) event area. Although the distances need to be calculated for all grid points in the domain, a fast algorithm known as the distance transform can be used so that all of these distances can be computed relatively quickly (Baddeley, 1992).

The Hausdorff metric is often used for comparing binary images, but it has some drawbacks, and numerous variations on Equation 6.16 have been proposed to circumvent them. For example, one frequently cited issue concerns its sensitivity to outliers because of the maximum in Equation 6.16. A simple modification of this is to use the partial Hausdorff distance (PHD), which replaces the maximum with the $k^{\text{th}}$ percentile. PHD, however, is not strictly a metric. Venugopal *et al.* (2005) propose a modification to the PHD that ensures it is a metric, called the forecast quality index (FQI), namely,

$$\text{FQI}(\hat{x}, x)$$
$$= \frac{\text{PHD}_k(\hat{x}, x)}{\dfrac{1}{n_x} \displaystyle\sum_{i=1}^{n_x} \text{PHD}_k(x, y_i)} \Bigg/ \frac{2\mu_{\hat{x}}\mu_x}{\mu_{\hat{x}}^2 + \mu_x^2} \frac{2\sigma_{\hat{x}}\sigma_x}{\sigma_{\hat{x}}^2 + \sigma_x^2},$$
$$(6.17)$$

where $\mu$ and $\sigma$ are the respective means and standard deviations of the field values (e.g. rainfall amounts) at non-zero pixels in each (original non-binary) field $\hat{x}$ and $x$, $n_x$ is the number of non-zero observed pixels, and the $y_i$ are surrogates of $x$ obtained by stochastic realizations from distributions having either the same probability distribution function or the same correlation structure as the observations.

Another modification of the Hausdorff metric was proposed by Baddeley (1992), and replaces the maximum in Equation 6.16 with an $L^p$ norm, in addition to applying a transformation $w$ to the shortest distances, $d(s, \cdot)$, to ensure that it is a metric. Specifically, to compare two binary fields $\hat{x}$ and $x$ for a given threshold defining sets $A$ (forecast event areas) and $B$ (observed event areas), the Baddeley $\Delta$ metric for forecast verification is given by

$$\Delta_w^p(A, B)$$

$$= \left[ \frac{1}{N} \sum_{i=1}^n |w\,(d(s_i, A)) - w\,(d(s_i, B))|^p \right]^{1/p}$$

$$= \left[ \frac{1}{N} \sum_{i=1}^n v_i^p \right]^{1/p} \qquad (6.18)$$

where $p$ is chosen *a priori*. If $p = 1$, then $\Delta$ gives the arithmetic average of $v$; in the limit, as $p$ approaches 0, $\Delta$ approaches the minimum $v_i$, and as $p$ approaches infinity $\Delta$ tends towards the maximum $v_i$, and is equivalent to the Hausdorff metric. Finally, $p = 2$ gives the familiar Euclidean norm for $v$. Gilleland *et al.* (2008) apply $\Delta$ with $p = 2$, along with an additional (unnecessary for forecast verification) constraint on $d(s_i, \cdot)$ based on the original form of $\Delta$ for image analysis, to features identified by the MODE procedure in order to automatically merge features within a field, and match the resulting features between forecast and observed fields. Gilleland (2011) demonstrates $\Delta$ with $p = 2$ on the Inter-Comparison Project (ICP) test cases as a forecast verification metric.

## 6.8.2 Field deformation

While metrics summarize the amount of similarity or difference in spatial patterns and/or intensities, it is also possible to employ methods that attempt to deform the shape of one field into another. Generally, it is only the spatial patterns of intensities that are deformed, and not the intensities themselves. Therefore, even if a perfect spatial deformation is obtained, the RMSE of the resulting deformed field will not be zero unless the intensities happen to match up perfectly. Even for perfect intensities, small-scale errors resulting from numerical approximations of the deformations will usually leave residual errors. Deforming an image can be accomplished by a variety of approaches. Keil and Craig (2007, 2009) use an optical flow procedure based on a pyramidal algorithm. The deformation seeks to minimize an amplitude-based quantity (e.g.

correlation coefficient, RMSE) at different scales within a fixed search environment. Specifically,

1. The two fields are upscaled by averaging $2^l$ pixels onto one grid point.
2. A displacement vector at each grid location is computed by translating one field within the range of $\pm 2$ grid points in all directions. The displacement that minimizes the squared difference within a local region centred on the grid point is chosen.
3. This vector field is applied to the original field to generate an intermediate deformation that accounts for large-scale motions (topmost level of the pyramid).
4. The intermediate field from step 3 is up-scaled by averaging $2^{l-1}$ grid points, and steps 2 and 3 are repeated, thereby correcting the previous deformation.
5. The final deformation is found by summing all of the previous vector fields.

Keil and Craig (2009) deform the forecast field into the observation field, and vice versa, then summarize the deformation with a displacement-amplitude score (DAS):

$$\text{DAS} = \frac{1}{D_{\max}(n_{obs} + n_{fcst})}(n_{obs}\bar{D}_{obs} + n_{fcst}\bar{D}_{fcst})$$

$$+ \frac{1}{I_{obs}(n_{obs} + n_{fcst})}(n_{obs}\bar{A}_{obs} + n_{fcst}\bar{A}_{fcst}). \qquad (6.19)$$

When morphing the forecast field into the observed field, $\bar{D}_{obs}$ is the mean displacement (deformation) error, and the amplitude error

$$\bar{A}_{obs} = \left[ \sum(\tilde{x} - x)^2 / n_{obs} \right]^{1/2}$$

where $\tilde{x}$ is the morphed forecast field and $n_{obs}$ is the number of points with non-zero error in the (morphed) forecast. $\bar{D}_{fcst}$ and $\bar{A}_{fcst}$ are the corresponding displacement and amplitude errors when the observed field is morphed into the forecast field. The weighted distance and amplitude errors are respectively normalized by $D_{max}$, the maximum distance across the domain, and $I_{obs}$, a characteristic intensity representing the typical amplitude of the observed field (e.g. derived from a large data sample).

Image warping is another way to deform one field into another. This approach assumes a model, and the deformation is determined through optimization of a likelihood function (Åberg *et al.*, 2005; Gilleland *et al.*, 2010b). The generalized model is given by

$$x(s) = \hat{x}\left(W_{s_1}(s), W_{s_2}(s)\right) + \text{error}(s), \qquad (6.20)$$

where $W$ maps grid points $s = (s_1, s_2)$ from $x$ to points in $\hat{x}$. The inverse warp, which exists and can be found numerically, defines the deformed forecast field.

For the thin-plate spline warp function, the entire warp, $W$, is defined by a set of *control points* (also known as tie points, or landmarks). Although these can be taken to be the entire set of points, typically only a relatively small number of points are necessary. While the movement of a few points defines the warp function, the resulting function is applied to all points in the domain to obtain the deformed image (or forecast in this context). More (fewer) control points lead to more (less) intricate deformations. Early verification applications identified individual features in the fields, but informative deformations are obtained using a regular grid of points without having to identify features.

To obtain the optimal warp, one obtains an objective function based on an assumed likelihood for the error term in Equation 6.20. The normal distribution function is usually assumed, and works well even if it is not a good assumption (Åberg *et al.*, 2005). To prevent the warp function from making non-physical and/or difficult to interpret transformations, it is also necessary to penalize the likelihood to guard against such deformations. The only parameters of interest are the forecast control points themselves (one chooses values for any nuisance parameters introduced by the assumed penalized likelihood *a priori*).

Once a deformed forecast field is found, a vector field describing the optimal warp is obtained. There are various ways to utilize this information. For example, traditional scores such as RMSE can be taken both before and after deformation to determine the amount of reduction in error. Complementary information about the amount of deformation, or bending energy, can give a good picture of how well a forecast performs (e.g. Gilleland et al., 2010b).

Figure 6.8 shows an example of the image warp procedure applied to an idealized geometric case. Clearly, the deformed forecast matches the observation very well. However, as can be seen in the middle panel of the bottom row, a lot of energy is required to deform the forecast to this extent. Accounting for the amount of deformation required along with the reduction in error is essential when attempting to summarize forecast performance succinctly.

Inspection of the vector field describing the optimal deformation allows for more diagnostic information. In particular, it can be clearly seen that the object is displaced too far east by the large vectors pulling the rain areas back to the west. The spatial extent bias is recognized by the smaller opposing vectors squeezing the forecast rain area into itself. Of course, these errors are obvious for such simple geometric shapes without having to resort to image warping. However, for more realistic fields, with more complicated vector fields, such information can be invaluable.

## 6.9 Comparison of approaches

With so many new methods proposed over a relatively short span of time, the need for guidance on what could be gleaned from them became a prominent issue. The Spatial Forecast Verification Methods Inter-Comparison Project (ICP, http://www.ral.ucar.edu/projects/icp) was established to get a better handle on their attributes. Much of what has been discussed in this chapter is a direct result from what has been learned in the ICP.

Test cases representing precipitation fields, including real cases, cases perturbed in known ways from real fields, and simple geometric cases were made available. As new methods are introduced, it is hoped that they will be tested on these same cases in order to facilitate future comparisons. Many of the new methods have been applied to these cases, and results have been reported in a special collection of papers in *Weather and Forecasting*, which includes a number of articles published in 2009 and 2010. Principal findings of the ICP are summarized

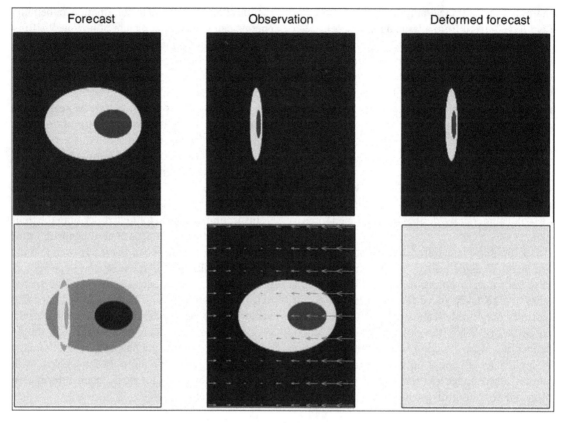

**Figure 6.8** Example application of image warping applied to one of the Inter-Comparison Project (ICP) geometric cases. Top row shows the forecast, observation and deformed forecast fields, respectively. Bottom row shows the error field between the forecast and the observation, the forecast with the image warp vectors superimposed, and the resulting error field between the deformed forecast and the observation

in Table 6.1. It should be noted that such a table does not represent the full range of methods, and there are many exceptions. For example, it ignores hybrid approaches, such as that of Lack *et al.* (2010), which inform about both structure and scale errors. However, it gives a quick overview of which methods could be used in unison to provide a more complete picture of forecast performance.

It should be noted that the spatial verification methods described in this chapter can be applied to the evaluation of any scalar field. Although spatial precipitation verification has received by far the greatest attention, these approaches are also being used to verify wind, cloud and lightning forecasts (e.g. Nachamkin, 2004; Casati and Wilson, 2007; Zingerle and Nurmi, 2008).

## 6.10 New approaches and applications: the future

A number of extensions to the new spatial verification approaches are desirable, for example, to include the time domain or to apply to ensemble forecasts. Some efforts are underway to develop these new capabilities. For example, object-based methods like MODE can be extended to include temporal as well as spatial attributes of objects by considering the objects to have three dimensions, where the third dimension is time. Attributes of the three-dimensional objects can then be extended to include time-related features, such as duration or lifetime of the objects, and the features of the three-dimensional objects can be compared just as

**Table 6.1** Summary table describing the broad categories of spatial verification methods, and which types of errors they inform about. 'Yes' means that this is the general purpose of the method, but 'No' does not mean that it is not possible to do (rather it has not been proposed, or is not fundamental to the approach). For example, the displacement methods can be applied at different scales, but this is not generally part of their foundation

| | Method | Scale-specific errors | Scales of useful skill | Structure errors | Location errors | Intensity errors | Hits, misses, false alarms, correct negatives |
|---|---|---|---|---|---|---|---|
| Filter | Traditional | No | No | No | No | Yes | Yes |
| | Neighbourhood | Yes | Yes | No | Sensitive, but no direct information | Yes | Yes |
| | Scale separation | Yes | Yes | No | Sensitive, but no direct information | Yes | Yes |
| Displacement | Field deformation | No | No | No | Yes | Yes | No |
| | Features-based | Indirectly | No | Yes | Yes | Yes | Yes, based on features rather than gridpoints |

is done with the two-dimensional objects. Davis *et al.* (2006b) implemented and demonstrated an early application of this concept.

Spatial verification methods are now starting to be applied to evaluating ensemble forecasts. In so far as an ensemble forecast is simply a collection of deterministic forecasts, any spatial method can be used to verify the individual ensemble members and rank them (e.g. Keil and Craig, 2007; Micheas *et al.*, 2007), or the distribution of spatial verification results used to characterize the quality of the forecast. The ensemble mean forecast can also be verified using spatial methods. Jung and Leutbecher (2008) applied a scale-dependent filter to ensemble forecasts to evaluate errors at planetary, synoptic and sub-synoptic scales.

Some methods are particularly well suited to ensemble verification. The neighbourhood methods of Atger (2001) and Marsigli *et al.* (2008) were specifically designed to evaluate and compare ensemble and deterministic forecasts. Zacharov and Rezacova (2009) recently adopted a neighbourhood approach to assessing the ensemble spread-skill relationship, using the fractions skill score as a measure for both spread and skill. The spread-skill relationship for spatial objects in ensemble forecasts was investigated by Gallus (2009) using the CRA and MODE techniques. It would be possible to verify the ensemble distribution of feature attributes using the same scores and diagnostics as are used for scalar distributions (see Chapters 7 and 8). Finally, the minimum spanning tree histogram and bounding box, described in Chapter 8, though not spatial methods per se, can be used to evaluate whether the observation is spatially located within the ensemble.

The work discussed in this section is just a beginning. To achieve the solidity required for operational use or common application in research, much additional development will be required. Nevertheless, these new approaches are promising and will lead to more fully useful spatial methods in the future.

## 6.11 Summary

Higher-resolution forecast models present new challenges in the determination of their perfor-

mance. First, it is necessary to obtain a meaningful observation field with which to verify them. Because observations are generally taken at points, or from sensors resolved at typically finer scales, care must be exercised in determining how to match them with forecasts. As noted in Section 6.1, one might downscale the forecast to specific observation locations, or else map the forecasts and observations onto a common grid. Most of this chapter concerns the latter choice, as a wealth of new approaches has been proposed for this situation.

Applying verification statistics on a grid point by grid point basis, as is customary, often yields results that are of questionable usefulness when applied to high- resolution forecasts (e.g. Mass *et al.*, 2002). The latter part of this chapter focused primarily on diagnostic spatial verification methods for handling this situation. The new methods generally fall into one of four categories: neighbourhood, scale separation, features-based and field deformation.

Neighbourhood methods, which apply smoothing filters, and scale separation methods that apply single-band pass filters, both inform directly about scales of useful skill, and because they usually rely on traditional verification statistics performed at different scales, they can provide information on intensity errors and occurrence of events (hits, misses and false alarms). Because the scale separation methods decompose the fields into independent scales, they have the potential to provide information about physically meaningful scales (e.g. large storm systems vs smaller convective activity). Either type of method will be sensitive to other types of errors, such as structure or location errors, but neither will inform directly about those types of errors.

The latter two categories, which can be grouped together as displacement methods, can be performed either on an entire field at once, as is done by field deformation methods, or individual features within a field can be found and various statistics about their attributes can be summarized, as is done by feature-based methods. In the latter case, it is necessary first to identify features of interest, and often determine whether non-contiguous features should be considered as part of the same cluster of features or not. Once such features are identified within a field, and clusters determined, then forecast and observed features must be matched. Neither approach informs directly about scales of useful skill, but both inform about location and intensity errors. Feature-based methods have the additional advantage of informing directly about structure errors, and they provide a holistic way to identify hits, misses and false alarms based on the occurrence and characteristics of the features themselves.

A logical next step would be to combine the advantages of filter and displacement approaches into hybrid schemes for verifying spatial forecasts. This would provide a fairly complete picture of forecast performance, and some such approaches have been proposed (e.g. Lack *et al.*, 2010). A challenge for many users will be achieving computational efficiency. While most of the methods described herein can be performed relatively quickly, when applied together the cost can multiply rapidly.

In summary, the important messages of this chapter are:

1. Extreme care must be taken in the process used to match spatial forecasts to observations; the results of the verification analysis will be highly dependent on this aspect of the verification.

2. Measures-oriented verification methods such as computation of RMSE, MAE, ETS, etc., can provide basic information about forecast performance for spatial forecasts, just as they can be used for point forecasts; however, interpretation of the results of these measures may be clouded by displacement errors that they don't directly measure.

3. Traditional verification approaches for spatial forecasts, such as the S1 score and the anomaly correlation, are useful for monitoring performance over time, particularly because these measures often were the only approaches used in the past and thus are the only measures available of historical forecast performance.

4. New spatial verification approaches can provide diagnostic information about forecast performance that cannot be obtained from traditional approaches, and these methods can account for errors such as displacement that commonly are of concern with high-resolution forecasting systems. Each of these methods is appropriate for considering particular types of verification questions.

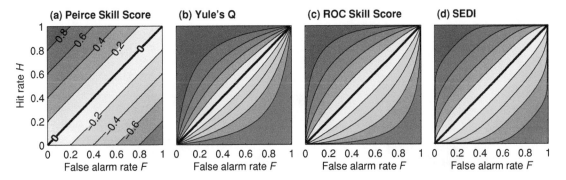

**Figure 3.2** Four base-rate independent performance measures plotted on ROC axes. All share the same colour scale as the Peirce Skill Score in panel (a). SEDI is the Symmetric Extremal Dependence Index

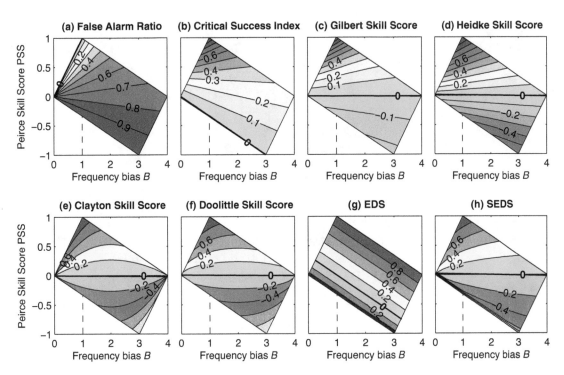

**Figure 3.3** Eight base-rate-dependent performance measures plotted on skill-bias axes for a base rate of 0.25. The vertical dashed lines indicate no bias. EDS, Extreme Dependency Score; SEDS, Symmetric Extreme Dependency Score

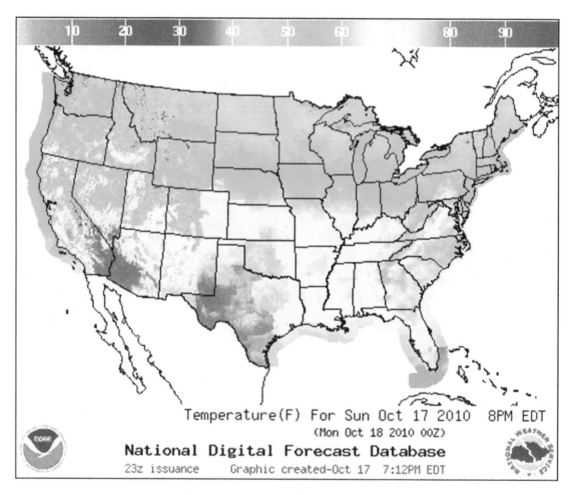

**Figure 6.1** Example of a spatial forecast

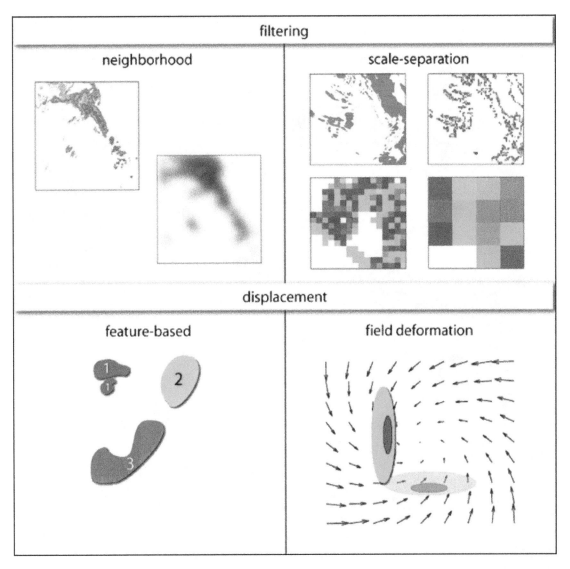

**Figure 6.4** Four types of spatial verification methods. From "Wavelets and field forecast verification" by Briggs and Levine, *Monthly Weather Review*, 125, 1329–1341, 1997. © American Meteorological Society. Reprinted with permission

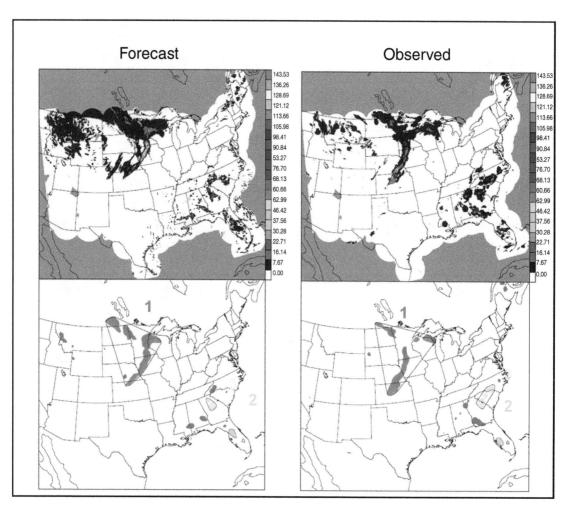

**Figure 6.7** Example of an application of the Method for Object-based Diagnostic Evaluation (MODE). See text for details

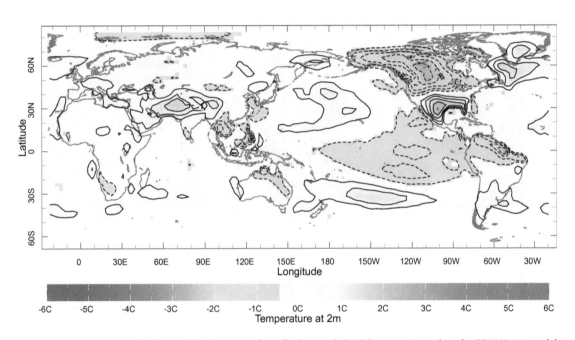

**Figure 11.1** Example of a 'deterministic' seasonal prediction made in February 2011 using the ECHAM 4.5 model (Roeckner *et al.* 1996). The prediction is expressed as the ensemble mean March–May 2011 temperature anomaly with respect to the model's 1971–2000 climatology

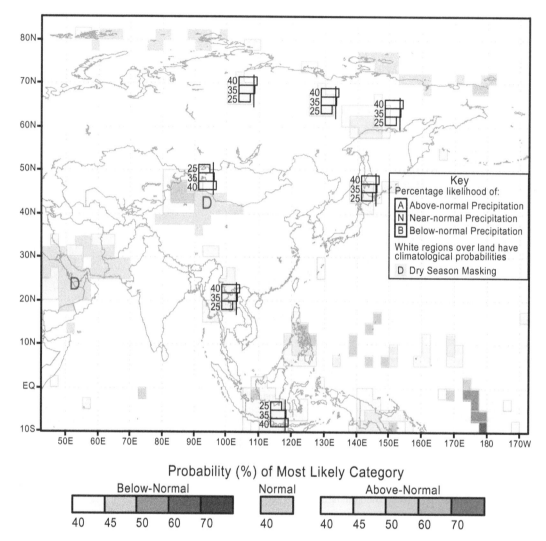

**Figure 11.2** Example of a 'probabilistic' seasonal prediction issued in February 2011 by the International Research Institute for Climate and Society. The prediction shows the probabilities that the March–May 2011 precipitation total will be in one of the three categories 'above-normal', 'normal' and 'below-normal', with these categories defined using the terciles of March–May totals between 1971 and 2000. The probabilities of the most likely category are shaded, but the probability for the normal category is constrained to a maximum of 40% based on prior verification analyses. Probabilities for all three categories are only shown for large areas and for areas with relatively sharp probabilities. The three horizontal bars are scaled by the corresponding forecast probability, and the thin vertical line indicates the climatological probability of 33%

# 7

# Probability forecasts

**Jochen Broecker**

*Max-Planck-Institute for the Physics of Complex Systems*

## 7.1  Introduction

In the widest sense, a forecast is a statement about the veracity of some proposition concerning the state of the world. In all but the most fortunate situations though, information on the state of the world is incomplete. Whatever the reason for the lack of information, forecasters should take into account and properly communicate this uncertainty in their statements. Based on these statements, forecast users should be able to decide upon which of the available actions is the most amenable for them, and what the likely consequences will be. As has been argued by various authors, forecasters should therefore issue their statements in the form of probabilities. For a discussion on why and to what extent forecasts should be cast as probabilities, the interested reader is referred to Good (1952), Thompson (1966), Murphy (1991b), de Finetti (1970), Brown (1970), and Blackwell and Girshick (1979) (a list that is not meant to be exhaustive); see also Murphy (1998) for an interesting historical overview.

In this chapter, after briefly explaining how probabilities can be employed to express forecasts, I will focus on how probabilistic forecasts can be *evaluated* using data. Section 7.2.1 contains a few remarks on the general concept of probability and on the notation used in this chapter. In Section 7.2.2, I discuss the concept of probabilistic forecasting schemes; these allow for evaluation using so-called scoring rules, to be explained in Section 7.3. The concept of scoring rules might seem a little arbitrary at first sight, but will be given justification through the concepts of resolution and reliability. These two forecast attributes are widely regarded as virtues for which the case can be made independent of scoring rules. Special attention will be given to binary problems, which have much in common with hypothesis testing. The relative operating characteristic, a device to evaluate binary forecasts, is revisited in Section 7.4.

Sections 7.2, 7.3 and 7.4 constitute the first and more theoretical part of this chapter; the practitioner though, to whom this book is intended to be a guide, will certainly be much interested in how forecasts could be evaluated given a sample of individual forecasts from a long-running forecasting scheme and corresponding verifications. These problems will be covered in the second part of this chapter, Sections 7.5 and 7.6. Several techniques for verifying probabilistic forecasts will be presented and illustrated with three examples.

*Forecast Verification: A Practitioner's Guide in Atmospheric Science*, Second Edition. Edited by Ian T. Jolliffe and David B. Stephenson.
© 2012 John Wiley & Sons, Ltd. Published 2012 by John Wiley & Sons, Ltd.

## 7.2  Probability theory

This section consists of two subsections. The first contains a compilation of probability concepts, mainly intended to fix notation and to point the reader to relevant references. The concept of conditional probability will be given special attention. In the second subsection, probabilistic forecasting schemes are introduced and discussed.

### 7.2.1  Basic concepts from probability theory

There is a plethora of books that provide an introduction to probability theory. The books of Mood *et al.* (1974) and Feller (1966) are relatively elementary, while Breiman (1973), Halmos (1974) and Loeve (1963) appeal to the more mathematically oriented reader. Savage (1972) and Jeffrey (2004) treat the subject from a personalistic point of view. For this chapter, we assume only a basic familiarity with the concept of probability, as for example detailed in Wilks (2006b, chapter 1), or in Sections 2.5 and 2.6 of this book.

In probability theory, one starts with considering a set $\Omega$ called *sample space*, subsets of which are called *events*. (In this chapter, italics indicate that an expression is to be considered a technical term.) The *probability* of an event $A$ will be denoted with $Pr(A)$.

Suppose we perform random experiments with outcomes in a set $\mathbb{K}$ having $K$ elements, called *categories*, which we simply identify with $\{1 \ldots K\}$, unless otherwise stated. In order to capture possible connections between different experiments, the individual experiments are usually considered parts of a single compound experiment with outcomes in the sample space $\Omega$. The outcome of an individual experiment thus becomes a function $X : \Omega \to \mathbb{K}$. Such a function is called a *random variable*. Later in this chapter, we will need to consider random variables with a continuous range of values. A rigorous treatment of such random variables requires some mathematical machinery that is beyond the scope of a 'Practitioner's Guide'; hence we will always argue as if $\Omega$ and $\mathbb{K}$ were finite.

A probability on $\Omega$ and a random variable $X$ give rise to a *distribution* of $X$, or rather, a distribution of

values of $X$ in $\mathbb{K}$. This distribution can be described in several ways. For discrete $\mathbb{K}$, we write

$$p_X(k) = Pr\{X = k\} \qquad \text{for all } k \in \mathbb{K}$$

for the *probability function* of $X$. (We will keep the subscript on $p_X$ in order to specify that this probability function belongs to the random variable $X$.)

If $\mathbb{K} =$ the real numbers, the *cumulative distribution function* is defined as

$$F_X(c) = Pr\{X < c\}.$$

The derivative $p_X(x) = F_X'(x)$ (if it exists) is referred to as the *density* of $X$.

The reader is probably aware of the fact that a probability gives rise to an integral for random variables, which is also referred to as the *expectation* of $X$. For a random variable $X$ that assumes a finite number of values (taken as $\{c_1, \ldots, c_K\}$ just for this definition), the expectation is defined as

$$E(X) = \sum_k c_k Pr\{X = c_k\}.$$

In terms of the cumulative distribution function and density of $X$, the expectation can be expressed in the following equivalent ways:

$$E(X) = \int_{\mathbb{R}} x \, \mathrm{d}F_X(x) = \int_{\mathbb{R}} x p_X(x) \mathrm{d}x \qquad (7.1)$$

The term 'expectation' can be misleading, and $E(X)$ should not be too literally interpreted as a 'value that is to be expected'. In general, $E(X)$ need not be equal to any of the possible values $\{c_1, \ldots, c_K\}$ of $X$, so there is no sense in which $E(X)$ is a typical outcome of $X$. Sometimes, the values of $X$ are not even quantities but merely labels of some categories, such as 'rain/hail/snow/sunshine'. In this case, the expectation ceases to have any meaning.

Very important in probabilistic forecasting is the concept of conditional probabilities and expectations. The probability $Pr(A)$ quantifies our *a priori* knowledge about specific event $A$ in the absence of any further information. Suppose now that, by performing an experiment (or equivalently, taking measurements), we learn that $X = c$ for some value

$c \in \mathbb{K}$. Based on that information, the probabilities for all events ought to change, depending on $c$. This new probability is called the *conditional probability* of the event $A$ given $X = c$, and is defined as

$$Pr[A|X = c] = \frac{Pr(A \cap \{X = c\})}{Pr\{X = c\}} \qquad (7.2)$$

It is important to keep in mind that $Pr[A|X = c]$ is a function of two arguments, namely the set $A$ and the number $c$. To all intents and purposes, the conditional probability $Pr[A|X = c]$ can be treated just like a standard probability depending on a parameter. That is, for fixed $c$, it obeys the laws of probability. In particular, for any other random variable $Y$, we can define the *conditional distribution* of $Y$ given $X$. We thus get conditional probability functions, cumulative distribution functions and conditional densities by applying the definitions for the unconditional version, *mutatis mutandis*. Equation 7.2 written in terms of probability functions gives the famous Bayes rule

$$p_{Y|X}(y|x) = \frac{p_{Y,X}(y, x)}{p_X(x)} \qquad (7.3)$$

which is in fact also true for densities. Heuristically, it is obtained by letting $Pr\{X = c\}$ go to zero in Equation 7.2. Using the conditional distribution function in Equation 7.1, we obtain the *conditional expectation* of $Y$ given $X$, defined thus

$$E[Y|X = c] = \int y\, p_{Y|X}(y|c)\mathrm{d}y$$

in terms of the conditional density.

This section will finish with a warning concerning the conditional expectation. Let $f(x, y)$ be some function, with $x, y \in \mathbb{K}$. Often, the notation $E_Y[f(X, Y)]$ is found, which is supposed to mean the expectation over $Y$, with $X$ being held constant. This notion becomes problematic if $X$ and $Y$ are dependent, since then keeping $X$ constant has side effects on $Y$. Then, $E_Y[f(X, Y)]$ can either mean to calculate $E[f(x, Y)]$ as a function of the non-random parameter $x$ and then setting $x = X$, or it can mean $E[f(X, Y)|X]$. In general, these notions coincide only if $X$ and $Y$ are independent. The conditional expectation can also be thought of as calculating $E[f(x, Y)]$ as a function of the non-random parameter $x$, but *taking into account* the side effects on the distribution of $Y$.

### 7.2.2 Probability forecasts, reliability and sufficiency

Let $Y$ denote the quantity to be forecast, commonly referred to as the *observation, predictant* or *target*. The observation $Y$ is modelled here as a random variable taking values in a set $\mathbb{K}$. As before, $\mathbb{K}$ is assumed to be a finite set of alternatives (e.g. 'rain/hail/snow/sunshine'), labelled $1 \ldots K$. Values of $Y$ (i.e. elements of $\mathbb{K}$) will be denoted by small lowercase letters like $k$ or $l$.

A *probability assignment* over $\mathbb{K}$ is a $K$-dimensional vector $p$ with non-negative entries so that $\sum_{k=1}^{K} p_k = 1$. Generic probability assignments will be denoted by $p$ and $q$. A *probabilistic forecasting scheme* is a random variable $\Gamma$, the values of which are probability assignments over $\mathbb{K}$. We can also think of $\Gamma$ as a vector of $K$ random variables $(\Gamma_1 \ldots \Gamma_K)$ with the property that

$$\Gamma_k \geq 0 \text{ for all } k, \qquad \sum_k \Gamma_k = 1$$

Since the probability assignments over $\mathbb{K}$ form a continuum, $\Gamma$ is, in general, a random variable with continuous range.

The reason for assuming $\Gamma$ to be random is that forecasting schemes usually process information available before and at forecast time. For example, if $\Gamma$ is a weather forecasting scheme with lead time 48 h, it will depend on weather information down to 48 h prior to when the observation $Y$ is obtained. Designing a forecasting scheme means modelling the relationship between this side information and the variable to be forecast. Taking both the observation as well as the forecast as random variables is often referred to as the distributions-oriented approach; see for example Murphy and Winkler (1987), Murphy (1993), Murphy (1996b), or Section 2.10 in this book. Section 3.2.2 considers the distributions-oriented approach in the context of binary events.

In the remainder of this section, we will discuss desirable attributes of probability forecasts. Clarifying what makes a 'good' forecast is obviously the first step towards evaluating them. The first property to be discussed here is *reliability*. (The reader is invited to compare the present reliability concept with that of Section 2.10 for deterministic forecasts.) On the condition that the forecasting scheme is equal to, say, the probability assignment $p$, the observation $Y$ should be distributed according to $p$. This means

$$Pr[Y = k | \Gamma = p] = p_k \qquad (7.4)$$

for all $k \in \mathbb{K}$. In particular, a reliable forecasting scheme can be written as a conditional probability. The reverse is also true: every conditional probability of $Y$ is a reliable forecast. For a proof of this statement, see Bröcker (2008).

In view of Equation (7.4), we will fix the notation $\pi_k^{(\Gamma)} = Pr[Y = k | \Gamma]$ for the conditional probability of the observation given the forecasting scheme. $\pi^{(\Gamma)}$ is a probabilistic forecasting scheme like $\Gamma$ itself. In terms of $\pi^{(\Gamma)}$ and $\Gamma$, the reliability condition (Equation 7.4) can be written simply as $\pi^{(\Gamma)} = \Gamma$. Since $\pi^{(\Gamma)}$ is reliable, it trivially holds that

$$\pi^{(\Gamma)} = \pi^{\left(\pi^{(\Gamma)}\right)}$$

In any case, $\pi^{(\Gamma)}$ is a conditional probability depending on $\Gamma$, independent of whether $\Gamma$ is reliable or not.

Reliability certainly is a desirable forecast property, but reliability alone does not make for a useful forecast; it was mentioned that any conditional probability (and, in particular, the unconditional probability distribution of $Y$) constitutes a reliable forecast. In order to distinguish between a more or less skilful forecast, we consider the concept of sufficiency. We will say that, with respect to $Y$, a random variable $\Gamma^{(2)}$ is *sufficient* for a random variable $\Gamma^{(1)}$ if

$$Pr\left[Y = k \,\middle|\, \Gamma^{(1)}, \Gamma^{(2)}\right] = Pr\left[Y = k \,\middle|\, \Gamma^{(2)}\right] \quad (7.5)$$

The interpretation is that, once $\Gamma^{(2)}$ is known, $\Gamma^{(1)}$ does not provide any additional information on $Y$. In the definition of sufficiency, both $\Gamma^{(1)}$ and $\Gamma^{(2)}$ need not be forecasting schemes, but can be arbitrary random variables.

## 7.3 Probabilistic scoring rules

In this section, the concept of scoring rules for probability forecasts will be discussed by which the success of such forecasting schemes can be measured and quantified. Scoring rules were already mentioned in Chapter 3 in a more general context. Here, we will consider scoring rules especially for probabilistic forecasts. Scoring rules can be thought of as a way to assign 'points' or 'rewards' to probability forecasts, providing quantitative indication of success in predicting the observation. Scoring rules measure the success of a single forecast; the overall score of a forecasting scheme is then taken as the average score over individual cases. In Section 7.4 the relative operating characteristic will be revisited, which, in contrast to scoring rules, cannot be written as an average score over individual instances.

### 7.3.1 Definition and properties of scoring rules

Consider a forecasting scheme $\Gamma$ and an observation $Y$. At some point, the forecasting scheme $\Gamma$ issues the forecast $p$ (a k-vector of probabilities for K categories), while at the same time, the observation $Y$ takes on the value $k$. A *scoring rule* (see, e.g., Matheson and Winkler, 1976; Gneiting and Raftery, 2007) is a function $S(p, k)$ that takes a probability assignment $p$ as its first argument and an element $k$ of $\mathbb{K}$ as its second argument. The value $S(p, k)$ can literally be interpreted as how well the forecast $p$ 'scored' when predicting the outcome $y$. The reader should be warned that different conventions exist in the literature as to whether a large or a small score indicates a good forecast. We will take a small score as indicating a good forecast.

One of the most commonly used scoring rules is the Brier score (Brier, 1950), which applies to the case where $k = 0$ or $1$ and reads as

$$S(p, k) = (p_1 - k)^2$$

In Section 7.3.2, further scoring rules are presented. Many scores come in several different versions applicable to different types of probability assignments, such as for densities and cumulative distribution functions; these are listed in Section 7.3.2 as well.

There are certain restrictions as to which scoring rules should be employed for evaluating probabilistic forecasts. We might, for example, contemplate a score of the form $S(p, k) = |p_1 - k|$, i.e. similar to the Brier score but with the absolute rather than the squared difference. We will see later explicitly why this score would lead to inconsistent results. In order to avoid such inconsistencies, scoring rules should be *strictly proper*, a property we want to discuss now (see Brown, 1970; Bröcker and Smith, 2007b; see also Sections 2.8 and 3.3). To define this concept, consider the *scoring function*, which for any two probability assignments $p$ and $q$ is defined as

$$s(p, q) = \sum_{k \in \mathbb{K}} S(p, k) q_k \qquad (7.6)$$

The scoring function is to be interpreted as the mathematical expectation of the score of a forecasting scheme that issues constant forecasts equal to $p$ for a random variable $Y$ that has in fact distribution $q$. We might reasonably demand that if the actual distribution of $Y$ is $q$, then any probability assignment different from $q$ should have a worse expected score than $q$ itself. This is the essence of the following definition. A score is called *proper* if the *divergence*

$$d(p, q) = s(p, q) - s(q, q) \qquad (7.7)$$

is non-negative, and it is called *strictly proper* if it is proper and $d(p, q) = 0$ implies $p = q$. The interpretation of $d(p, q)$ as a measure of discrepancy between $p$ and $q$ is obviously meaningful only if the scoring rule is strictly proper. It is important to note that $d(p, q)$ is, in general, not a metric, as it is neither symmetric nor does it fulfil the triangle inequality. We will exploit strict propriety extensively in Section 7.3.3, where it is shown that strictly proper scoring rules indeed quantify positive forecasting attributes such as reliability and sufficiency. A few remarks on improper scoring rules will follow at the end of this subsection. From now on, the term 'scoring rule' is assumed to mean 'strictly proper scoring rule' unless otherwise stated.

We will now discuss a few general aspects of scoring rules and how to construct them in a systematic way. The quantity

$$e(q) = s(q, q) \qquad (7.8)$$

is called the *entropy* associated with the scoring rule $S$. That this nomenclature is justified will be seen later when we consider specific examples of scoring rules (and their associated entropies). Taking the minimum over $p$ in Equation (7.7), we obtain (due to propriety and with Equation 7.8)

$$e(q) = \min_p s(p, q) \qquad (7.9)$$

From the definition in Equation 7.6, we see that $s(p, q)$ is linear in $q$. Therefore, Equation 7.9 demonstrates that the entropy $e(q)$ is a minimum over linear functions and hence concave (Rockafellar, 1970). From Equation 7.8 we gather that $s(p, q)$ is the linear function that is tangent to the entropy $e$ at $p$. These arguments can be reversed, which yields a recipe for constructing a scoring rule from a given concave (but otherwise arbitrary) function $e(q)$: Set

$$s(p, q) = e(p) + \sum_k \frac{\partial e}{\partial p_k}(p) \cdot (q_k - p_k) \quad (7.10)$$

Then $s(p, q)$ is the linear function that is tangent to $e$ at $p$. Thus, $s(p, q)$ provides a scoring function with corresponding scoring rule $S(p, k) = s(p, e_k)$ (here, $e_k$ is the $k^{\text{th}}$ unit vector).

We conclude this section by showing that the mentioned improper scoring rule $S(p, k) = |p_1 - k|$ for the case $k = 0$ or $1$ yields inconsistent results. This discussion might serve as a further justification of requiring that scoring rules be strictly proper. First, note that the scoring rule might alternatively be written as $S(p, k) = 1 - p_k$; versions of this scoring rule have indeed been proposed in the literature. The divergence of this score is given by $d(p, q) = p_1 - q_1 + 2(p_1 - q_1)q_1$. The minimum of the divergence as a function of $p$ for fixed $q$ should obtain at $q$ if this score were proper. However, the minimum obtains at $p_1 = 0$, respectively $p_1 = 1$ for $q < 0.5$, respectively $q > 0.5$. This means that to score well, a forecaster would *always* issue forecasts of either $p_1 = 0$ or $1$. Hence, this score encourages overconfident forecasts. In other words, with this score a good probability forecast that is not overconfident might look poor. A thorough discussion of the inconsistencies brought about by using improper scoring rules can be found in Brown (1970) and Bröcker and Smith (2007b).

### 7.3.2 Commonly used scoring rules

All presented scoring rules are strictly proper. The precise definitions of the scoring rules might differ from author to author.

- **Quadratic Score**   The scoring rule is given by

$$S(p, k) = \sum_j (p_j - \delta_{jk})^2$$

where we use $\delta_{jk} = 1$ if $j = k$ and $\delta_{jk} = 0$ otherwise. For $\mathbb{K} = \{0, 1\}$, this score reduces to twice the Brier score. The divergence and entropy read as

$$d(p, q) = \sum_j (p_j - q_j)^2, \qquad e(p) = 1 - \sum_j p_j^2$$

The Quadratic scoring rule can be written as

$$S(p, k) = \sum_j p_j^2 - 2p_k + 1$$

and often the constant offset of 1 is omitted. Moreover, this variant of the quadratic scoring rule has a version in terms of densities given as

$$S(p, x) = \int p(z)^2 dz - 2p(x).$$

The divergence and entropy are, respectively,

$$d(p, q) = \int (p(x) - q(x))^2 dx,$$

$$e(p) = \int -p(x)^2 dx.$$

- **Ignorance Score** (or logarithmic score)   This scoring rule is given by

$$S(p, k) = -\log p_k$$

The divergence and entropy read as

$$d(p, q) = \sum_j -\log \left( \frac{p_j}{q_j} \right) q_j,$$

$$e(p) = \sum_j -\log \left( p_j \right) p_j,$$

respectively. The Ignorance Score has a version in terms of densities. As with the Quadratic scoring rule, the corresponding expressions are given by replacing summation with integration, and probability assignments with densities. An interesting property of the Ignorance Score is that $S(p, k)$ depends only on $p_k$ (and not on any $p_{j \neq k}$), i.e. only on the probability that is actually verified. This property is called *locality*. It can be shown that if $K \geq 3$, any strictly proper local score must be an affine function of the Ignorance score – this fact has been rediscovered many times; see e.g. Bernardo (1979) and Savage (1971), who attributes it to Gleason.

- **Pseudospherical Score**   This scoring rule is given by

$$S(p, k) = -\frac{p_k^{\alpha-1}}{\|p\|_\alpha^{\alpha-1}}$$

with some $\alpha > 1$ and $\|p\|_\alpha = [\sum_l p_l^\alpha]^{1/\alpha}$. The propriety of this score follows from Hölder's inequality. The divergence and entropy read as

$$d(p, q) = \|q\|_\alpha - \frac{\sum_j q_j p_j^{\alpha-1}}{\|p\|_\alpha^{\alpha-1}}, \qquad e(p) = -\|p\|_\alpha$$

The Pseudospherical scoring rule has a version in terms of densities. Again, the corresponding expressions are given by replacing summation with integration, and probability assignments with densities.

- **Ranked Probability Score**   This scoring rule is given by

$$S(p, k) = \sum_j \left( \sum_{i=1}^j p_i - H(j - k) \right)^2$$

Here, the function $H(x)$, the Heaviside function, is 1 if $x \leq 0$ and zero otherwise. Also this score has a version for continuous observations, called the *Continuous Ranked Probability Score* (CRPS). To define the CRPS, we replace summation with integration, and probability assignments with densities. Noting that $\int_{z \leq x} p(z) dz = F(x)$ is the cumulative distribution function, we arrive at

$$S(F, x) = \int_z (F(z) - H(z - x))^2 dz$$

In terms of references for scoring rules, see Epstein (1969) and Murphy (1971) for a discussion of the Ranked Probability Score. Scoring rules for continuous variables are discussed in Matheson and Winkler (1976) and Gneiting and Raftery (2007); for applications of these scoring rules in a meteorological context, see Raftery *et al.* (2005) and Bröcker and Smith (2008).

As to why e($p$) is called the entropy, it is clear that this nomenclature is justified for the Ignorance Score, since in that case e($p$) agrees with what is known as the entropy both in statistical physics and information theory. In general, the term 'entropy' is associated with a measure of uncertainty inherent to a distribution. In our situation, it seems reasonable to call the distribution that assigns equal probability of $1/K$ to all elements of $\mathbb{K}$ the 'most uncertain' distribution. A simple calculation will show that this distribution indeed maximizes the entropy for all aforementioned scoring rules.

### 7.3.3 Decomposition of scoring rules

A scoring rule $S(p, k)$ provides a means to evaluate a probabilistic forecasting scheme individually for each instant in time. Certainly, to get statistically meaningful results, we should be interested in the *average score* of a forecast over many pairs of forecasts and corresponding observations. For an infinitely large data set, the average score would be equal to the mathematical expectation $E(S(\Gamma, Y))$. (Since $\Gamma$ is random, the expectation affects both $\Gamma$ and $Y$.) The aim of this subsection is to discuss a decomposition of the expectation $E(S(\Gamma, Y))$ of the score. This result is of theoretical interest, as it provides a justification of the scoring rule methodology. The corresponding result for finite data sets will

be discussed in Section 7.5.4. The decomposition is presented here as three different statements in Equations 7.11 to 7.13. The first and third statements, applied to the Brier score, amount to a very well known decomposition, apparently due to Murphy (1973). We will abstain from giving mathematical proofs. The interested reader is referred to the classical paper by Murphy (1973) for the Brier score, to Hersbach (2000) for the CRPS, and to Bröcker (2009) for the general case.

To formulate the first decomposition, the probability of $Y$ given $\Gamma$ is needed, which was already considered in Section 7.2.2 and denoted with $\pi^{(\Gamma)}$. With this, the first decomposition reads as

$$E(S(\Gamma, Y)) = E\left(S(\pi^{(\Gamma)}, Y)\right) + E\left(d(\Gamma, \pi^{(\Gamma)})\right)$$
$$(7.11)$$

The second term on the right-hand side in Equation 7.11 is positive definite, and referred to as the *reliability term*. It describes the deviation of $\Gamma$ from $\pi^{(\Gamma)}$. Recalling that $\Gamma = \pi^{(\Gamma)}$ indicates a reliable forecast, the interpretation of the reliability term as the average violation of reliability becomes obvious. Hence, Equation 7.11 essentially says that $\pi^{(\Gamma)}$ achieves a better expected score than $\Gamma$, confirming our intuitive understanding that reliability is a virtuous forecast property. The first term in Equation 7.11 is referred to as the *potential score* (the terminology follows Hersbach, 2000). The potential score will be subject to the other two decompositions.

Our second decomposition involves the concept of sufficiency. Let $\Gamma^{(1)}, \Gamma^{(2)}$ be two forecasting schemes, and suppose that $\Gamma^{(1)}$ is sufficient for $\Gamma^{(2)}$. Our second decomposition is

$$E(S(\pi^{(2)}, Y)) = E(S(\pi^{(1)}, Y)) + Ed(\pi^{(2)}, \pi^{(1)})$$
$$(7.12)$$

This decomposition quantifies the difference in potential score between two forecasting systems where one is sufficient for the other. Again, note that d($\ldots$) is positive definite, whence $\Gamma^{(1)}$ has a better potential score than $\Gamma^{(2)}$. In summary, Equation 7.12 says that: if $\Gamma^{(1)}$ is sufficient for $\Gamma^{(2)}$, then $\pi^{(\Gamma^{(1)})}$ achieves a better expected score than $\pi^{(\Gamma^{(2)})}$ confirming our intuitive understanding that $\Gamma^{(1)}$ contains more information than $\Gamma^{(2)}$.

Our third decomposition considers the (non-random) forecasting scheme $\bar{\pi} = Pr(Y = k)$. Equivalently, $\bar{\pi}$ is the expectation of $\pi^{(\Gamma)}$,

$$\bar{\pi} = E\pi^{(\Gamma)}$$

In meteorology, $\bar{\pi}$ is often referred to as the *climatology* of $Y$. Our third decomposition is

$$E(S(\pi^{(\Gamma)}, Y)) = e(\bar{\pi}) - Ed(\bar{\pi}, \pi^{(\Gamma)}) \quad (7.13)$$

(This is a consequence of Equation 7.12 and the fact that any forecasting scheme $\Gamma$ is sufficient for $\bar{\pi}$.) This relation gives a concise description of the potential score in terms of the fundamental uncertainty $e(\bar{\pi})$ of $Y$ and the positive definite *resolution term* $Ed(\bar{\pi}, \pi^{(\Gamma)})$.

A few more words on the interpretation of Equation 7.13 might be in order. The entropy of the climatology $e(\bar{\pi})$ can be seen as the expectation of the score of the climatology as a forecasting scheme. Hence, the entropy quantifies the ability of the climatology to forecast random draws from itself. The resolution term describes, roughly speaking, the average deviation of $\pi^{(\Gamma)}$ from its expectation $\bar{\pi}$; it can therefore be interpreted as a form of 'variance' of $\pi^{(\Gamma)}$. The resolution term is indeed given by the standard variance of $\pi^{(\Gamma)}$ in case of the Brier score. It might seem counterintuitive at first that the larger this 'variance', the better the score. Consider an event that happens with a 50% chance. Then $p = 0.5$ is a reliable forecast that has zero resolution with respect to any score. Consider another forecast that says either $p = 0.1$ or $p = 0.9$, and that is also reliable. Necessarily, the forecasts $p = 0.1$ and $p = 0.9$ must both occur with frequency 0.5. The latter forecast is clearly more informative than the former; if it says '0.1', we know for sure that the event is unlikely, while '0.9' is a reliable indicator of a likely event. The fact that the second forecast is more informative is actually the reason for its larger variability.

The decompositions show that proper scoring rules respect the notions of reliability and sufficiency, which have been motivated independently from the the concept of scoring rules. Indeed, reliability improves the score, and if a forecasting scheme is sufficient the former will have a better potential score than the latter forecast. In general though, using different scoring rules will generally lead to a different ranking of a given set of forecasting schemes. In particular, unreliable forecasting schemes might be ranked differently by different scoring rules, even if one is sufficient for the other. One scoring rule might strongly penalize the lack of reliability of one forecast, while another scoring rule might be more sensitive to lack of resolution of other forecasts. Furthermore, there is no general rule for how a score will rank forecast schemes (reliable or unreliable) as long as none of the two forecast schemes is sufficient for the other. In summary, the actual ranking of such forecasts will depend on the particular scoring rule employed.

## 7.4 The relative operating characteristic (ROC)

In this section, we will consider an observation $Y$ taking only the values 0 or 1 (binary problem). Probability forecasts for such problems can be described by a single number $\Gamma = p_1$ since necessarily $p_0 = 1 - p_1$. For this reason, probability forecasts for binary problems are amenable to evaluation with the receiver operating characteristic (ROC), discussed in Section 3.3. We will briefly discuss what such an evaluation would portend about $\Gamma$ as a probability forecast.

First we recall the basic definitions in relation to ROCs. Suppose there is a random variable $W$, which will henceforth be referred to as the *decision variable*. In Chapter 3, $W$ is called weight of evidence. The idea is that a large value of $W$ is indicative of the event $Y = 1$, while a small $W$ is indicative of the event $Y = 0$. That is, $W$ exceeding a certain threshold $w$ could be interpreted as signalling an impending event. When evaluating probability forecasts by means of the ROC, then $\Gamma$ will play the role of $W$. We will keep writing $W$ in this section and mention explicitly when $W$ is to be interpreted as a probability forecast.

We define the *hit rate* as

$$H(w) = Pr(W \geq w | Y = 1) \quad (7.14)$$

and the *false alarm rate* as

$$F(w) = Pr(W \geq w | Y = 0) \quad (7.15)$$

Alternative names for the hit rate are rate of true positives or the power of the test $W \geq w$. Alternative names for the false alarm rate are rate of false positives or the size of the test $W \geq w$. The ROC consists of a plot of $H$ versus $F$, with $w$ acting as a parameter along the curve. It follows readily from the definitions that both $H$ and $F$ are monotonically decreasing functions of $w$ with limits 0 for increasing $w$ and 1 for decreasing $w$. Hence, the ROC curve is a monotonically *increasing* arc connecting the points $(0, 0)$ and $(1, 1)$. The exact shape of the ROC generally depends on the statistics of $Y$ and $W$. A typical ROC is shown in Figure 7.1.

An important question is when should the decision variable $W$ be considered better than another decision variable $W'$ (for a given observation $Y$). We will refer to $W$ as being *not superior* to $W'$, if for all false alarm rates $F$ between 0 and 1, the corresponding hit rates $H$ of $W$ are equal to or less than the corresponding hit rates $H'$ of $W'$. In other words, the ROC curve corresponding to $W$ is nowhere above that of $W'$.

A probability forecast (for binary problems) is essentially a decision variable that in addition one would like to be reliable. How does reliability matter for the ROC? The answer is that in order to be *optimal*, it is sufficient, but not necessary, that the decision variable is a reliable probability forecast. (Here, by 'optimal' we mean that no decision variable of the form $f(W)$, which would arise, e.g., through post-processing, can ever be superior to $W$.) Put differently, post-processing an already reliable probability forecast will at best result in the same ROC; but an optimal decision variable need not be a reliable probability forecast.

To see why these statements are true, we first note that the ROCs for some decision variables $W$ and $f(W)$ will be the same if $f$ is a monotonically increasing function. This can be shown using the definitions of hit rates and false alarm rates. Alternatively, one can argue that a monotonic transformation of $W$ merely amounts to a relabelling of the threshold values; the patterns of warnings and non-warnings remain exactly the same. We can conclude from this discussion that if $W$ is a reliable probability forecast and $f$ is a monotonic transformation, then $f(W)$ will in general lose reliability but still have the same ROC.

We now discuss why a reliable probability forecast is an instance of an optimal decision variable. Let $W$ be some decision variable, and consider the *likelihood ratio*

$$\lambda(w) = \frac{p_{W|Y}(w, 1)}{p_{W|Y}(w, 0)} \qquad (7.16)$$

(Here, $p_{W|Y}(w, y)$ is the conditional density of $W$ given $Y$; in Chapter 3 of this book, the notations $f_0(w) = p_{W|Y}(w, 0)$ and $f_1(w) = p_{W|Y}(w, 1)$ are used.) Applying the fundamental Neyman–Pearson lemma (Mood *et al.*, 1974), it is possible to show that the decision variable $\lambda(W)$ is optimal in the sense that no decision variable of the form $f(W)$ can possibly be superior to $\lambda(W)$. However, if $W$ is a reliable probability forecast, then by definition $W = Pr(Y = 1|W)$. Using the Bayes rule, we can write

$$W = Pr(Y = 1|W) = \frac{\lambda(W)}{\left(\lambda(W) + \dfrac{1 - \bar{\pi}}{\bar{\pi}}\right)}$$

where $\bar{\pi} = Pr(Y = 1)$. This relation shows that $W$ is a monotonically increasing function of $\lambda(W)$.

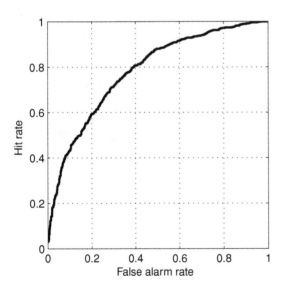

**Figure 7.1** The empirical relative operating characteristic (ROC) curve for the forecasts of Example A at a lead time of 8 days. The presence of skill in the forecast is manifest in the fact that the ROC curve shows a significant bend towards the north-western corner of the plot

Therefore, the decision functions $W$ and $\lambda(W)$ give the same ROC, and we have shown that a reliable probability forecast $W$ is another instance of an optimal decision variable.

As a final remark, we note that $\lambda(w)$ is the slope of the ROC curve of $W$ at threshold value $w$ (see Atger, 2004). We can conclude that replacing the decision variable $W$ with the slope of the ROC curve at $w = W$ is the best we can do in terms of improving the ROC through transforming or 'calibrating' $W$. If the ROC curve happens to be concave, then the slope (as a function of $w$) is monotonically increasing. Since monotonically increasing transformations of the decision variable do not change the ROC, we see that a concave ROC is optimal in that it cannot be improved any further by a transformation of the decision variable. On the other hand, a ROC curve that is not concave is not yet optimal, and therefore $W$ cannot be a reliable probability forecast. This might serve as a preliminary reliability test.

## 7.5 Evaluation of probabilistic forecasting systems from data

Sections 7.5 and 7.6 form the second part of this chapter, where we will be concerned with the question of how to assess the performance of a probabilistic forecasting scheme given instances of forecasts and corresponding observations. An *archive* $T$ is a series of pairs of forecasts and corresponding observations, that is, $T = \{(p(n), y(n)), n = 1 \ldots N\}$, where $p(n)$ are realizations of the probability forecast $\Gamma$ and $y(n)$ are realizations of the observation $Y$. The components of $p(n)$ are still written as subscripts, i.e. $p(n) = (p_1(n) \ldots p_K(n))$, and we have

$$p_k(n) \geq 0 \text{ for all } k, n, \qquad \sum_k p_k(n) = 1 \text{ for all } n.$$

### 7.5.1  Three examples

Several evaluation methods will be demonstrated by application to three examples of probabilistic forecasting systems. Hopefully, the examples are reasonably similar to problems the reader might encounter in practice, yet simple enough to not obscure the general ideas. For easy reference, the three examples will be associated with three imaginary forecasters referred to as A, B and C. Although the specific problems are different, all three forecasters used the following raw data. In terms of measurements, we use 2-metre temperature data from London's Heathrow Airport weather station, taken daily at 12.00 UTC. In terms of forecasts, dynamical weather forecasts from the European Centre for Medium Range Weather Forecasts (ECMWF) for 2-metre temperature are used. The ECMWF maintains an ensemble prediction system with 50 members. The ensemble members are independent runs of a global weather model, each generated with slightly perturbed initial conditions (there is also an unperturbed run, the control, which is not used here). In addition, the ECMWF issues a single run called a high-resolution forecast, generated with the same model but with roughly double the spatial resolution. Forecasts were available from 1 January 2001 until 31 December 2005, from the then operational ECMWF prediction system, featuring lead times from 1 to 10 days and a spatial resolution (for the ensembles) of about 80 km. (For more information, the reader is referred to Persson and Grazzini, 2005.) Only the forecast information relevant for London Heathrow is used. We will focus on lead times of 5 and 8 days only, for which we have respectively $N = 1814$ and $N = 1810$ daily values. All data were verified at noon. We will use the following notation: The station data will be denoted by $x(n), n = 1 \ldots N$. The ensemble is written as $\hat{x}(n), n = 1 \ldots N$, where each $\hat{x}(n)$ has 50 entries, denoted as $\hat{x}_k(n), k = 1 \ldots 50$. The high-resolution forecast is written as $h(n), n = 1 \ldots N$.

All three forecasters are interested in forecasting errors in the high-resolution forecast, i.e. making statements about $\delta(n) = x(n) - h(n)$. To this end, the ensemble will be employed. The details of the individual schemes and objectives are different though, as will be explained presently. It is necessary to keep in mind that although the basic idea in all three forecasting problems is the same, they are very different in detail, in particular concerning their statistics.

### Example A: binary forecasts
Forecaster A is only interested in whether the actual temperature will be lower or higher than predicted

by the high-resolution forecast. That is, her observations are given by

$$y(n) = \begin{cases} 1 & \text{if } x(n) \geq h(n), \\ 0 & \text{else.} \end{cases}$$

She uses the ensemble to build her probability forecasts by just counting the number of ensemble members, which turn out to be higher than the high-resolution forecast:

$$p_1(n) = \frac{\#\{k; x_k(n) \geq h(n)\} + \frac{1}{2}}{51},$$

and of course, $p_0(n) = 1 - p_1(n)$. The way in which forecaster A calculates her forecasts is essentially based on the assumption that the ensemble comprises a collection of equally likely scenarios of the atmosphere's future development – this is clearly an idealizing assumption. At least, as a safeguard against too overconfident forecasts in either direction, forecaster A adds $\frac{1}{2}$ in the numerator and 1 in the denominator.

There exist considerably more sophisticated ways to form (binary or multi-categorical) probability forecasts from either ensembles or other forecast information, which need not have any probabilistic character. Just to give the reader a feel of how such an approach might look and without going into the details, we mention logistic regression. Logistic regression assumes a model of the form

$$p_1 = \frac{\exp(x\beta')}{1 + \exp(x\beta')}$$

where $\beta$ are the coefficients, and $x$ is some variable providing forecast information. The coefficients $\beta$ can be determined by optimizing the performance of this scheme, for example as measured by the empirical score (see Section 5.3). More detailed discussions of logistic regression can be found in McCullagh and Nelder (1989) and Hastie *et al.* (2001). For applications of logistic regression (and alternatives) in meteorological contexts, see for example Crosby *et al.* (1995), Wilks and Hamill (2007), Primo *et al.* (2009) and Bröcker (2009).

## Example B: three-category forecasts

Forecaster B is interested in three rather than two categories: whether the actual observation will be higher than, about the same as, or lower than the high-resolution forecast. That is, his observations are given by

$$y(n) = \begin{cases} 1 & \text{if } x(n) \leq h(n) - \Delta_1, \\ 2 & \text{if } h(n) - \Delta_1 \leq x(n) < h(n) + \Delta_2, \\ 3 & \text{if } h(n) + \Delta_2 < x(n). \end{cases}$$

The two thresholds $\Delta_1, \Delta_2$ are chosen so that the three categories occur with about equal relative frequency. That is, the climatological probability of each category comes out to be about $1/3$. Forecaster B uses the ensemble to build his probability forecast very much like forecaster A. The forecast probability of each category is equal to the relative number of ensemble members in that category.

## Example C: continuous forecasts

Forecaster C is interested in continuous probability forecasts for the deviation $\delta(n)$ between station measurements and the high-resolution forecast. Using the ensemble, she builds densities as follows. She computes the ensemble mean and standard deviation

$$\mu(n) = \frac{1}{50} \sum_k x_k(n),$$

$$\sigma^2(n) = \frac{1}{49} \sum_k (x_k(n) - \mu(n))^2,$$

and then forms her forecast density

$$p(x, n) = \frac{1}{\sqrt{2\pi}\sigma} e^{-\frac{1}{2}\left(\frac{x - \mu(n)}{\sigma(n)}\right)^2}$$

Again, more sophisticated methods to transform ensembles into forecast densities exist. A variety of approaches have been proposed to transform ensembles into probability density functions; see for example Roulston and Smith (2003), Gneiting *et al.* (2005), Jewson (2003), Raftery *et al.* (2005) and Bröcker and Smith (2008). A more detailed account of this topic can be found in Chapter 8 of this book.

### 7.5.2 The empirical ROC

The first performance measure we consider is the ROC. As the ROC can only be applied to binary problems, we discuss results for forecaster A only. The following formulae assume that the archive $T$ has been sorted in ascending order along $p_1(n)$. With this step applied, the hit rates and false alarm rates can be estimated by approximating probabilities in the definitions (Equations 7.14 and 7.15) with observed frequencies. This yields

$$H(n) = 1 - \frac{\sum\limits_{m=1}^{n} y(m)}{N_1},$$

$$F(n) = 1 - \frac{\sum\limits_{m=1}^{n} 1 - y(m)}{N_0},$$

with

$$N_1 = \sum_{m=1}^{N} y(m), \qquad N_0 = N - N_1.$$

Here $H(n)$ and $F(n)$ are shorthands for $H(p_1(n))$ and $F(p_1(n))$, respectively. The ROC curve for forecaster A is shown in Figure 7.1.

Note that the forecast $p_1(n)$ appears only implicitly in these formulae through the initial sorting of the data. This again demonstrates that the entire ROC analysis remains unaffected by any monotonic transformation of $p_1(n)$, as such a transformation would not change the ordering.

A very common summary statistic for the ROC is the *area under the curve* (AUC). Ideally, the AUC is 1, while an AUC of $1/2$ indicates zero association between the forecast and the observation. By approximating the integral with a sum, we obtain the following estimate of the AUC (the second equality follows after some algebra):

$$\text{AUC} = -\sum_{n=1}^{n} H(n) \cdot (F(n) - F(n-1))$$

$$= \frac{1}{N_0 N_1} \left( \sum_{n=1}^{N} n y(n) - \frac{N_1(N_1+1)}{2} \right)$$

The term in brackets is the Mann–Whitney U-statistic (see, e.g., Mood *et al.*, 1974, Chapter 11,

section 5.5). Translating known facts about the U-statistic gives that in case of zero association between forecasts and observations, i.e. if they are independent, the AUC has a normal distribution, with mean and standard deviation given by

$$\mu = \frac{1}{2}$$

$$\sigma^2 = \frac{1}{12} \left( \frac{1}{N_0} + \frac{1}{N_1} + \frac{1}{N_0 N_1} \right)$$

Thus in case of zero association, the *p*-value,

$$1 - \Phi \left( \frac{\text{AUC} - \mu}{\sigma} \right)$$

has uniform distribution, where $\Phi$ is the standard normal cumulative distribution function.

The AUC for forecaster A is 0.731, with a *p*-value of essentially zero. Hence, such an AUC would be exceedingly unlikely if there were no association between forecasts and observations.

### 7.5.3 The empirical score as a measure of performance

Motivated by the discussion on scoring rules, we would like to quantify the performance of the presented forecasting schemes by means of the expected score $E(S(\Gamma, Y))$ – a quantity that is not available operationally, though. Given an archive $T$ of forecast-observation pairs, the performance of the forecast system – in other words, the expected score – can be estimated by the *empirical score*

$$\hat{S} = \frac{1}{N} \sum_{n=1}^{N} S(p(n), y(n))$$

The empirical score evaluates the average performance of the forecast system over all samples in the archive.

Frequently, one is interested not in the absolute but the relative score with respect to some reference forecast. Such a relative score is often referred to as *skill score*. More precisely, we consider

$$\hat{S}K = \frac{\hat{S} - \hat{S}_0}{\hat{S}_\infty - \hat{S}_0}$$

**Table 7.1** Absolute and relative (skill) scores of forecasts A, B and C, using the quadratic scoring rule. The uncertainty information represents plus or minus the standard deviation. The column headings $\hat{S}$, $\hat{S}_0$ and $\hat{S}K$ refer to the empirical score, the score of the reference forecast (climatology), and the skill score, respectively; see Section 7.5.3 for exact definitions

| Forecast | Lead time (days) | $\hat{S}$ | $\hat{S}_0$ | $\hat{S}K$ |
|---|---|---|---|---|
| A | 5 | $0.214 \pm 0.0052$ | 1/4 | $0.14 \pm 0.021$ |
|   | 8 | $0.192 \pm 0.005$ | 1/4 | $0.23 \pm 0.02$ |
| B | 5 | $0.595 \pm 0.0098$ | 2/3 | $0.12 \pm 0.015$ |
|   | 8 | $0.577 \pm 0.0097$ | 2/3 | $0.13 \pm 0.015$ |
| C | 5 | $-0.141 \pm 0.004$ | $-0.122 \pm 0.0013$ | $0.16 \pm 0.029$ |
|   | 8 | $-0.109 \pm 0.0028$ | $-0.0887 \pm 0.00091$ | $0.23 \pm 0.028$ |

Here, $\hat{S}_0$ is the score of a (usually low skill) reference forecast, while $\hat{S}_\infty$ is the best (i.e. minimum) attainable score. A popular reference forecast is the climatology. For all scoring rules considered here, the minimum attainable score is zero, so that we can write

$$\hat{S}K = 1 - \frac{\hat{S}}{\hat{S}_0}$$

The various scores of all three examples, A, B and C, are reported in Table 7.1.

The quadratic score was used throughout (which amounts to the Brier score in case of forecaster A). The figures give the mean value plus or minus the standard deviation. The standard deviation for the skill score can be computed by propagation of errors (see, e.g., Meyer, 1975).

A perhaps surprising observation in connection with Table 7.1 is that the scores of forecaster A and forecaster B become better with longer lead times (although the increase is barely significant). In view of the fact that we are forecasting errors in the high-resolution forecast (and not the temperatures themselves), this observation can be seen as a further justification of the ensemble methodology. Having said this, the score of forecaster C becomes worse with increasing lead time. Note, however, that in the case of forecaster C, comparison of results across different lead times is problematic, as the climatological distribution of the observations also changes; this does not occur with forecaster A and forecaster B. (We repeat that although the data behind all three examples are the same, the respective definitions of

the observation $y(n)$ differ.) Since the score $S_0$ of the climatology becomes worse as well, the relative (skill) score of forecaster C indeed shows improvement with increasing lead time. Whether the absolute or relative performance is relevant depends on the problem specification. We learn from this that comparing forecasting problems with different climatological distributions requires due care, and that both absolute and relative scores need to be looked at, as they may show different behaviour.

### 7.5.4 Decomposition of the empirical score

As we have seen in Section 7.3.3, the mathematical expectation of the score admits a decomposition into various terms. A similar decomposition holds for the empirical score, to be discussed now. To explain this decomposition, let $T$ again be the data archive of forecast-observation pairs. We assume that the forecasting scheme $\Gamma$ only issues forecasts from a finite set $\{q^{(1)} \ldots q^{(D)}\}$ of probability assignments. Clearly, this assumption is not fulfilled in example C, which features a continuous range of forecasts. Even if the range of forecasts is finite, the number of possible forecast values might still be too large for the subsequent analysis to be applicable. In such situations, the forecasts first have to be distributed among suitably chosen bins and then averaged, with the in-bin averages being the new set of forecasts $\{q^{(1)} \ldots q^{(D)}\}$. This procedure of *binning* (or stratifying) forecasts is important also for reliability analysis, which will be the subject of Section 7.6. For this reason, Section 7.5.5 has been devoted to binning of forecasts. We proceed here assuming this step has been applied. To get a decomposition of the original forecasts, the relation (Equation 7.18) below needs to be modified (see Stephenson *et al.*, 2008b).

Let $N_{kd}$ be the number of instances in the data set where $p(n) = q^{(d)}$ and at the same time $y(n) = k$. The numbers $N_{kd}$ form what is known as a $K \times D$ *contingency table* (cf. Chapter 4 of this book). Further, let $N_{\bullet d} = \sum_k N_{kd}$ be the number of instances in the data set where $p(n) = q^{(d)}$, i.e. we sum over the rows of the contingency table. Obviously, $\sum_d N_{\bullet d} = N$, the total number of instances in the data set. Now, define for $d = 1 \ldots D$ and $k = 1 \ldots K$ the *relative observed frequencies*

$$\bar{o}_k^{(d)} = \frac{N_{kd}}{N_{\bullet d}} \qquad (7.17)$$

For every $d$, the vector $\bar{o}^{(d)} = (\bar{o}_1^{(d)}, \ldots, \bar{o}_K^{(d)})$ is a probability assignment. Moreover, it should be seen as an approximation to $\pi^{(\Gamma)}$, the conditional probability of $Y$ given $\Gamma$. Indeed, if we let the number of instances go to infinity and assume the right ergodicity properties, then

$$\bar{o}_k^{(d)} \to \frac{Pr(Y = k, \Gamma = q^{(d)})}{Pr(\Gamma = q^{(d)})}$$
$$= Pr(Y = k | \Gamma = q^{(d)})$$

Finally, we set

$$\bar{o}_k = \frac{N_{k\bullet}}{N}$$

Note that $\bar{o}_k$ is the average of $\bar{o}_k^{(d)}$ over $d$. On the other hand, $\bar{o} = (\bar{o}_1, \ldots, \bar{o}_K)$ is an approximation of the climatological probability distribution of $Y$, namely

$$\bar{o}_k \to Pr(Y = k)$$

The right-hand side was called $\bar{\pi}$ in Section 7.3.3.

In terms of these objects, the decomposition

$$\frac{1}{N} \sum_n S(p(n), y(n)) = e(\bar{\pi}) - \frac{1}{N} \sum_d N_{\bullet d} d(\bar{o}, \bar{o}^{(d)})$$
$$+ \frac{1}{N} \sum_d N_{\bullet d} d(q^{(d)}, \bar{o}^{(d)})$$
$$(7.18)$$

holds. This decomposition should be considered as an empirical version (i.e. summation over samples replacing mathematical expectations) of Equations 7.11 and 7.13.

In view of the decomposition (Equation 7.18), we could entertain the following thought. We define a new probabilistic forecasting scheme $\Phi$ as follows: whenever $\Gamma = q^{(d)}$ for some $d$, we set $\Phi = \bar{o}^{(d)}$. The decomposition (Equation 7.18) shows that the empirical score of $\Phi$ is always better than that of $\Gamma$, since the reliability term of $\Phi$ is zero. This seems to suggest a foolproof way of improving forecast skill: we simply adopt $\Phi$ as our new forecasting scheme. This strategy would *always* increase skill, even if the original forecast were in fact reliable! How can this be? Roughly speaking, this comes about because we 'recalibrate' not only the systematic deviations from reliability, but also those arising merely through sampling variations and because our sample size is finite.

Another way to say this is that the recalibrated forecast $\Phi$ is evaluated 'in sample', which means that the same data were used to both build and evaluate $\Phi$. The forecast $\Phi$ looks better than it actually would be if we evaluated it on instances of the data set that were not used to generate $\Phi$. In other words, the resolution (second) term in Equation 7.18 is an overly optimistic estimate of the performance of $\Phi$, while the reliability (third) term is a too pessimistic estimate of the reliability of the original forecasting scheme $\Gamma$. This needs to be accounted for when these terms are interpreted and used. How this is done with the reliability term will be explained in Section 7.6 in connection with reliability analysis. In the next subsection, we will consider more conservative ways to estimate the resolution term, which will be employed to determine appropriate binnings of the forecasts.

### 7.5.5 Binning forecasts and the leave-one-out error

The difficulty in analysing the decomposition (Equation 7.18) also becomes manifest if we try to use the resolution term to determine an appropriate binning of a forecast with a continuous range. It is easy to see that the resolution term improves the more bins we use. Indeed, if we chose the bins so that each bin contains exactly one forecast ($p(n)$, say), then the corresponding observed frequency would be built upon a single observation $y(n)$ and thus be equal to $\delta_{k,y(n)}$. Thus, the recalibrated forecast $\Phi$ would be perfect!

Clearly though, a thus recalibrated forecast would be utterly useless in practice. How can we choose the bins so that on the one hand a reasonable amount of forecast falls into each bin, but on the other hand we have enough bins in order to obtain some resolution? Ideally, we should evaluate the resolution term on data that were not used to compute the observed frequencies. But this would require twice the amount of data.

Nonetheless, several approaches exist to simulate the evaluation on independent data, requiring only limited amounts of additional computation. One such approach is the leave-one-out calculation. The leave-one-out score can be written as

$$\hat{S}_{\text{LOO}} = \frac{1}{N} \sum_n S(\bar{o}_{\text{LOO}}(n), y(n))$$

where $\bar{o}_{\text{LOO}}(n)$ is computed like a standard relative observed frequency but with the $n^{\text{th}}$ sample $(p(n), y(n))$ being left out of the data set. This can also be written as follows. Letting

$$\bar{o}_k^{(d,l)} = \frac{N_{k,d} - \delta_{kl}}{N_{\bullet d} - 1} \quad for\ k, l = 1 \ldots K, d = 1 \ldots D$$

we have

$$\hat{S}_{\text{LOO}} = \frac{1}{N} \sum_{l,d} S(\bar{o}^{(d,l)}, e_l) N_{l,d}$$

with $e_l$ being the $l^{\text{th}}$ unit vector. In Figure 7.2, both the standard resolution term (marked with circles) as well as the leave-one-out resolution term (marked with diamonds) are shown as a function of the number of bins for example A.

This plot readily demonstrates the discussed effects. Note that the forecasts in example A are in fact not continuous but feature quantum jumps of $1/51$, as there are only 50 ensemble members. Hence it does not make sense to check more than 51 bins here. Nonetheless, even within this range, there is already a marked difference between the two resolution terms. The resolution of the observed frequen-

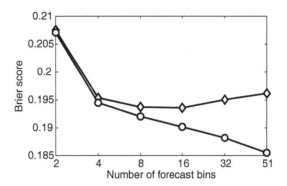

**Figure 7.2**  The Brier score for recalibrated forecasts of Example A at a lead time of 8 days (i.e. the resolution term of the original forecast's Brier score). The score is plotted against the number of bins employed to calculate the observed frequencies. The resolution term was calculated both with and without the leave-one-out technique (diamonds and circles, respectively). The leave-one-out score shows a clear minimum, while the in-sample score improves with ever-increasing number of bins

cies appears to improve with increasing number of bins, when evaluated in sample, while the out-of-sample performance starts to deteriorate if more than about 16 bins are used. For subsequent analyses, we will make a safe choice and use 10 bins.

A similar analysis can be carried out for the other two forecast schemes. The obvious question here is how to bin the forecasts, as these are multi-dimensional probability vectors (in the case of forecaster B) or even density functions (in the case of forecaster C), which cannot be ordered. A workable approach is to choose some characteristic of the probability forecasts, for example their standard deviation $\sigma$ (in the case of forecaster C) or their logarithmic entropy,

$$\sum_{k=1}^{3} -\log(p_k) p_k$$

in the case of forecaster B. In general, such a characteristic is simply a function $f(p)$ of the probability forecasts. Having selected a function $f(p)$ and a series of values $f_0 < \ldots < f_D$ in the range of $f(p)$, we say that the probability forecast $p(n)$ belongs to bin $d$ if $f_{d-1} < f(p(n)) < f_d$. (We define $f_0 = -\infty$, $f_D = \infty$ in order that the binning encompasses all forecasts.) This way of binning forecasts will be referred to as *stratifying* or *binning forecasts along* $f(p)$. Appropriate choices for $f(p)$ are problem dependent; two possible choices will be discussed in Section 7.6 in connection with examples B and C.

## 7.6 Testing reliability

In this section, we focus on statistical tests for reliability of a forecasting scheme. Recalling the definition in Equation 7.4, the reliability of a forecasting scheme $\Gamma$ means that

$$Pr(Y = k|\Gamma = p) = p_k \qquad \text{for } k = 1 \dots K. \tag{7.19}$$

This is a relation between $K - 1$ independent functions, as summing over $k$ on both sides gives 1. For binary forecasts where $\mathbb{K} = \{0, 1\}$, the condition (Equation 7.19) reads as

$$Pr(Y = 1|\Gamma = (1 - p, p)) = p \tag{7.20}$$

The problem here is that the conditional probabilities on the left-hand sides in the reliability condition (Equations 7.19 or 7.20) are not available operationally. If they were, the entire machinery of calculating observed frequencies as outlined in the previous section would be unnecessary. Consequently, these conditional probabilities have to be estimated from the available data archive, and the reliability condition has to be assessed using statistical hypothesis testing. Even if the forecasting scheme were reliable, we would not expect the approximated conditional probabilities to obey the reliability condition (Equation 7.19) exactly, due to sampling variations and model bias. It therefore needs to be analysed whether the observed deviations are consistent with those that would be expected for a reliable forecast. If they are, then the hypothesis of reliability cannot be denied based on the given data.

Common reliability tests approximate the conditional probabilities with observed frequencies. To this end, the forecasts again need to be distributed among suitably chosen bins, and the observations

grouped among suitably chosen categories. Then a reliability test amounts to a test on the distribution of the contingency table $N_{kd}$. We will therefore adopt the same setup as in Section 7.5.3 and investigate reliability in terms of the contingency table. To summarize, a reliability test might proceed as follows:

1. Determine a number of bins $B_1 \dots B_D$ and stratify the forecasts $p(n)$ among these bins. Likewise, distribute the observations among some categories, if necessary.
2. Compute the entries $N_{kd}$ of the contingency table as the number of instances in the data set where $p(n) \in B_d$ and $y(n) = k$.
3. Letting $N_{\bullet d} = \sum_k N_{kd}$ the number of instances in the data set where $p(n) \in B_d$, define the observed frequencies as in Equation 7.17.
4. Compare the observed frequencies $\bar{o}^{(d)}$ with the in-bin average:

$$p^{(d)} = \frac{1}{N_{\bullet d}} \sum_{p(n) \in B_d} p(n). \tag{7.21}$$

The last point in particular merits further discussion. We will carry out this programme for the three considered examples, discussing in particular how to 'compare' observed frequencies with in-bin averaged forecast probabilities.

### 7.6.1 Reliability analysis for forecast A: the reliability diagram

How to bin binary forecasts was discussed in Section 7.5.5, including finding an appropriate number of bins. A classical tool to compare observed frequencies with forecast probabilities is the reliability diagram, which consists of a plot of $\bar{o}_1^{(d)}$ versus $p_1^{(d)}$ as a function of $d$. The graph should be close to the diagonal. The reliability diagram for forecast A is presented in Figure 7.3.

Due to sampling variations, reliability diagrams typically do not align perfectly with the diagonal, even if the forecast is truly reliable. Therefore, any reliability diagram should provide additional guidance as to how close to the diagonal the graph is

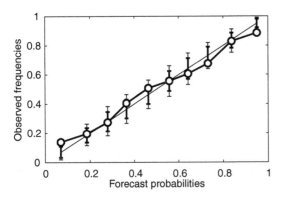

**Figure 7.3** A reliability diagram for forecasts of Example A at a lead time of 8 days. Standard confidence bars are shown (thick line), and Bonferroni corrected bars are added (thin line). The reliability of the most extreme points is hard to assess

to be expected if the forecast were in fact reliable. For this purpose, confidence bars have been plotted about the diagonal in Figure 7.3, which ought to be interpreted as follows. If the forecast were reliable, then each individual point would have a 5% chance of falling above and 5% chance of falling below the interval marked by the thick bars. In other words, the bars represent symmetric confidence intervals with coverage probability of 0.9. If the individual points on the graph were independent, then the chance of finding the entire reliability curve within the bars would be $0.9^{10} = 0.34$. By plotting additional bars with larger coverage probability, this latter figure can be increased to 0.9. More precisely, drawing additional bars (shown with thin lines in Figure 7.3) with a coverage probability of $\sqrt[10]{0.9} = 0.9895$, we get a 0.9 chance of finding the entire plot within the thin bars. This widening of the bars is known as the Bonferroni correction. Note that the intervals are in fact not independent, and so the Bonferroni correction gives a conservative estimate of the simultaneous confidence interval.

The bars are constructed as follows (see Bröcker and Smith, 2007a, for a detailed discussion). Regarding $N_{\bullet d}$ as fixed, the $N_{1d}$ should follow a binomial distribution. Let $B(m, g, M)$ denote the probability of less than $m$ successes in $M$ Bernoulli trials with a probability of success $g$.

Then $B\left(\bar{o}_1^{(d)} N_{\bullet d}, p_1^{(d)}, N_{\bullet d}\right)$ gives the probability that, for a reliable forecast, an observed frequency smaller than $\bar{o}_1^{(d)}$ occurs. In other words, the interval

$$I_d = \left\{x; \ 0.05 \leq B\left(x N_{\bullet d}, p_1^{(d)}, N_{\bullet d}\right) \leq 0.95\right\}$$

contains the observed frequency $\bar{o}_1^{(d)}$ with a 90% chance, with an equal chance of 5% for being either below or above that interval. Therefore, $I_d$ constitutes the interval marked by the consistency bar.

Figure 7.3 represents an instance of the often encountered situation that some of the observed frequencies are on the verge of the consistency bars, especially at the end points. In particular if the number of instances is large the region of interest is compressed along the diagonal. The relevant information might get obscured by graphics symbols, and much of the plot area does not contain any information. Another presentation of reliability diagrams, which makes more efficient use of the entire plot, was presented in Bröcker and Smith (2007a). Instead of the observed frequencies, it was suggested to plot the transformed values

$$\mu_d = B\left(\bar{o}_1^{(d)} N_{\bullet d}, p_1^{(d)}, N_{\bullet d}\right)$$

From the discussion above, we gather that for reliable forecasts, the $\mu_d$ are expected to be smaller than a given number $1 - \alpha$ with a probability $1 - \alpha$. In other words, if the forecast were reliable, and if we could repeat the reliability test an infinite number of times with new data each time, then we would find a fraction $\alpha$ of all runs exhibiting a $\mu_d$ smaller than $\alpha$. Again in other words, if we marked an interval of length $\alpha$ on the y-axis, any $\mu_d$ would fall into this interval with a probability $\alpha$. Therefore, the $\mu_d$ provide direct quantitative information as to whether the deviations from reliable behaviour are systematic or merely random and are thus easier to interpret than the relative observed frequencies used in standard reliability diagrams. In particular, the interpretation of the $\mu_d$ plot is always exactly the same, independent of the size of the data set. A plot of the $\mu_d, d = 1 \ldots D$ is shown in Figure 7.4.

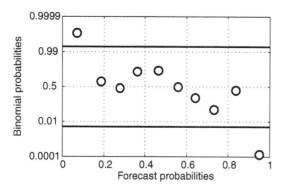

**Figure 7.4** Reliability diagram as in Figure 7.2, but on probability paper. The interesting regions are displayed much better than in a standard reliability diagram. The most extreme points are seen to be inconsistent with reliability

The readability of the $\mu_d$ diagram is further improved by scaling the ordinate by the logit-transformation,

$$\log\left(\frac{\mu}{1-\mu}\right)$$

This has the effect of displaying the small probabilities $0.1, 0.01, 0.001, \ldots$ as well as the large probabilities $0.9, 0.99, 0.999, \ldots$ equidistantly. For a $\mu_d$ diagram, the $N_{\bullet d}$ should not be less than about 8 and be roughly equal for different $d$, as otherwise significant variations are ignored.

In Figure 7.4, the probabilities on the $y$-axis refer to individual points on the plot, not to the entire diagram. That is, the probability of finding all dots in the band between 0.01 and 0.99, for example, is not 0.98, but only $0.98^{10} = 0.82$. To get bands with prescribed coverage probabilities for the entire diagram, again a Bonferroni correction can be applied. In Figure 7.4, the two horizontal solid lines mark the Bonferroni corrected 0.05 and 0.95 probabilities; i.e. we expect to find all dots inside the two lines with 90% probability. It is seen that the most extreme forecast probabilities show deviations from reliable behaviour. The lowest probability is too high (overestimating the actual frequency of events), while the highest probability is too low. This would be hard to tell from the traditional reliability diagram as in Figure 7.3. The *critical probability* is defined as 1 minus the probability encom-

passed by the two most extreme observed frequencies (i.e. $1 - (\mu_{max} - \mu_{min})$). In the present case, the critical probability is 0.0086. We can conclude that, if the forecast were reliable, it is very unlikely we would see something like Figure 7.4.

### 7.6.2 Reliability analysis for forecast B: the chi-squared test

According to our recipe, we first need to determine suitable bins and distribute the forecasts among them. In the case of forecaster B, the probabilities are not ordinal variables but vectors. As already discussed in Section 7.5.5, the forecasts can be binned along some function $f(p)$. For forecaster B's forecasts, we will use the function

$$f(p) = \sum_{k=1}^{3} v_k p_k$$

with $v = [0, 1/3, 2/3]$. Qualitatively, this function is large if the forecast puts most probability mass on the category 'high', while it is small if most mass is on the category 'low'. The leave-one-out analysis presented in Section 7.5.5 suggested around 12 bins, but to simplify the graphical presentation, only four bins were used for this example. The in-bin averages of the forecast probabilities are shown in Figure 7.5 as dark bars in the foreground. The corresponding observed frequencies are represented by light-grey bars in the background. Visually, there appears to be fair agreement between forecast probabilities and observed frequencies.

We will now consider more quantitative ways to test the agreement between forecast probabilities and observed frequencies. The null hypothesis of reliability is essentially a statement about the distribution of the contingency table, and by comparing this distribution with the actual numbers $N_{kd}$, we can decide whether the hypothesis is consistent with the data or not. The full distribution of the contingency table is multinomial with some parameter $g_{kd}$, but unfortunately the null hypothesis does not specify the distribution of the contingency table completely – it does not make any statement about the marginals $g_{\bullet d}$. What comes to our rescue here is a well-known fact about the distribution of generalized likelihood ratios; see, for example,

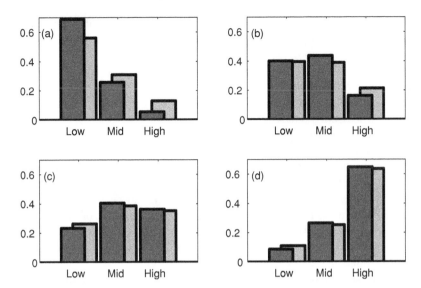

**Figure 7.5** Forecast probabilities (dark bars) and histograms of observed frequencies (light bars) for forecasts of Example B at a lead time of 8 days. The forecasts were binned as explained in Section 7.6.2. Forecasts and observed frequencies show a good qualitative agreement. For a more quantitative analysis using the $\chi^2$-test, see Section 7.6.2

theorem 7, chapter IX in Mood *et al.* (1974), and also Chapter 4 of this book. Consider the quantity

$$R = \sum_{k,d} -\log\left(\frac{p_k^{(d)}}{N_{kd}/N_{\bullet d}}\right) N_{kd}/N$$

henceforth referred to as the *R-statistic*. It follows from the above-mentioned theorem that $2N \cdot R$ is asymptotically of chi-squared ($\chi^2$) distribution with $D(K-1)$ degrees of freedom. Hence, rather than $R$ itself, we consider the $p$-values $1 - P_{\chi^2}(2NR)$, where $P_{\chi^2}$ is the cumulative $\chi^2$-distribution function with $D(K-1)$ degrees of freedom. Under the null hypothesis, these $p$-values should have a uniform distribution on $[0, 1]$. By rejecting the null hypothesis if the $p$-value is smaller than some $\alpha$ and accepting it otherwise, we obtain a test for reliability of size $\alpha$. The precise interpretation is that, by this strategy, a forecast that is in fact reliable will be rejected as unreliable with a chance of $\alpha$.

There is an interesting connection between the $R$-statistic and the Ignorance Score. Using the relative observed frequencies, we can write

$$R = \frac{1}{N} \sum_d \mathrm{d}\left(p^{(d)}, \bar{o}^{(d)}\right) N_{\bullet d}$$

using the divergence d for the Ignorance Score. This shows that $R$ is in fact the reliability term in the decomposition (Equation 7.18) for the Ignorance Score. As discussed already in Section 7.5.4, the conclusion that the forecasting scheme $\Gamma$ is unreliable whenever the reliability term is positive is not justified, since this term is always positive. The conclusion *is* justified, though, if $R$ is not merely positive but unusually large, i.e. larger than would normally be expected if $\Gamma$ were in fact reliable. What should be considered an unusually large value for $R$ is quantified by the $\chi^2$-distribution.

The $R$-statistic, albeit being a useful statistical test for reliability, does not provide detailed information. If a forecast lacks reliability, one might like to know more as to where reliability fails. A possible approach is to look at individual $R$-statistics for each $d = 1 \ldots D$. Let

$$R_d = \sum_k -\log\left(\frac{p_k^{(d)}}{\pi_k^{(d)}}\right) \pi_k^{(d)} = \mathrm{d}\left(p^{(d)}, \bar{o}^{(d)}\right)$$

Assuming that $N_{\bullet d}$ is fixed, $2N_{\bullet d}R_d$ has a $\chi^2$-distribution with $K - 1$ degrees of freedom. Therefore, $P_{\chi^2}(2N_{\bullet d}R_d)$ is uniformly distributed on $[0, 1]$, where again $P_{\chi^2}$ is the cumulative $\chi^2$-distribution function, now with $K - 1$ degrees of freedom. For the present example, these values turn out to be 0, 0.0117, 0.3225 and 0.2504 for the four categories, respectively. Given commonly chosen significance levels of $\alpha$ such as 0.05 or 0.01, we would reject reliability for the forecasts of categories (1, 2) or of category 1, respectively.

A problem of the discussed goodness-of-fit test (and, in fact, of any other goodness-of-fit test) is that it is always possible to construct histograms that will pass the test perfectly, despite exhibiting obvious pathologies that are unlikely to arise by mere randomness. For example, histograms often display a clear trend or are convex. A trend upwards indicates that higher categories are assigned forecast probabilities that are too small, or in other words, they verify too often. In the case of ensemble forecasts, this indicates under-forecasting of the ensembles. Convex histograms can arise for two reasons. Either the ensembles systematically exhibit too small a spread, and thus the extreme ranks verify too often. Or the histogram in fact confounds two forecast bins, one containing over-forecasting and one containing under-forecasting ensembles. Since the $R$-statistic is invariant against reordering of the categories, however, forecasts with convex or tilted histograms might pass the goodness-of-fit test undetected. There are numerous other statistics suitable for goodness-of-fit tests that, other than Pearson's classical statistic, are sensitive to the ordering of the categories (Elmore, 2005; Jolliffe and Primo, 2008), such as the Cramér–von Mises statistic; see also Section 8.3.1 of this book. However, apart from the $R$-statistic it is not clear if any of the common goodness-of-fit statistics correspond to the reliability term of a proper score.

### 7.6.3 Reliability analysis for forecast C: the PIT

Example C deserves special attention as here both the forecasts as well as the observations occupy a continuous range. A workable approach to analyse such forecasts is first to categorize the observations (and bin the forecasts), and subsequently apply techniques for discrete multi-categorical forecasts, as explained in Section 7.6.2. This analysis is greatly facilitated by first applying a *probability integral transform (PIT)* to the observations. Let $p(x, n), n = 1 \ldots N$ be the forecasts written as den-

sities. The PIT of the observations $y(n)$ leads to new observations $z(n)$ defined as

$$z(n) = \int_{-\infty}^{y(n)} p(x, n) \mathrm{d}x$$

In words, the PIT of $y(n)$ is the forecast probability mass below $y(n)$. The pertinent fact about the PIT is that the forecast distributions for the $z(n)$ are uniform distributions on the interval [0, 1]. Therefore, reliability just means that the $z(n)$ should indeed be uniformly distributed on the interval [0, 1] as well, for arbitrary binning of the forecasts (see Tay and Wallis, 2000; Gneiting *et al.*, 2005, for application of the PIT in a meteorological context and further references).

With these facts in mind, we proceed to apply our reliability recipe. We bin the forecasts along $\mu(n)$ (see Section 7.5.1, Example C), the mean error between ensemble members and high-resolution forecasts. The number of bins could be determined using a leave-one-out analysis, which, in the present context, gives an optimal value of about eight bins. Again for illustrative purposes, though, we will use only four bins. These bins were chosen to be equally populated. The (PITed) observations $z(n)$ were categorized by rounding to the first digit. Thus, our contingency table in point 2 of the recipe contains $10 \times 4$ entries, which should be all roughly equal. The observed frequencies are shown in Figure 7.6 as grey bars.

The corresponding probability forecasts are all equal to $1/10$ and are shown as a horizontal dashed line. For all forecast bins, the extreme categories are too heavily populated, indicating that the forecasts underestimate the variability in the observation. A possible reason might be the normal forecast densities, which are probably inadequate for the given problem. Furthermore, the forecasts show a clear conditional bias. In forecast bin 1, the forecasts have a tendency to be too low, while in forecast bin 2, the forecasts are too high, i.e. the signal in the forecasts is too strong. If we had aggregated all forecasts into one bin, this would have resulted in a U-shaped histogram, confounding the two distinct phenomena of conditional bias and insufficient forecast spread; see Hamill (2001) for a thorough discussion of the interpretation of rank histograms.

As in the other two examples, the likeness of observed frequencies and forecast probabilities can

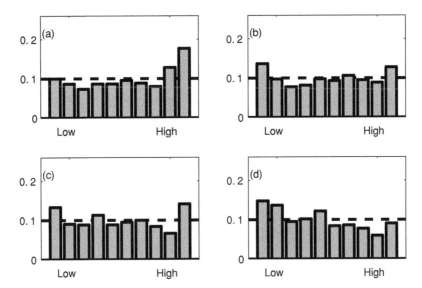

**Figure 7.6** PIT histograms for forecasts of Example C at a lead time of 8 days. The forecasts were binned along the mean value $\mu(n)$ of the forecast. The histograms indicate an insufficient spread as well as a conditional bias of the forecasts. This is confirmed by a more quantitative analysis using the $\chi^2$-test (see Section 7.6.3)

be checked with a $\chi^2$-test. In the present example, the $p$-values for the four bins turn out to be 0.0001, 0.1156, 0.0136 and 0.0004, indicating a statistically significant departure from reliability.

## Acknowledgements

The views expressed in this chapter (for which I carry sole responsibility) were shaped through discussions with many people, of whom I would like to mention Johannes Bröcker, Liam Clarke, Leo Granger, Renate Hagedorn, Sarah Hallerberg, Kevin Judd, Holger Kantz, Devin Kilminster, Francisco Doblas Reyes, Mark Roulston, Stefan Siegert, Leonard A. Smith, Dave Stainforth and Antje Weisheimer. Both station data as well as ensemble forecasts were kindly provided by the European Centre for Medium Range Weather Forecasts. The Centre for the Analysis of Time Series, London School of Economics, provided further support through a visiting research fellowship.

# 8

# Ensemble forecasts

**Andreas P. Weigel**

*Federal Office of Meteorology and Climatology MeteoSwiss, Zurich, Switzerland*

## 8.1 Introduction

Only a few differential equations are needed to describe the dynamics of the atmosphere. These equations are well known, and their numerical integration constitutes the basic principle of numerical weather prediction (NWP). Thus, NWP may at first sight appear like an entirely deterministic problem, but in fact it is not. Firstly, the initial state of the atmosphere, from which the equations are integrated forward in time, cannot be determined in unlimited accuracy. Secondly, even the smallest uncertainties in the initial conditions have a tendency to propagate to larger scales and to act as quickly growing noise terms (e.g. Lorenz, 1993). In their combination, these two aspects introduce the element of chance and probability to NWP, with the uncertainties growing as longer timescales are considered. Therefore, rather than integrating a single trajectory from a best guess of the initial state, it is more consistent with the true nature of atmospheric flow to consider the evolution of the entire uncertainty range of possible initial states. This mathematical problem, which is formally described by the Liouville equation, can in practice not be solved analytically due to the high dimensionality of the system (Ehrendorfer, 2006). However, it can be approached pragmatically by Monte Carlo techniques (Leith, 1974; Stephenson and Doblas-Reyes, 2000), which leads to the concept of ensemble forecasting. The idea is to integrate a model repeatedly from a set of perturbed initial states of the atmosphere (the 'initial condition ensemble'), and then to analyse how the resulting set of forecast trajectories (the 'forecast ensemble') propagates through phase-space. If the initial condition ensemble truly characterizes the underlying initial condition uncertainty (see Kalnay *et al.*, 2006 for an overview of ensemble generation techniques), and if the atmospheric dynamics are well represented by the model, then the forecast ensemble should, at any lead time, mark the flow-dependent distribution of possible outcomes. In practice, of course, neither are the models perfect, nor is it trivial to construct a representative initial condition ensemble. Consequently, ensemble forecasts can be subject to systematic biases and errors, implying major challenges in their interpretation. Nevertheless, ensemble forecasting has become a matter of routine in NWP with innumerable successful applications.

As longer timescales of several weeks to seasons are considered, prediction uncertainty is increasingly dominated by uncertainties in relevant boundary conditions, such as sea-surface temperatures. By including these uncertainties in the ensemble generation and modelling process, it has been possible to extend the concept of ensemble forecasting to climate timescales, allowing, for example, the computation of seasonal ensemble forecasts (see Chapter 11 in this book; Stockdale *et al.*, 1998). Further extensions of the concept include the perturbation of model parameters, with the aim to sample parameter uncertainty (e.g. Doblas-Reyes *et al.*, 2009), and the combination of the output of several models to a multi-model ensemble, with the aim to sample model uncertainty (e.g. Palmer *et al.*, 2004; Weigel *et al.*, 2008a).

How can the quality of ensemble forecasts be assessed? Despite their probabilistic motivation, ensemble forecasts are *a priori* not probabilistic forecasts, but rather 'only' finite sets of deterministic forecast realizations. To derive probabilistic forecasts from the ensemble members, further assumptions concerning their statistical properties are required. The question as to how the quality of ensemble forecasts can be assessed therefore depends on how they are interpreted, i.e. whether they are seen as finite random samples from underlying forecast distributions, or whether they have been converted into probability distributions and are interpreted probabilistically. For both interpretations specific metrics exist.

Those metrics that are designed for the 'sample interpretation' aim at characterizing the collective behaviour and statistical properties of the ensemble members. They do not require that some form of probability distribution has been derived prior to verification, but often represent the first step towards a probabilistic interpretation to follow. An overview of commonly applied tests of this kind is presented in Section 8.3. On the other hand, once a probabilistic forecast has been derived from an ensemble, verification is in principle straightforward by applying standard probabilistic skill metrics, as discussed in Chapter 7. One important aspect in this context is that the skill scores obtained then not only measure model quality, but also the appropriateness of the assumptions made when converting the ensemble into a probability distribution. This and other aspects of probabilistic ensemble verification are discussed in Section 8.4. The focus is thereby not on a repetition of the skill metrics and discussions provided in Chapter 7, but rather on issues of specific relevance for ensemble forecasts, such as the effect of ensemble size on skill. Many of the tests and metrics presented in Sections 8.3 and 8.4 are illustrated with examples, and I therefore start, in Section 8.2, with a short description of the data used for these examples. The chapter closes with a summary in Section 8.5.

## 8.2 Example data

The examples presented in this chapter are based on three data sources: (i) medium-range ensemble forecasts produced by the European Centre for Medium Range Weather Forecasts (ECMWF) Ensemble Prediction System (EPS; Palmer *et al.*, 2007); (ii) seasonal ensemble forecasts produced by the ECMWF seasonal prediction System 3 (Anderson *et al.*, 2007); and (iii) synthetic ensemble forecasts from a Gaussian toy model.

For the examples based on medium-range EPS data, forecasts of 500 hPa geopotential height with lead times of 1 to 15 days have been considered. The example forecasts have been issued in 12-hour intervals during the period December 2009 until February 2010 and have been evaluated over the northern hemisphere extratropics (20–90°N). There are 51 ensemble members. Verification is against the ECMWF operational analysis.

For the examples based on seasonal System 3 data, hindcasts of mean near-surface (2 m) temperature, averaged over the months December to February, have been used. If not stated otherwise, the Niño3.4 region (5°S–5°N, 120–170°W) is used as the target domain. Data have been obtained from the ENSEMBLES project database (Royer *et al.*, 2009; Weisheimer *et al.*, 2009). All hindcasts have been started from 1 November initial conditions and cover the period 1960–2001. There are nine ensemble members. If systematic model biases have been removed prior to verification, the hindcasts are referred to as 'bias-corrected', and if additional corrections have been applied to improve ensemble spread, they are referred to as 'recalibrated'. Otherwise, the hindcasts are referred to as 'raw'. For recalibration, the climate-conserving recalibration (CCR) technique of Weigel *et al.* (2009) has been applied. Verification is against data from the

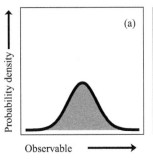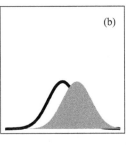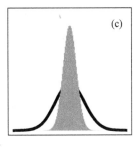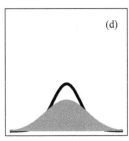

**Figure 8.1** Illustration of reliable and unreliable ensemble forecasts. For a given initial condition uncertainty, consider the 'true' probability distribution of possible outcomes to be expected after a given amount of time (black solid line). Assume that an ensemble prediction system is applied to estimate this probability distribution (indicated by the grey shaded areas). If the forecast is reliable (a), the ensemble members should be statistically indistinguishable from truth and be sampled from the true distribution of possible outcomes. For the toy model of Section 8.2, this corresponds to a parameter setting $\beta = 0$ and $\gamma = 1$. If the forecasts are subject to a systematic bias (b), the ensemble distribution is shifted with respect to the distribution of possible outcomes (toy model configuration: $\beta > 0$ and $\gamma = 1$). Underdispersive ensemble forecasts (c) fail to sample the full range of uncertainties and are too sharp ($\beta = 0$ and $\gamma < 1$), while overdispersive forecasts (d) overestimate the true uncertainty range ($\beta = 0$ and $\gamma > 1$)

40-years ECMWF Re-Analysis (ERA-40) dataset (Uppala *et al.*, 2005).

The toy-model applied in this chapter is a Gaussian stochastic generator of forecast-observation pairs fulfilling preset conditions with respect to forecast skill and ensemble properties. A forecast-observation pair is constructed in a two-step process: Firstly, a 'predictable signal' $s$ is sampled from a normal distribution with mean 0 and variance 1. Secondly, an $m$-member ensemble forecast $\hat{x} = (\hat{x}_1, \hat{x}_2, \ldots, \hat{x}_m)$ and a verifying observation $x$ are constructed upon $s$ according to

$$x = s + \varepsilon_x$$
$$\hat{x}_i = s + \varepsilon_i + \beta$$
$$\text{where} \quad \varepsilon_x \sim N(0, \alpha)$$
$$\varepsilon_i \sim N(0, \gamma\alpha)$$

$\sim N(\mu, \sigma)$ here stands for 'a random number drawn from a normal distribution with mean $\mu$ and variance $\sigma^2$'; $\varepsilon_x$ represents the unpredictable and chaotic components of the observations, while $\varepsilon_i$ is the perturbation of the $i^{\text{th}}$ ensemble member. There are three free parameters $\alpha$, $\beta$ and $\gamma$. Parameter $\alpha$ controls the signal-to-noise ratio; without loss of generality, $\alpha$ is set to 1. $\beta$ is the systematic bias of the forecasts, and $\gamma$ controls the dispersion char-

acteristics of the ensemble. If $\beta = 0$ and $\gamma = 1$, then $\hat{x}_i$ and $x$ are sampled from the same underlying distribution, and the forecasts are *reliable* (illustrated in Figure 8.1a). If $\beta > 0$ (respectively $\beta < 0$), the forecasts are positively (negatively) biased (Figure 8.1b). If $\gamma < 1$, the forecasts are *underdispersive* (Figure 8.1c), and if $\gamma > 1$, they are *overdispersive* (Figure 8.1d). A more in-depth discussion of toy models of this kind can be found in Weigel and Bowler (2009).

## 8.3 Ensembles interpreted as discrete samples

As is discussed at several places in this book (e.g. Sections 2.10, 7.3.3 and 11.3), prediction skill is a multifaceted quantity that needs to be characterized by several 'skill attributes'. Here we look at how these attributes are to be interpreted in the context of ensemble forecasts, and how they can be measured if the ensembles are construed as discrete samples from an underlying probability distribution.

One of the most important skill attributes of *probabilistic* forecasts is 'reliability', which characterizes the degree to which the forecast probabilities are consistent with the relative frequencies of the observed outcomes (see Section 7.6).

In the context of *ensemble forecasts*, reliability implies that the ensemble members and observed outcomes are consistently sampled from the same underlying probability distributions, i.e. that they are statistically indistinguishable from each other. This property is sometimes also referred to as 'ensemble consistency' (e.g. Wilks, 2006b). In the following two subsections, an overview of commonly applied tests of ensemble reliability is provided. A distinction is hereby made between tests for one-dimensional forecasts of one variable at one location (Section 8.3.1), and multidimensional forecasts that are jointly issued at several locations and/or for several variables (Section 8.3.2).

'Resolution' and 'discrimination' are two further central attributes of probabilistic prediction skill. Both attributes are related in that they measure whether there is an association in the sense of a correlation between what is predicted and what is observed. Resolution takes the perspective of the forecasts, quantifying the degree to which the observed outcomes change as the forecasts change. Discrimination takes the perspective of the observations, asking whether different observed outcomes can be correctly discriminated by the forecasts. More detail on the difference between resolution and discrimination is given in Section 2.10. Resolution and discrimination of ensemble forecasts are typically assessed by means of standard probabilistic approaches, requiring that the ensembles have been converted into probabilistic forecasts prior to verification. To the author's knowledge, not many tests of discrimination or resolution have been reported that are directly applicable to ensembles interpreted as samples from a probability distribution. One exception, a recently introduced test for discrimination, is presented in Section 8.3.3.

### 8.3.1    Reliability of ensemble forecasts

*Mean squared error and mean ensemble variance*

This section starts with the description of an easy-to-implement criterion for ensemble reliability that is frequently applied to obtain first insight into the dispersion characteristics of an ensemble prediction system (e.g. Stephenson and Doblas-Reyes, 2000). This criterion states that it is a necessary condition

for ensemble reliability that the mean squared error (MSE) of the ensemble mean forecasts is identical to the mean intra-ensemble variance, apart from an ensemble size-dependent scaling factor. How can this criterion be understood?

Consider a set of $n$ ensemble forecasts $\hat{x}_1, \ldots, \hat{x}_n$ and corresponding observations $x_1, \ldots, x_n$. Assume there are $m$ ensemble members for each forecast, with $\hat{x}_{t,i}$ being the $i^{\text{th}}$ ensemble member of the $t^{\text{th}}$ forecast. Let $\bar{\hat{x}}_t$ be the sample mean of the $t^{\text{th}}$ ensemble forecast, i.e.

$$\bar{\hat{x}}_t = \frac{1}{m} \sum_{i=1}^{m} \hat{x}_{t,i}$$

and let $\mu_t$ be the expectation of the underlying (and usually unknown) probability distribution from which $\hat{x}_t$ is sampled. Finally, let $s_t^2$ be the sample variance of the ensemble members of $\hat{x}_t$:

$$s_t^2 = \frac{1}{m-1} \sum_{i=1}^{m} (\hat{x}_{t,i} - \bar{\hat{x}}_t)^2$$

In a reliable ensemble prediction system the observations and all $m$ ensemble members should be statistically indistinguishable from each other. This implies that for all $t = 1, 2, \ldots, n$ the observation $x_t$ and all ensemble members of $\hat{x}_t$ should be samples from the same underlying distribution. From this it follows that for any ensemble member $i = 1, 2, \ldots, m$

$$E_t \left[ (x_t - \mu_t)^2 \right] = E_t \left[ \left( \hat{x}_{t,i} - \mu_t \right)^2 \right]$$

and thus

$$E_t \left[ (x_t - \mu_t)^2 \right] = E_t \left[ \frac{1}{m} \sum_{i=1}^{m} \left( \hat{x}_{t,i} - \mu_t \right)^2 \right] \quad (8.1)$$

with $E_t$ being the expectation over $t$. The term in squared brackets on the right-hand side of Equation 8.1 is the variance of forecast $\hat{x}_t$, which can be estimated without bias by $s_t^2$. Using this in the right-hand side of Equation 8.1, and using

$$E_t \left[ (x_t - \bar{\hat{x}}_t)^2 \right] = E_t \left[ \left( x_t - \mu_t + \mu_t - \bar{\hat{x}}_t \right)^2 \right]$$
$$= E_t \left[ (x_t - \mu_t)^2 \right] + E_t \left[ \left( \bar{\hat{x}}_t - \mu_t \right)^2 \right]$$

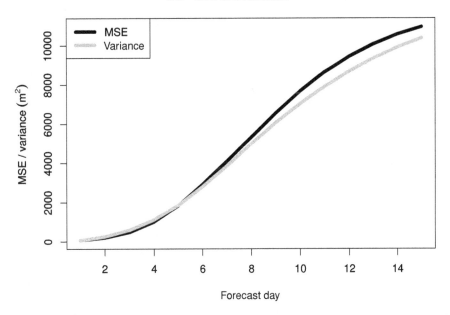

**Figure 8.2** Mean squared error (MSE, black line) of 500 hPa geopotential height mean forecasts as a function of lead time based on data from the European Centre for Medium Range Weather Forecasts (ECMWF) Ensemble Prediction System (EPS) ensemble, along with the average intra-ensemble variance (grey line). The values shown have been computed for and averaged over the northern hemisphere extratropics (20–90°N) and the period December 2009 until February 2010. The squared errors are with respect to the operational analysis. The variances have been scaled by a factor of 1.02 to account for the effect of finite ensemble size. The ECMWF is acknowledged for providing these data

in the left-hand side of Equation 8.1, one obtains

$$E_t\left[(x_t - \bar{\hat{x}}_t)^2\right] - E_t\left[(\bar{\hat{x}}_t - \mu_t)^2\right] = E_t\left[s_t^2\right]$$

(8.2)

Based on the expression for the variance of a sample mean, it can be shown that

$$E_t\left[(\bar{\hat{x}}_t - \mu_t)^2\right] = \frac{E_t\left[s_t^2\right]}{m}$$

Applying this in Equation 8.2 yields the aforementioned reliability criterion:

$$E_t\left[(x_t - \bar{\hat{x}}_t)^2\right] = \frac{m+1}{m}E_t\left[s_t^2\right]$$

(8.3)

The left-hand side of Equation 8.3 corresponds to the MSE of the ensemble mean forecasts, and the right-hand side is the average intra-ensemble sample variance, multiplied by an ensemble-size-

dependent inflation factor $(m + 1)/m$. The inflation factor can be neglected for large ensemble sizes, when there is (almost) identity between the ensemble mean MSE and the average ensemble variance.

As an example consider Figure 8.2, which is based on an evaluation of medium-range forecasts of 500 hPa geopotential height, averaged over the northern hemisphere. Data have been produced by the ECMWF EPS (see Section 8.2). Displayed are the MSE of the ensemble mean forecasts and the average intra-ensemble variance as a function of lead time. The variances have been inflated with a factor of 1.02 to account for the effect of finite ensemble size (there are 51 ensemble members). The plot shows that for lead times of up to about 6 days the identity of Equation 8.3 is well satisfied. For longer lead times, however, the MSE grows faster than the average ensemble variance, implying that the forecasts become increasingly unreliable.

Note that Equation 8.3 is a necessary but not sufficient condition of ensemble reliability. In other words, if Equation 8.3 is *not* satisfied, one can conclude that the forecasts are unreliable; however, if Equation 8.3 is satisfied, this does not yet imply that the forecasts are reliable; further tests are then necessary to demonstrate reliability. For instance, in the special case of all distributions being normal, it can be shown that it is a sufficient criterion of ensemble reliability if, in addition to Equation 8.3, the climatological variance of the observations is identical to that of the ensemble members (Johnson and Bowler, 2009; Weigel *et al.*, 2009).

### Rank histograms

The rank histogram, also known as analysis rank histogram, Talagrand diagram or binned probability ensemble, has been proposed independently by Anderson (1996), Hamill and Colucci (1997) and Talagrand *et al.* (1998), and is probably one of the most widely used tests for ensemble reliability. The basis of this test is an analysis of how the observed outcomes rank with respect to the corresponding ensemble members. From that, insight can be gained into the average dispersion characteristics of the forecasts. How is a rank histogram constructed?

Consider an $m$-member ensemble forecast $\hat{x}_t = (\hat{x}_{t,1}, \ldots, \hat{x}_{t,m})$ and a corresponding observation $x_t$. Assume that $x_t$ exceeds $M$ of the $m$ ensemble members ($M \le m$) and is exceeded by the remaining ($m - M$) members. The rank of $x_t$ with respect to the ensemble members of $\hat{x}_t$ is then given by $r_t = M + 1$. If $x_t$ is smaller than all ensemble members, then the observation has rank $r_t = 1$, and if $x_t$ exceeds all ensemble members, then $r_t = m + 1$. For a set of $n$ forecast-observation pairs, the rank histogram is constructed by determining the ranks of all observations, then counting how often each of the $m + 1$ possible rank values is taken, and finally displaying these numbers in the form of a histogram.

If an ensemble prediction system is reliable, then the ensemble members and observations should be statistically indistinguishable. It would then be equally likely for $x_t$ to take any of the rank values $r_t = 1, 2, \ldots, m + 1$. Consequently, a reliable ensemble prediction system would, apart from deviations due to sampling variability, yield a flat rank histogram with an expected number of $n/(m + 1)$ counts per rank category. This is illustrated in Figure 8.3a, where a rank histogram is shown that has been constructed from 10 000 reliable toy model

forecasts and corresponding observations. On the other hand, rank histograms significantly deviating from flatness imply that the forecasts are unreliable. In this case, the histogram shape may give hints about the nature of the forecast deficiencies. For instance, if the forecasts are subject to a systematic unconditional positive bias, there is an enhanced probability that an observation is exceeded by the majority of the ensemble members, leading to an overpopulation of the lower ranks and hence a negatively sloped histogram (Figure 8.3b). In fact, if the bias is large enough, the observations would at some point be systematically exceeded by all ensemble members, leaving all rank categories empty apart from the lowest one. Similarly, a negative unconditional bias yields positively sloped histograms (Figure 8.3c). If both the lowest and highest ranks are overpopulated and the central ranks depleted (Figure 8.3d), this indicates ensemble underdispersion: since the spread is on average too low, there is an enhanced probability that an observation is not captured by the ensemble and is assigned rank 1 or $m + 1$. Conversely, ensemble forecasts that are overdispersive capture the observation too often and result in a peaked histogram (Figure 8.3e).

An important aspect in the interpretation of rank histograms is the distinction between random deviations from flatness due to sampling uncertainty, and systematic deviations from flatness due to specific forecast deficiencies.

A common approach to assess this problem is the $\chi^2$ goodness-of-fit test, with the null hypothesis being that the rank histogram is uniform. If $n$ is the number of sets of forecast-observation pairs, $m$ the ensemble size, $n_i$ the number of counts in the $i^{\text{th}}$ rank category, and $e = n/(m + 1)$ the number of counts that would be expected in each rank category were the histogram perfectly flat, then the test statistic $T$ for the $\chi^2$ goodness-of-fit test is given by

$$T = \sum_{i=1}^{m+1} \frac{(n_i - e)^2}{e} \qquad (8.4)$$

Since $T$ follows approximately a $\chi^2$-distribution with $m$ degrees of freedom under the null hypothesis, this test is easy to use for assessing significance. As an example, consider the rank

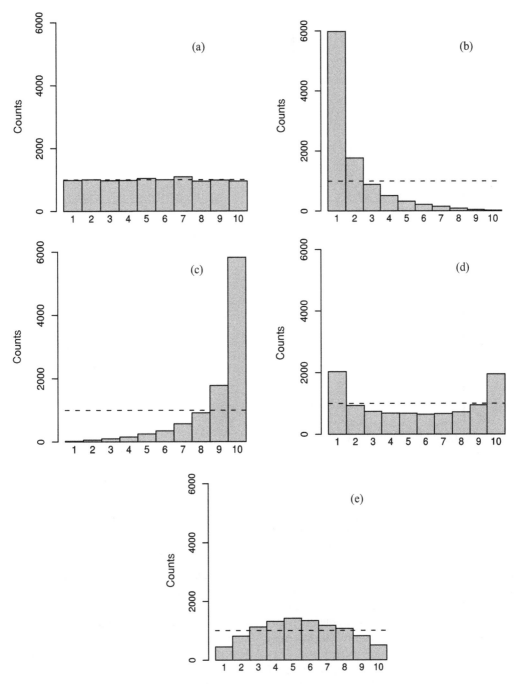

**Figure 8.3** Rank histograms for synthetic toy model forecasts with different dispersion characteristics. Each histogram is based on 10 000 synthetic forecast-observation pairs generated with the toy model of Section 8.2. Ensemble size is nine. (a) Reliable forecasts as illustrated in Figure 8.1a (toy model configuration: $\alpha = 1$, $\beta = 0$, $\gamma = 1$). (b) Forecasts subject to a systematic positive bias ($\alpha = 1$, $\beta = 1.5$, $\gamma = 1$) as in Figure 8.1b. (c) Forecasts subject to a systematic negative bias ($\alpha = 1$, $\beta = -1.5$, $\gamma = 1$). (d) Underdispersive forecasts ($\alpha = 1$, $\beta = 0$, $\gamma = 0.6$) as in Figure 8.1c. (e) Overdispersive forecasts ($\alpha = 1$, $\beta = 0$, $\gamma = 1.5$) as in Figure 8.1d. The dashed horizontal lines indicate the number of counts per category that would be expected were the forecasts perfectly reliable

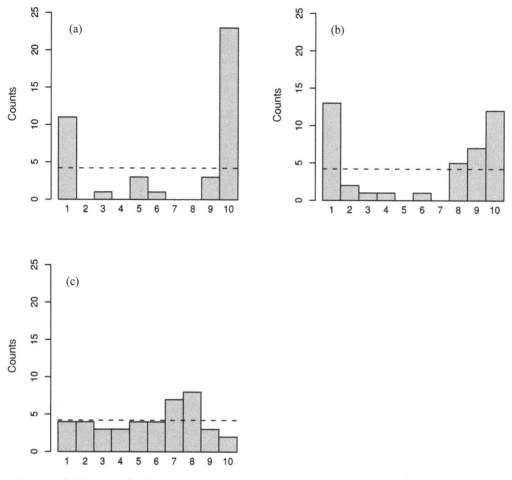

**Figure 8.4**   Rank histograms for European Centre for Medium Range Weather Forecasts (ECMWF) System 3 seasonal forecasts of average 2 m temperature in the Niño3.4 region as described in Section 8.2. (a) Raw ensemble output; (b) bias-corrected forecasts; and (c) recalibrated forecasts. The dashed horizontal lines indicate the number of counts per category that would be expected were the forecasts perfectly reliable

histograms obtained from seasonal ensemble forecasts (ensemble size 9) for the prediction context described in Section 8.2. The histograms have been constructed once from 'raw' ensemble forecasts (Figure 8.4a), once from bias-corrected forecasts (Figure 8.4b), and once from recalibrated forecasts (Figure 8.4c). The 99th percentile value of the null distribution, a $\chi^2$-distribution with nine degrees of freedom, is 21.7. The rank histogram of Figure 8.4a yields $T = 117.5$. Its shape indicates both underdispersion and a systematic negative bias: underdispersion because the outer ranks are overpopulated, and a systematic negative bias because the

highest rank is more populated than the lowest rank. The histogram of Figure 8.4b, which has the systematic bias removed but still indicates underdispersion, has $T = 51.8$; and the histogram of the recalibrated forecasts of Figure 8.4c has $T = 7.5$. Consequently, the null hypothesis of the rank histogram being uniform can be rejected at the 1% level for the raw and bias-corrected forecasts, but not for the recalibrated forecasts. Note that here it has been assumed that the $n$ sets of forecast-observation pairs are independent from each other. If this assumption cannot be justified, for example if highly correlated forecasts at neighbouring gridpoints or consecutive days are jointly considered in the rank histogram, the critical $\chi^2$-values need to be adjusted to account for serial dependence (Wilks, 2004). Otherwise there would be an enhanced danger of misinterpretation (Marzban *et al.*, 2010).

While the $\chi^2$ goodness-of-fit test is widely used and easy to implement and interpret, it has been criticized for its lack of power for small sample sizes. Moreover, the test statistic $T$ only considers the sum of the departures from uniformity, but is insensitive to any order that may exist for these departures, so that sloped, peaked or U-shaped histograms may remain undetected. Elmore (2005) has therefore proposed to use tests from the Cramér–von Mises family of statistics (summarized in Choulakian *et al.*, 1994) as an alternative, since they are order-sensitive and more powerful than the $\chi^2$ goodness-of-fit test. Among the tests in this family, the Watson test (Watson, 1961) is particularly sensitive to U-shaped and peaked histograms, while the Anderson–Darling test (Anderson and Darling, 1952) is very sensitive to sloped histograms. Another strategy to deal with the shortcomings of the $\chi^2$ goodness-of-fit test has been proposed by Jolliffe and Primo (2008). The authors suggest tests based on decompositions of the $\chi^2$ test statistic into components that correspond to specific alternatives, for example sloped or U-shaped histograms. This technique is argued to be more flexible than the Cramér–von Mises tests because the decomposition can be individually tailored to the question of interest. Moreover, since their approach is entirely based on $\chi^2$-distributions, the assessment of significance and computation of $p$-values is relatively easy.

So far it has tacitly been assumed that there are no ties between observed outcomes and the corresponding ensemble members, implying that the ranks of the observations are always well defined. However, in reality this assumption is not necessarily satisfied. For instance, in the context of ensemble forecasts of precipitation it often happens that both the observed outcome and some ensemble members take the value 0. The rank of the observation is then no longer uniquely determined. Hamill and Colucci (1998) have dodged this problem by calculating random small deviates to be added to the tied forecasts and the observation, a technique sometimes referred to as 'dithering' (e.g. Casati *et al.*, 2004). If the magnitudes of these deviates are sufficiently small, an artificial ranking is introduced without substantially affecting further calculations. Alternatively, one could simply assign a random rank value between $(M + 1)$ and $(M + M_{tied} + 1)$ to the observation, with $M$ being the number of ensemble members exceeded by the observation, and $M_{tied}$ being the number of ensemble members tied with the observation. This approach has, for example, been applied by Sloughter *et al.* (2007) for the verification of precipitation forecasts.

### Conditional exceedance probabilities (CEP)

Similarly to the reliability criterion of Equation 8.3, the flatness of rank histograms is a necessary but not sufficient condition of ensemble reliability. Hamill (2001), for instance, has shown examples of ensemble forecasts that are subject to different conditional biases and thus unreliable, but nevertheless yield flat rank histograms. In fact, it is a common source of misinterpretation that ensemble deficiencies, such as underdispersion or bias, are assumed to be 'stationary' and unconditional, while they actually may vary from case to case and be conditional on the specific flow characteristics. Consequently, rank histograms should always be interpreted with caution and accompanied by other tests further characterizing the ensemble. One such test could be based on the analysis of conditional exceedance probabilities (CEPs), an approach that has been suggested by Mason *et al.* (2007) to detect and quantify conditional biases.

Consider an ensemble forecast $\hat{x}$ and the corresponding observed outcome $x$. Let there be $m$ ensemble members, and let $\hat{x}_{(i)}$ be the $i^{th}$ ensemble member, with the ensemble members being *sorted in ascending order* (the subscript index has been put in brackets to indicate order statistics). As has been stated above, ensemble reliability implies that the observed outcome and the corresponding ensemble members are samples from the same flow-dependent underlying probability distribution. Consequently, the probability that the verifying observation exceeds a given percentile of the forecast distribution should be independent of the physical value of that percentile. For instance, the probability that the median ensemble member of a temperature forecast is exceeded by the verifying observation should always be 50%, regardless whether the median forecast is, say, $-5°C$ or $26°C$. More generally, if $\hat{x}_{(i)}$ is assumed to mark the $[i/(m+1)]^{th}$ percentile of the forecast distribution, reliability implies that the conditional exceedance probability $CEP = P(x > \hat{x}_{(i)}|\hat{x}_{(i)})$ should always be

$$1 - \left(\frac{i}{m+1}\right),$$

i.e. independent of the value of $\hat{x}_{(i)}$. This is what is tested by the approach of Mason *et al.* (2007): For a given ensemble member rank $i$ and a given set of forecast-observation pairs, a parametric CEP-curve $P(X > \hat{X}_{(i)}|\hat{X}_{(i)})$ is estimated and then compared to the ideal CEP, which is a horizontal line at

$$1 - \left(\frac{i}{m+1}\right)$$

Mason *et al.* (2007) suggest applying a generalized linear model with binomial errors and a logit link function to obtain

$$P\left(X > \hat{X}_{(i)}\big|\hat{X}_{(i)}\right) = \frac{1}{1 + \exp\left(-\beta_{0,i} - \beta_{1,i}\hat{X}_{(i)}\right)}$$

(8.5)

with parameters $\beta_{0,i}$ and $\beta_{1,i}$ to be estimated from the generalized linear regression.

Figure 8.5 illustrates the CEP-approach for the seasonal forecast examples described in Section 8.2 and Figure 8.4. The black line is the climatological (i.e. unconditional) probability that an observation exceeds a specific value, while the grey lines are the CEPs obtained for the ensemble medians using Equation 8.5. Figure 8.5a is for raw forecasts. If the forecasts were perfectly reliable, the CEP should be a horizontal line at 0.5, since the median is considered (indicated by the dashed line). However, this is clearly not the case in this example, where a strong conditional bias is apparent. If high temperatures are predicted, the forecasts are negatively biased and very likely to be exceeded by the observations. Conversely, if low temperatures are predicted, the forecasts are subject to a positive bias and mostly exceed the observation. This behaviour is an indication that

the model climatology has lower interannual variability than the observations. Removing the systematic (negative) bias of the forecasts shifts the entire CEP-curve downwards towards lower exceedance probabilities, but does not change its shape (Figure 8.5b). After a more sophisticated recalibration, which also corrects for systematic errors in interannual variability, the CEP is strongly improved, but still not perfectly flat (Figure 8.5c). The remaining deviation from flatness can, at least partially, be attributed to sampling errors in the quantile estimates of the forecast distributions. Since the exceedance probabilities are not independent from the quantile estimates, such sampling errors lead to sloped CEP curves, even if the ensemble forecasts are perfectly reliable (Bröcker *et al.*, 2011). This implies that the assumption that the ensemble members mark equidistant quantiles of an underlying forecast distribution is not appropriate in this context. As a strategy to overcome this deficiency, Mason *et al.* (2011) have proposed to split each ensemble forecast into two halves so that quantiles and exceedance probabilities can be calculated independently from each other.

### Spread-skill relationships

Spread-skill relationships (sometimes also referred to as 'spread-error' or 'accuracy-spread' relationships) refer to a family of tests that have been proposed for assessing a specific aspect of reliability, namely the information contained in ensemble spread. The basic question these tests seek to answer is whether there is an association between the predicted uncertainty of a forecast, and the accuracy of the forecast. Predicted uncertainty is thereby typically measured by the ensemble spread, and forecast accuracy by the absolute or squared error of the ensemble mean. It has been argued that such a spread-skill relationship should exist if the ensemble forecasts are reliable and truly sample the underlying flow-dependent probability distribution of possible outcomes. However, many studies analysing spread-skill relationships have only found weak correlations (e.g. Whitaker and Loughe, 1998; Atger, 1999; Hamill *et al.*, 2004). To some degree this may be due to the fact that any spread-skill relationship that may exist can by construction hardly be detected if the case-to-case variability of spread

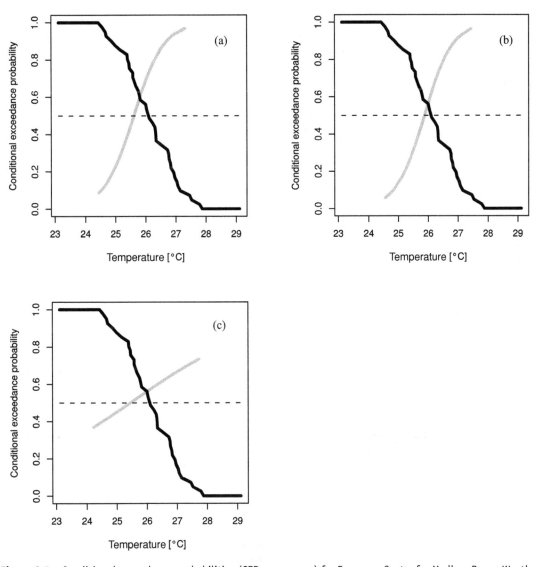

**Figure 8.5**   Conditional exceedance probabilities (CEP, grey curve) for European Centre for Medium Range Weather Forecasts (ECMWF) System 3 seasonal forecasts of average 2 m temperature in the Niño3.4 region, as described in Section 8.2 and shown in Figure 8.4. Here only the ensemble medians are considered. (a) Raw ensemble output; (b) bias-corrected forecasts; and (c) recalibrated forecasts. The black curve is the climatological, i.e. unconditional, probability that an observation exceeds a specific value. The dashed line indicates the CEP that would be expected were the prediction system perfectly reliable

is low, regardless of whether or not the forecasts are reliable (Whitaker and Loughe, 1998). However, there is also a deeper misconception in how spread-skill relationships are typically interpreted. If an ensemble forecast is reliable, then its spread does not quantify the expected error, but rather the error *uncertainty*, i.e. the error *variance*. This implies that

any standard regression between spread and forecast error is by construction difficult to interpret, regardless of how the 'error' is defined (Mason and Stephenson, 2008).

A possible alternative strategy to assess whether there is useful information in the case-to-case variability of ensemble spread has been sketched by

Mason and Stephenson (2008). The authors suggest fitting distributions to the ensemble forecasts, once with a fixed variance over the entire verification set, and once with the variance being estimated individually for each forecast on the basis of ensemble spread. By comparing the skill obtained for the two resulting sets of probabilistic forecast data, the contribution of variable ensemble spread to prediction skill can be estimated.

### 8.3.2 Multidimensional reliability

The tests that have been discussed so far are a priori only applicable to one-dimensional point forecasts, i.e. forecasts of one variable at one location. However, for many applications multidimensional forecasts are needed, i.e. joint forecasts of several variables at multiple locations. This then obviously also requires information on the collective reliability of such multidimensional forecasts. It is thereby not sufficient to evaluate the individual forecast dimensions separately from each other. For instance, a model may on average yield good temperature and precipitation forecasts, but the simultaneous forecasts of these two variables may be inconsistent and unrealistic. In the following, three approaches of assessing multidimensional ensemble reliability are discussed: (i) minimum spanning tree histograms, (ii) multivariate rank histograms, and (iii) bounding boxes.

*Minimum spanning tree histograms*
The minimum spanning tree histogram as a tool for assessing multidimensional reliability has been proposed by Smith (2001) and then further discussed by Smith and Hansen (2004), Wilks (2004) and Gombos *et al.* (2007). The principle is as follows.

Consider an ensemble of $m$ $k$-dimensional vector forecasts; $k$ could, for example, be the number of variables or locations simultaneously predicted. Each of the $m$ ensemble members can then be interpreted as a point in a $k$-dimensional space. For instance, Figure 8.6a shows an example of a nine-member seasonal ensemble forecast of temperature, simultaneously issued for the Niño3.4 region and a region in southern Africa (ignore the connecting lines for the moment). The two-dimensional en-

semble members of this joint forecast are displayed as dots in the two-dimensional prediction space. A spanning tree for this forecast can be formally constructed by connecting all forecast points with line segments such that no closed loops are generated. If $l$ is the sum of the lengths of all line segments, the minimum spanning tree (MST) is defined as that spanning tree which minimizes $l$. The length of the MST is thus a well-defined metric characterizing the collective 'closeness' of a set of points in a $k$-dimensional space. The thin lines in Figure 8.6a show the MST for this specific example.

As has been stated repeatedly in this chapter, the ensemble members of a reliable prediction system should be statistically indistinguishable from truth. This implies that on average the $k$-dimensional 'distance' of the observations from any of the corresponding ensemble members should be similar to the average mutual distance of the ensemble members from each other. Consequently, the MST length of a reliable ensemble forecast should on average not be significantly affected if a random ensemble member is replaced by the verifying observation. This is the reliability criterion that is tested by MST histograms. Let $l_0$ be the length of the MST of our $k$-dimensional $m$-member ensemble forecast, let $l_i$ be the length of the MST that is obtained if the $i^{th}$ ensemble member is replaced by truth (see Figures 8.6b and 8.6c for examples), and let $r$ be the rank of $l_0$ within the set $\{l_0, l_1, \ldots, l_m\}$. For a reliable forecast, $l_0$ should not be systematically larger or smaller than any $l_i \in \{l_1, \ldots, l_m\}$; i.e. it should be equally likely for $l_0$ to take any of the rank values $r = 1, \ldots, m + 1$. This can be assessed by determining $r$ for all forecast-observation pairs and then displaying the values obtained in the form of a histogram, the so-called MST histogram. If the forecasts are reliable, the MST histogram should be uniform. This can be tested by a $\chi^2$ goodness-of-fit test (Wilks, 2004) or tests from the Cramér–von Mises family (Gombos *et al.*, 2007), as discussed in Section 8.3.1. Note that, similarly to the one-dimensional rank histogram, MST histogram flatness is a necessary but not sufficient criterion of reliability. Also note that, despite some similarities, MST histograms generally cannot be considered as a multidimensional generalization of rank histograms and need to be interpreted differently. For instance, underdispersive ensemble

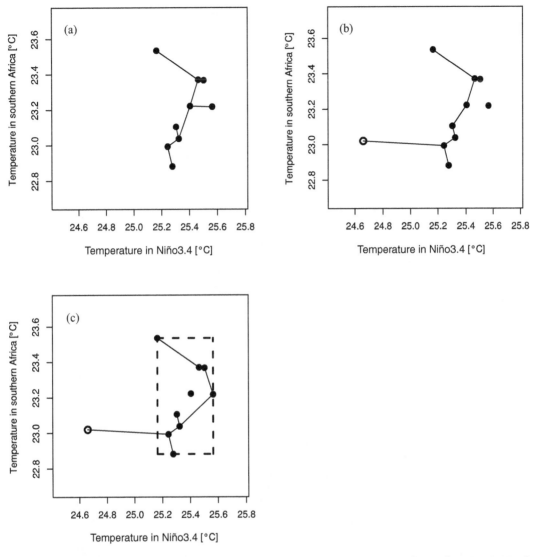

**Figure 8.6** Minimum spanning tree (MST) for a two-dimensional European Centre for Medium Range Weather Forecasts (ECMWF) System 3 seasonal ensemble forecast of 2 m temperature, jointly issued for the Niño3.4 region and southern Africa (20–35°S, 10–40°E). More details on the prediction context are provided in Section 8.2. The nine ensemble members are shown as black dots, the verifying observation as an unfilled circle. (a) The MST spanned by the ensemble members only; (b) and (c) MSTs obtained by replacing one (arbitrary) ensemble member with the observation. The dashed box in (c) is the two-dimensional bounding box of this ensemble forecast

forecasts, which are associated with U-shaped rank histograms, typically yield negatively sloped MST histograms.

Figure 8.7 shows examples of MST histograms obtained from seasonal temperature forecasts that have been jointly issued for the Niñõ3.4 region and southern Africa. Figure 8.7a is for raw en- semble forecasts. The pronounced overpopulation of the first bin is due to the combined effect of systematic bias and underdispersion. Applying the $\chi^2$ goodness-of-fit test of Equation 8.4 yields $T = 322.3$ (a value of only 21.7 would be required to reject the null hypothesis of uniformity at the 1% level). By removing the systematic bias, the

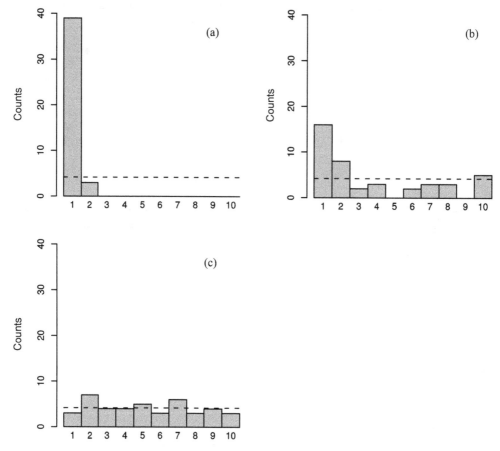

**Figure 8.7** Minimum spanning tree (MST) histograms for two-dimensional European Centre for Medium Range Weather Forecasts (ECMWF) System 3 seasonal ensemble forecasts of 2 m temperature, jointly issued for the Niño3.4 region and southern Africa (20–35°S, 10–40°E). More details on the prediction context are provided in Section 8.2. (a) Raw ensemble output; (b) bias-corrected forecasts; and (c) recalibrated forecasts. The dashed horizontal lines indicate the number of counts per category that would be expected were the forecasts perfectly reliable

MST histogram of Figure 8.7b is obtained, which is much closer to uniformity ($T = 48.5$) but still negatively sloped. Only if the forecasts are additionally recalibrated (Figure 8.7c) does it follow that the null distribution cannot be rejected at the 1% level ($T = 4.2$).

In the examples discussed above, the standard Euclidean $L_2$-norm has been used to determine the lengths of the line segments between the forecast points. Note that this distance metric is not always the most appropriate choice. Problems may arise, for example, if one or several of the forecast vector dimensions have significantly larger variance than the remaining ones. The subspace spanned by those high-variance components then essentially dominates the MST lengths and in consequence

the shape of the histograms. To avoid this implicit weighting of the dimensions, Wilks (2004) suggests standardizing the variances of each forecast dimension prior to calculating the MSTs. Further inconsistencies may arise if some forecast dimensions are highly correlated with each other, as may, for example, be the case if the dimensions represent neighbouring gridpoints. For such cases the Mahalanobis norm has been suggested as an alternative to the Euclidean norm (Stephenson, 1997; Wilks, 2004; Gombos *et al.*, 2007) since it accounts for covariance and is scale-invariant, thus measuring the 'statistical' distance between two points rather than the 'geometrical' distance.

Finally, note that MST lengths are not the only way to characterize the 'closeness' of multidimensional ensemble forecasts. For instance, Stephenson and Doblas-Reyes (2000) suggest the application of optimum projection methods, also known as 'multidimensional scaling' (MDS), to display multidimensional ensemble forecasts on two-dimensional maps for assessing their dispersion characteristics. Since MDS is based on the mutual distances of the ensemble members from each other, it represents a natural way to visualize multidimensional ensemble closeness.

### Multivariate rank histograms

Another approach to validate multidimensional reliability is given by multivariate rank (MVR) histograms (Gneiting et al., 2008). While the previously discussed MST histograms assess how the mutual 'closeness' of the ensemble members relates to their distance from the observations, MVR histograms assess how the observations *rank* with respect to the individual ensemble members. MVR histograms therefore represent a generalization of the scalar rank histogram discussed in Section 8.3.1. The key challenge in the construction of MVR histograms is that one needs to come up with a suitable definition of how to rank multidimensional vectors. Once this has been done, MVR histograms are constructed in the same way as scalar histograms, i.e. by counting how often each of the possible rank values is taken by the observations.

To determine the rank of a $k$-dimensional observation $x$ with respect to an ensemble of $m$ $k$-dimensional vector forecasts $\hat{x}_1, \ldots, \hat{x}_m$, Gneiting et al. (2008) propose the following procedure:

*Step 1:* Define $v_o := x$ and $v_i := \hat{x}_i$ $(i = 1, \ldots, m)$. Let $v_{0,l}$ denote the $l^{\text{th}}$ dimension of the observation and $v_{i,l}$ the $l^{\text{th}}$ dimension of the $i^{\text{th}}$ ensemble member.

*Step 2:* Determine the 'pre-rank' $\rho_i$ of each vector $v_i$ $(i = 0, \ldots, m)$:

$$\rho_i = 1 + \sum_{\substack{j=0 \\ j \neq i}}^{m} q_{i,j} \quad \text{with}$$

$$q_{i,j} = \begin{cases} 1 & \text{if} \quad v_{i,l} \geq v_{j,l} \\ & \text{for all} \quad l \in \{1, \ldots, k\} \\ 0 & \text{otherwise} \end{cases}$$

In the case of two-dimensional vectors, which can be displayed as points in the two-dimensional prediction space, the pre-rank of a specific vector $v_i$ can be interpreted as the number of points contained in the box to its lower left, including $v_i$ itself. This is illustrated in Figure 8.8a–c for the forecast example of Figure 8.6. Note that there are several ties in the pre-ranks.

*Step 3:* Determine the multivariate rank $r$. If the pre-rank of the observation, $\rho_0$, is not tied, then $r$ is given by $\rho_0$. If there are ties, these are resolved at random. For instance, in the example of Figure 8.8, two ensemble members have the same pre-rank as the observation, namely 1. The value of $r$ is then a random integer between 1 and 3.

Repeating these steps for all forecast-observation pairs and displaying the multivariate ranks in the form of a histogram yields the MVR histogram. The shape of MVR histograms is interpreted in the same way as their one-dimensional counterparts of Section 8.3.1. In particular, flat MVR histograms are associated with reliable forecasts, while overconfidence manifests itself in U-shaped MVR histograms. For the same reasons as discussed above in the context of MST histograms, it may be useful to standardize and rotate the forecast vectors by a Mahalanobis transform (Stephenson, 1997; Wilks, 2006b) prior to calculating the MVR histogram.

### Bounding boxes

All tests discussed so far are based on the reliability criterion that the ensemble members should be statistically indistinguishable from the observations. A less stringent criterion is applied in the bounding box approach, which has been suggested and discussed by Weisheimer et al. (2005), and by Judd et al. (2007). In this approach, an observed outcome is already considered as being consistent with the corresponding ensemble forecast, if it falls into the 'bounding box' (BB) spanned by the ensemble forecast. In the one-dimensional case, the BB is simply the interval defined by the smallest and the largest one of the ensemble members. For a $k$-dimensional forecast, the BB is defined by the minimum and maximum value of each ensemble component. The dashed rectangle shown in Figure 8.6c illustrates the BB of the two-dimensional example discussed in

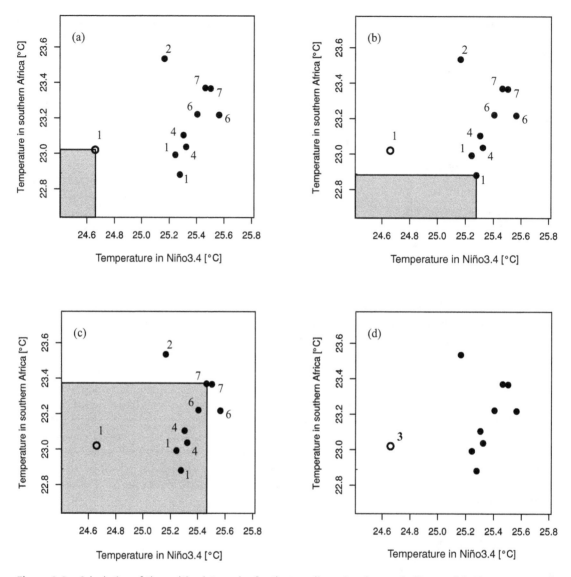

**Figure 8.8** Calculation of the multivariate rank $r$ for the two-dimensional example discussed in Figure 8.6. Panels (a)–(c) show the nine ensemble members (filled circles) and the observation (unfilled circle) together with their associated pre-ranks, which are obtained by counting how many points are contained in the box to the lower left of each vector, including the vector itself. Three exemplary boxes are displayed. The multivariate rank $r$ is shown in (d). Since the pre-rank of the observation is 1 and is tied with the pre-rank of two ensemble members, $r$ needs to be randomized over the set $\{1, 2, 3\}$. Here the random choice has been $r = 3$

the subsection on MST histograms. The basic argument of the BB approach is that a reliable ensemble prediction system should capture the observations more frequently than forecasts that are biased or underdispersive. In fact, if the forecasts are reliable, and if the $k$ dimensions are independent from each other (which is usually not the case), then the expected proportion of observations captured by the BBs is given by

$$\left( \frac{m - 1}{m + 1} \right)^{k}$$

with $m$ being the number of ensemble members. Thus, by analysing BB capture rates, some basic information on the consistency and applicability of a set of ensemble forecasts can be obtained without requiring any distributional assumptions to be made. BBs have the additional advantage of being easy to compute for any number of dimensions, making them particularly useful for assessing high-dimensional ensemble forecasts. Moreover, due to their intuitive interpretation, they are easy to communicate, also to non-experts. However, there is the obvious disadvantage that capture rates can always be artificially enhanced by simply inflating the ensemble forecasts to unrealistically large spread values. In other words, overdispersion is not penalized. Moreover, being defined by the minimum and maximum values of an ensemble, BBs are unduly affected by outliers and may fail in characterizing the bulk ensemble properties appropriately. Thus, it is not recommended to use BBs instead of other verification metrics, but rather as a complement to them.

Finally, note that bounding boxes may also be useful to analyse the consistency of ensemble forecasts through time. Model deficiencies could, for example, be detected by tracking the evolution of a BB through time and identifying those periods when the verifying observation falls out of the BB, i.e. is not 'shadowed' any more (Smith, 2001).

### 8.3.3   Discrimination

Next to reliability, discrimination is another important attribute of prediction skill. Discrimination measures whether forecasts differ when their corresponding observations differ (see also Sections 2.10 and 11.3.1). For example, do forecasts for days that were wet indicate more (or less) rainfall than forecasts for days that were dry? Despite being one of the most fundamental skill attributes, not many tests for discrimination (or the related skill attribute of *resolution*) have been published that are directly applicable to ensemble forecasts. Rather, discrimination is usually assessed probabilistically, i.e. by applying a suitable probabilistic skill metric (e.g. the area under the ROC curve; see Section 7.5.2) after the ensemble forecasts have been transformed into probability distributions. In the following, a

non-probabilistic measure of ensemble discrimination is presented: the 'generalized discrimination score', $D$, sometimes also referred to as the 'Two Alternatives Forced Choice' (2AFC) score.

### The generalized discrimination score for ensembles

The generalized discrimination score, $D$, was introduced by Mason and Weigel (2009) as a generic verification framework to measure discrimination. Formulations of $D$ have been derived for most types of forecast and observation data, ranging from single yes-no forecasts for binary outcomes to probabilistic forecasts of outcomes measured on a continuous scale (see also Section 11.3.1). In many cases, $D$ is equivalent to tests that are already known under a different name. For instance, if probabilistic forecasts of binary outcomes are considered, $D$ is equivalent to the trapezoidal area under the ROC curve (Section 7.5.2). Regardless of which verification context is considered, $D$ is always based on the same simple principle and seeks to answer the following question: Given any two differing observations, what is the probability that these observations can be correctly discriminated (i.e. ranked) by the corresponding forecasts? If the forecasts do not contain any useful information, then the probability that the forecasts correctly discriminate any two observations is 50%, which corresponds to random guessing. Consequently, the expected value of $D$ for a set of such forecasts is 0.5. The more successfully the forecasts are able to discriminate the observations, the closer the score is to 1. On the other hand, forecasts that consistently rank the observations in the wrong way would yield $D = 0$.

How can $D$ be formulated for ensemble forecasts? The procedure has been described in Weigel and Mason (2011) and is as illustrated in Figure 8.9. Firstly, all possible sets of two forecast-observation pairs are constructed from the verification data. Then, for each of these sets, the question is asked whether the forecasts can be used successfully to rank the observations. The proportion of times that this question is correctly answered yields $D$. This procedure requires a definition of how to rank ensemble forecasts. Indeed, when assessing whether two observations $x_s$ and $x_t$ are correctly discriminated by the ensemble forecasts $\hat{x}_s$ and $\hat{x}_t$, one

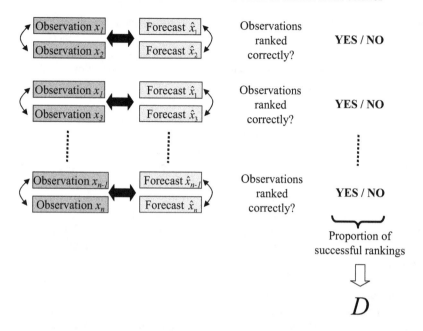

**Figure 8.9** Conceptual illustration of the calculation of the generalized discrimination score *D*. Adapted from Weigel and Mason (2011)

needs to be able to say whether $\hat{x}_s > \hat{x}_t$ or $\hat{x}_s < \hat{x}_t$. Weigel and Mason (2011) propose the following definition: $\hat{x}_s$ is called 'larger' than $\hat{x}_t$ if the probability that a random ensemble member from $\hat{x}_s$ is larger than a random ensemble member from $\hat{x}_t$ exceeds 50%. Applying this criterion for all sets of two forecast-observation pairs, and assuming that the observations are measured on a continuous (or-dinal) scale, leads to the following formulation of *D* for ensemble forecasts: Consider a set of *n* observations $x_1, \ldots, x_n$ and corresponding ensemble forecasts $\hat{x}_1, \ldots, \hat{x}_n$. Let there be *m* ensemble members, and let $\hat{x}_{t,i}$ be the $i^{\text{th}}$ ensemble member of the $t^{\text{th}}$ forecast. *D* is then given by

$$D = \frac{1}{2}\left(\tau_{\hat{R},x} + 1\right) \qquad (8.6)$$

with $\tau_{\hat{R},x}$ being Kendall's rank correlation coefficient (Sheskin, 2007) between the *n* observations and an *n*-element vector $\hat{R} = (\hat{R}_1, \ldots, \hat{R}_n)$, which corresponds to the ranks of the ensemble forecasts. The components of $\hat{R}$ can be computed with the following algorithm (see Weigel and Mason, 2011,

for details):

$$\hat{R}_s = 1 + \sum_{\substack{t=1 \\ t \neq s}}^{n} \hat{u}_{s,t} \quad \text{with}$$

$$\hat{u}_{s,t} = \begin{cases} 1 & if \quad \sum_{i=1}^{m} \hat{r}_{s,t,i} > m(m+0.5) \\ 0.5 & if \quad \sum_{i=1}^{m} \hat{r}_{s,t,i} = m(m+0.5) \\ 0 & if \quad \sum_{i=1}^{m} \hat{r}_{s,t,i} < m(m+0.5) \end{cases}$$

$$(8.7)$$

with $\hat{r}_{s,t,i}$ being the rank of $\hat{x}_{s,i}$ with respect to the set of pooled ensemble members $\{\hat{x}_{s,1}, \hat{x}_{s,2}, \ldots, \hat{x}_{s,m}, \hat{x}_{t,1}, \hat{x}_{t,2}, \ldots, \hat{x}_{t,m}\}$, if sorted in ascending order. Note that Equation 8.6 only holds for observations measured on a continuous scale. Expressions for binary and categorical observations are given in Weigel and Mason (2011).

While this expression may at first sight appear 'bulky', it is straightforward to implement and has a simple and intuitive interpretation. For the previously discussed seasonal forecasts of mean temperature in the Niño3.4 region (e.g. Figure 8.4a), one obtains $D = 0.92$. This means that in 92% of the cases any two observed seasonal mean temperatures would have been correctly discriminated *a priori* by the corresponding ensemble forecasts. Note that this score is invariant to bias corrections and recalibration, since the forecasts and observations are reduced to an ordered scale. As such, $D$ can be considered as a measure of the potential usefulness rather than the actual value of the forecasts.

## 8.4   Ensembles interpreted as probabilistic forecasts

The tests presented in the last section have in common that they interpret ensembles formally as finite sets of deterministic forecast realizations. However, once the ensemble members have been converted into some form of probability distribution, standard probabilistic skill metrics, as discussed in Chapter 7, can also be used for verification. While it may be argued that finding an appropriate probabilistic ensemble interpretation is a problem on its own and not related to verification, both tasks are in fact closely interlinked, because any probabilistic skill estimate of an ensemble prediction system inadvertently not only measures model quality but also

the appropriateness of the ensemble interpretation method applied. This section therefore starts with a brief discussion of common approaches of probabilistic ensemble interpretation (Section 8.4.1). After that, a short summary of probabilistic skill scores frequently applied for ensemble verification is presented (Section 8.4.2). In Section 8.4.3 the impact of finite ensemble size on skill is discussed.

### 8.4.1   Probabilistic interpretation of ensembles

It is very common for ensemble forecasts to be issued as probabilities for a binary or categorical event. These probabilities are often estimated by simply taking the proportion of ensemble members predicting the event ('frequentist interpretation'). Implicitly, the underlying probabilistic interpretation is that of a set of discrete 'delta distributions' marked by the ensemble members, as illustrated in Figure 8.10a. (A delta distribution is a mathematical construct that can be interpreted as a distribution with an infinitely sharp peak bounding a finite area.) However, even if the ensembles are reliable, probabilities derived from such a frequentist ensemble interpretation can only yield reasonable probabilistic estimates if many ensemble members are available and if the event considered does not correspond to a climatologically rare event.

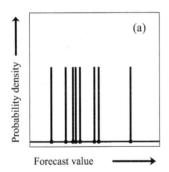
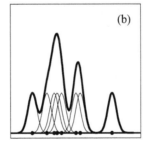
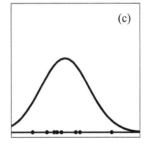

**Figure 8.10**  Approaches of probabilistic ensemble interpretation. (a) The frequentist interpretation, where the underlying probabilistic paradigm is that of a set of discrete delta peaks defined by the ensemble members (shown as black dots). In this interpretation, the probability for a specific event to happen is estimated by the proportion of ensemble members predicting that event. (b) The method of kernel dressing, where the individual ensemble members are dressed with a suitable kernel distribution (thin lines) characterizing the sampling uncertainty of the ensemble members. The sum of these kernels yields the forecast distribution (heavy black line). (c) The method of distribution fitting, where suitable parametric distributions are fitted to the ensemble members

Better probabilistic forecasts can be obtained by 'dressing' the ensemble members with so-called kernel functions, i.e. by replacing the discrete delta-distributions by suitable continuous probability distributions characterizing the uncertainty of each ensemble member. The forecast distribution is then given by the (non-parametric) sum of the dressed ensemble members, as illustrated in Figure 8.10b. The choice of kernel very much depends on the prediction context. For unbounded variables, such as temperature, usually Gaussian kernels are applied (e.g. Roulston and Smith, 2003; Wang and Bishop, 2005), while for precipitation Gamma kernels have been suggested (Peel and Wilson, 2008). A key challenge is thereby the definition of a kernel standard deviation, or bandwidth. For Gaussian kernels, often a bandwidth of $b = (4/3)^{1/5} m^{-1/5} \sigma_m$ is applied (Silverman, 1986), with $m$ being the ensemble size and $\sigma_m$ being the intra-ensemble standard deviation, but other choices may be more appropriate, particularly if the ensemble distributions are skewed (Peel and Wilson, 2008).

A third frequently applied approach of probabilistic ensemble interpretation is the fit of suitable parametric distributions to the ensemble forecast, as illustrated in Figure 8.10c. If the ensemble members really can be interpreted as independent random samples from an underlying forecast distribution, and if justified assumptions can be made concerning the shape of this distribution, then this approach can yield significant further skill improvement. For forecasts of temperature and geopotential height, often Gaussian distributions are fitted (e.g. Wilson et al., 1999), while for precipitation again Gamma distributions are usually more appropriate (e.g. Hamill and Colucci, 1998). To assess potential bifurcations and multimodalities in the ensemble forecasts, multivariate Gaussian distributions (Stephenson and Doblas-Reyes, 2000) and Gaussian mixture models (Wilks, 2002) have been successfully applied.

However, regardless of whether the forecast probabilities have been obtained by ensemble member counting, by kernel dressing or by distribution fitting approaches, even a perfectly estimated forecast distribution, due to the effect of model error, usually does not correspond to the true distribution of possible outcomes. This problem can be partially addressed, and prediction skill enhanced, by including the error-statistics of past forecasts into the ensemble interpretation. Such approaches are sometimes referred to as statistical post-processing, recalibration or ensemble MOS (for further discussion on this see, e.g., Bröcker and Smith, 2008; Gneiting et al., 2005; Hamill and Whitaker, 2006; Primo et al., 2009; Roulston and Smith, 2003; Stephenson et al., 2005; Tippett et al., 2007; Weigel et al., 2009; Wilks and Hamill, 2007).

### 8.4.2 Probabilistic skill metrics applied to ensembles

Once a probabilistic forecast has been formulated from an ensemble, regardless by which method, standard probabilistic skill scores can be applied. The choice of score is thereby mainly determined by the prediction context (e.g. whether forecasts of binary, categorical or continuous outcomes are issued) and by the skill attribute of interest (e.g. reliability or resolution). If the ensemble forecasts are issued as discrete probabilities for a binary event, reliability and resolution are often assessed with the reliability and resolution components of the Brier score decomposition (see Section 7.5.4; Murphy, 1973), or with reliability diagrams (Section 7.6.1; Wilks, 2006b). The forecast attribute of discrimination, on the other hand, can be assessed with the trapezoidal area under the ROC curve (Sections 7.4 and 7.5.2; Mason, 1982a).

Common summary measures of ensemble prediction skill are the (half) Brier score (BS; Brier, 1950) for binary events; the ranked probability score (RPS; Epstein, 1969) and the ignorance score (IS; Roulston and Smith, 2002) for categorical events; and the continuous ranked probability score (CRPS; Hersbach, 2000) as well as the proper linear score (Bröcker and Smith, 2007b) for outcomes measured on a continuous scale. All these summary measures share the property of being strictly proper (see Sections 7.3.1 and 7.3.2), meaning that skill cannot be 'artificially' enhanced by hedging the forecasts towards other values against the forecaster's true belief. For a more in-depth discussion on the properties of these scores, and of scoring rules in general, the reader is referred to Chapter 7. Here only a short summary of the most frequently used summary scores in the context of probabilistic ensemble verification is provided, namely BS, RPS, CRPS and IS.

## Brier Score

The prediction context of the Brier Score (BS) is that of discrete probability forecasts issued for binary outcomes ('event' and 'non-event'). Let $\hat{p}_t$ denote the probability assigned to the event by the $t^{th}$ forecast, and define $y_t = 1$ (respectively $y_t = 0$) if the $t^{th}$ observation corresponds to an event (respectively, non-event). The (half) Brier score (BS) is then given by:

$$BS = \frac{1}{n} \sum_{t=1}^{n} (\hat{p}_t - y_t)^2 \qquad (8.8)$$

The BS is often formulated as a skill score by relating it to the score obtained from a reference forecast strategy. If climatology is used as a reference, and if $c$ is the climatological probability of the event, the Brier skill score (BSS) is given by:

$$BSS = 1 - \frac{BS}{BS_{Cl}} \quad \text{with} \quad BS_{Cl} = \frac{1}{n} \sum_{t=1}^{n} (c - y_t)^2 \qquad (8.9)$$

Positive values of the BSS indicate forecast benefit with respect to the climatological forecast. Being a strictly proper scoring rule, the BS can be decomposed into three components measuring reliability, resolution and uncertainty ('uncertainty' is the expectation value of the score if climatology was used as a forecasting strategy). For more details on the Brier Score decomposition see Sections 7.3.3 and 7.5.4; see also Murphy, 1973; Wilks, 2006b). Note that the practical implementation of this decomposition often requires that forecast probabilities are binned into a small set of probability intervals. This requires two extra components to be included in the decomposition (Stephenson *et al.*, 2008b).

## Ranked probability score

The ranked probability score (RPS) can be interpreted as a multicategorical generalization of the BS. It is applicable to discrete probabilistic forecasts issued for categorical events. In the following,

$K$ denotes the number of categories, and $c_k$ denotes the climatological probability that the observed outcome is in category $k$. For a set of $n$ forecast-observation pairs, $\hat{p}_{t,k}$ is the probability assigned by the $t^{th}$ forecast to the $k^{th}$ category. Further, we define $y_{t,k} = 1$ if the $t^{th}$ observation is in category $k$, and $y_{t,k} = 0$ otherwise. Finally, let $\hat{P}_{t,k}$ and $Y_{t,k}$ denote the $k^{th}$ component of the $t^{th}$ cumulative forecast and observation vectors, and $C_k$ the $k^{th}$ category of the cumulative climate distribution; i.e.

$$\hat{P}_{t,k} = \sum_{l=1}^{k} \hat{p}_{t,l}, \quad \text{and} \quad Y_{t,k} = \sum_{l=1}^{k} y_{t,l},$$

$$\text{and } C_k = \sum_{l=1}^{k} c_l.$$

Using this notation, the RPS is given by

$$RPS = \frac{1}{n} \sum_{t=1}^{n} \sum_{k=1}^{K} (\hat{P}_{t,k} - Y_{t,k})^2 \qquad (8.10)$$

Like the BS, the RPS is often formulated as a skill score to assess forecast benefit with respect to a reference strategy. If climatology is used as a reference, the ranked probability skill score (RPSS) is given by:

$$RPSS = 1 - \frac{RPS}{RPS_{Cl}} \quad \text{with}$$

$$RPS_{Cl} = \frac{1}{n} \sum_{t=1}^{n} \sum_{k=1}^{K} (C_k - Y_{t,k})^2 \qquad (8.11)$$

Note that also for the RPS, formulations exist of a decomposition into three terms representing reliability, resolution and uncertainty (Murphy, 1972; Candille and Talagrand, 2005).

## Continuous ranked probability score

The continuous ranked probability score (CRPS) is defined as the integrated squared difference between the cumulative forecast and observation distributions; it has several appealing characteristics, apart from being proper. Firstly, it is defined on a continuous scale and therefore does not require reduction of the ensemble forecasts to discrete probabilities of binary or categorical events as is necessary for the BS and RPS. Secondly, it can be interpreted as an integral over all possible BS values and allows, in analogy to the BS, the decomposition into terms representing reliability, resolution and uncertainty (for formulations of this decomposition see Hersbach, 2000; Candille and Talagrand, 2005). Thirdly, the deterministic limit of the CRPS is identical to the mean absolute error and thus has a clear interpretation. For the special situation of ensemble forecasts that are interpreted as a set of delta distributions as illustrated in Figure 8.10a ('frequentist' interpretation), Hersbach (2000) has derived an expression for the CRPS that is easy to implement, as follows.

Consider a set of $n$ ensemble forecasts $\hat{x}_1, \ldots, \hat{x}_n$ and corresponding observations $x_1, \ldots, x_n$. Let there be $m$ ensemble members, and let $\hat{x}_{t,(i)}$ be the $i^{th}$ ensemble member of the $t^{th}$ forecast with the $m$ ensemble members being sorted in ascending order (the brackets around the subscript index $i$ indicate order statistics). Further, define $\hat{x}_{t,(0)} = -\infty$ and $\hat{x}_{t,(m+1)} = \infty$. The CRPS is then given by:

$$CRPS = \frac{1}{n} \sum_{t=1}^{n} \left[ \sum_{i=1}^{m} \alpha_{t,i} \left( \frac{i}{m} \right)^2 \right.$$
$$\left. + \sum_{i=0}^{m-1} \beta_{t,i} \left( 1 - \frac{i}{m} \right)^2 \right] \quad (8.12)$$

where

$$\alpha_{t,i} = \begin{cases} 0 & \text{if} & x_t \leq \hat{x}_{t,(i)} \\ x_t - \hat{x}_{t,(i)} & \text{if} & \hat{x}_{t,(i)} < x_t \leq \hat{x}_{t,(i+1)} \\ \hat{x}_{t,(i+1)} - \hat{x}_{t,(i)} & \text{if} & \hat{x}_{t,(i+1)} < x_t \end{cases}$$

and

$$\beta_{t,i} = \begin{cases} \hat{x}_{t,(i+1)} - \hat{x}_{t,(i)} & \text{if} & x_t \leq \hat{x}_{t,(i)} \\ \hat{x}_{t,(i+1)} - x_t & \text{if} & \hat{x}_{t,(i)} < x_t \leq \hat{x}_{t,(i+1)} \\ 0 & \text{if} & \hat{x}_{t,(i+1)} < x_t \end{cases}$$

For the seasonal prediction example used above (e.g. Figure 8.4), one obtains CRPS = 0.26°C for the raw ensemble forecasts, 0.23°C for the bias-corrected forecasts, and 0.17°C for the recalibrated forecasts. Note that the CRPS has the same unit as forecasts and observations, highlighting its interpretability as a probabilistic generalization of the mean absolute error.

## Ignorance score

The ignorance score (IS), sometimes also referred to as the logarithmic score, is applicable to discrete probabilistic forecasts issued for categorical events, just as the RPS. Again, $K$ denotes the number of categories, and $\hat{p}_{t,k}$ is the probability assigned to the $k^{th}$ category by the $t^{th}$ forecast. By $k^*(t)$ we denote the category the $t^{th}$ observation has fallen into. The ignorance score is then given by

$$IS = -\frac{1}{n} \sum_{t=1}^{n} \log_2 \left( \hat{p}_{t,k^*(t)} \right) \quad (8.13)$$

Like the other skill metrics described above, the ignorance score is strictly proper and can therefore be decomposed into reliability, resolution and uncertainty (Weijs et al., 2010). Apart from that, the IS has some interesting properties that set it apart from the RPS. Firstly, it has a clear information theoretical interpretation in that it measures the average information deficit (in bits) of a user, who is in possession of a forecast, but does not yet know the true outcome (Roulston and Smith, 2002). Secondly, it is a *local* score, meaning that it only considers the probability assigned to the observed category and ignores the probabilities assigned to categories that have not been observed. Whether or not this attribute of *locality* is a desirable property is subject to ongoing discussion (see Section 2.8; see also Mason, 2008; Benedetti, 2010) and depends, among others, on the application context. Thirdly, the IS becomes infinite in cases when probability 0 has been assigned to the observed category. This may be considered as an unwanted property of the IS, particularly in the context of ensemble forecasting where zero probability is sometimes obtained if ensemble size is small and if probabilities are estimated by the proportion of ensemble members predicting an event. On the other hand, Weijs et al. (2010) have pointed out that 'it is constructive to give an infinite penalty to a forecaster who issues a wrong forecast that was supposed to be certain. This is fair because the value that a user would be willing to risk when trusting such a forecast is also infinite.' To avoid infinite IS values, one should therefore in the first place rethink the method of probabilistic ensemble interpretation applied. If one nevertheless wants to stick to a frequentist ensemble interpretation, a simple 'quick fix' to avoid zero probabilities could be to distribute an additional virtual ensemble member over all forecast categories (Roulston and Smith, 2002).

### 8.4.3 Effect of ensemble size on skill

In Section 8.4.1 it has been stated that the skill of probabilistic ensemble forecasts depends (i) on the quality of the ensemble prediction system itself, and (ii) on the method that is applied to convert the finite set of discrete ensemble members into a probability distribution. In the following, a third factor is discussed, namely ensemble size. Indeed, if the moments of a forecast distribution are to be estimated from $m$ ensemble members, then the accuracy of these estimates decreases as $m$ is reduced. For instance, the ensemble mean as an estimator for the central tendency of the underlying forecast distribution (standard deviation $\sigma$) is associated with a standard error of

$$\frac{\sigma}{\sqrt{m}}$$

Hence, even if the ensemble members by themselves are statistically indistinguishable from truth and thus perfectly reliable, probabilistic forecasts derived from them are inherently unreliable due to sampling uncertainty. This intrinsic unreliability is increased, and resolution is decreased, as ensemble size gets smaller (Richardson, 2001; Ferro, 2007a; Weigel *et al.*, 2007a; Ferro *et al.*, 2008). Having a direct impact on the precision with which a forecast distribution is estimated, large ensembles can therefore be particularly important if extreme events are to be forecast. Of course, if a forecast distribution is heavily biased or overconfident, even infinite ensemble size would not yield a 'good' forecast.

Most probabilistic skill metrics, particularly those that are sensitive to reliability, reveal a pronounced dependency on ensemble size. This can be seen, for example, in Figure 8.11 (to be discussed in more detail later), where the ranked probability skill score (RPSS) for synthetic toy model forecasts is shown as a function of ensemble size (black solid lines). For all three toy model configurations considered, skill drops as ensemble size is reduced.

Generally, this ensemble size dependency is a reasonable property in that it reflects the reduced forecast quality due to enhanced sampling uncertainty. However, there are verification questions when one would like to have the ensemble dependency quantified and removed from a skill metric. For example, this is the case if several models with unequal ensemble size are to be compared to identify the better model; or if the expected prediction skill of large-sized forecast ensembles is to be estimated on the basis of small-sized hindcast ensembles; or if one wants to determine the potential prediction skill which could be reached if the ensemble size were infinite.

For most probabilistic skill metrics the impact of ensemble size has not yet been explicitly quantified. An exception is the Brier (skill) score and the rank probability (skill) score, where two related approaches exist, which will be presented in the following. The first approach ('RPSS$_D$') reduces the ensemble size dependency of the RPSS by adjusting the reference forecast strategy; the second approach ('unbiased estimator') directly formulates an unbiased estimator for the effect of ensemble size on the RPS. Both approaches are based on a simple 'frequentist' ensemble interpretation, i.e. forecast probabilities are estimated by taking the fraction of ensemble members predicting an event.

The notation applied in the following is the same as introduced above in the subsections on the Brier and ranked probability scores (Section 8.4.2).

### The RPSS$_D$

The conceptual basis of this approach was introduced by Müller *et al.* (2005). It is motivated from the notion that an ensemble prediction system that produces better than random forecasts may nevertheless yield negative RPSS values, if ensemble size $m$ is small and climatology is chosen as a reference strategy. This is, for example, visible in Figure 8.11a for $m < 5$ (black solid line). Müller *et al.* (2005) argue that this is because the RPSS is based on an 'unfair' comparison between two forecast strategies of different reliability: on the one hand the ensemble forecasts, which, due to their finite ensemble size, are intrinsically unreliable; and on the other hand the climatological reference, which is perfectly reliable. As a 'fairer' reference, Müller *et al.* (2005)

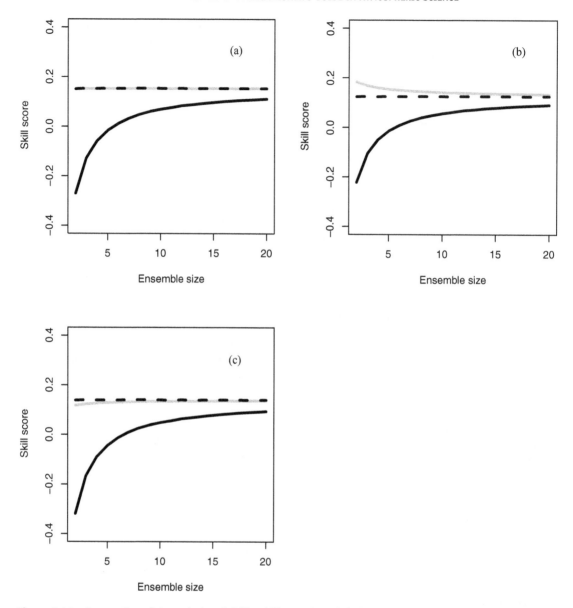

**Figure 8.11** Expectation of the ranked probability skill score (RPSS) (solid black line), the RPSS$_D$ (solid grey line), and the unbiased estimator of Equation 8.18 (black dashed line), as functions of ensemble size. The expected skill has been estimated from 10 000 score values, each of which is based on 50 sets of synthetic forecast-observation pairs generated with the toy model of Section 8.2. Three equiprobable categories are considered: (a) is for reliable forecasts (toy model configuration: $\alpha = 1$, $\beta = 0$, $\gamma = 1$) as illustrated in Figure 8.1a; (b) is for underdispersive forecasts ($\alpha = 1$, $\beta = 0$, $\gamma = 0.6$) as in Figure 8.1c; and (c) is for overdispersive forecasts ($\alpha = 1$, $\beta = 0$, $\gamma = 1.5$) as in Figure 8.1d

suggest applying $m$-member random forecasts sampled from climatology, since their intrinsic unreliability is similarly affected by ensemble size as that of the ensemble forecasts, thus reducing the overall ensemble size dependency of the RPSS. For the limit of large $m$, this modified reference strategy converges to the conventional climatological reference, which can be interpreted as a random sample from climatology with infinitely many ensemble members.

On the basis of this new reference strategy, Weigel *et al.* (2007a) have formulated a revised RPSS, the so-called $RPSS_D$, which can be computed by adding a correction term $D$ to the conventional climatological reference score:

$$RPSS_D = 1 - \frac{RPS_m}{RPS_{Cl} + D} \qquad (8.14)$$

with

$$D = \frac{1}{m} \sum_{k=1}^{K} \sum_{l=1}^{k} \left[ c_l \left( 1 - c_l - 2 \sum_{i=l+1}^{k} c_i \right) \right].$$

The index $m$ of $RPS_m$ is used here to indicate that the forecasts are based on $m$ ensemble members. If the $K$ forecast categories are equiprobable, the correction term $D$ simplifies to

$$D = (K^2 - 1)/(6Km).$$

For the binary event/non-event situation of the BSS, the modified reference strategy yields

$$BSS_D = 1 - \frac{BS_m}{BS_{Cl} + \frac{1}{m}c(1 - c)} \qquad (8.15)$$

In Figure 8.11, the expected $RPSS_D$ is shown (grey solid lines) as a function of ensemble size for (a) reliable toy model forecasts, (b) underdispersive forecasts, and (c) overdispersive forecasts. In all three cases the $RPSS_D$ reduces the ensemble size significantly with respect to the RPSS. However, only if the forecasts are reliable (Figure 8.11a) is the ensemble size dependency entirely removed (see also Tippett, 2008).

Note that generalizations of the $RPSS_D$ exist for situations when not all verification samples have the same ensemble size (Weigel *et al.*, 2008b), and when the output of several ensemble prediction systems is combined to a weighted multi-model ensemble (Weigel *et al.*, 2007b).

### Unbiased estimator

Under the assumption that the members of an ensemble forecast are 'exchangeable', i.e. statistically indistinguishable from each other (but not necessarily from truth), Ferro *et al.* (2008) have derived an

unbiased estimator for the effect of ensemble size on the RPS. Let $RPS_m$ denote the RPS obtained from an $m$-member ensemble prediction system, and let $E(RPS_M)$ denote the RPS that would be expected had the ensemble prediction system $M$ instead of $m$ members. An unbiased estimator of $E(RPS_M)$ is then given by

$$E(RPS_M) \cong RPS_m - \frac{M - m}{M(m - 1)n}$$

$$\times \sum_{t=1}^{n} \sum_{k=1}^{K} \hat{P}_{t,k} \left( 1 - \hat{P}_{t,k} \right) \qquad (8.16)$$

where '$\cong$' means *is estimated without bias from*. For the binary prediction context of the BS, Equation 8.16 simplifies to

$$E(BS_M) \cong BS_m - \frac{M - m}{M(m - 1)n} \sum_{t=1}^{n} \hat{p}_t \left( 1 - \hat{p}_t \right) \qquad (8.17)$$

Note that Ferro *et al.* (2008) have also derived a similar expression for the CRPS.

Estimates of the scores to be expected for infinite ensemble size can be easily obtained by taking the limit $M \to \infty$ in Equations 8.16 and 8.17. From this, it is straightforward to formulate an unbiased estimator of the RPSS for infinite ensemble size:

$$E(RPSS_\infty) = 1 - \frac{E(RPS_\infty)}{RPS_{Cl}} \cong 1 - \frac{RPS_m}{RPS_{Cl}}$$

$$+ \frac{\sum_{t=1}^{n} \sum_{k=1}^{K} \hat{P}_{t,k} \left( 1 - \hat{P}_{t,k} \right)}{RPS_{Cl}(m - 1)n} \qquad (8.18)$$

For the toy model forecasts analysed in Figure 8.11, Equation 8.18 has been applied to determine $E(RPSS_\infty)$ as a function of ensemble size (dashed lines). The plots confirm that the ensemble size dependency is entirely removed, regardless of whether the forecasts are reliable, underdispersive or overdispersive. This is different from the $RPSS_D$ of Equation 8.14, which is an unbiased estimator of $E(RPSS_\infty)$ only if the forecasts are reliable (Tippett, 2008).

## 8.5  Summary

This chapter has provided an overview of approaches to assess the quality of ensemble forecasts. *A priori*, ensembles are 'only' finite sets of deterministic forecast realizations that have been started from different initial conditions and/or are subject to different boundary conditions, and that are thought to represent samples from an underlying flow-dependent forecast probability distribution. In practice, ensembles are usually interpreted and applied as probabilistic forecasts, necessarily involving further statistical assumptions. These two levels of interpretation are reflected in the skill metrics commonly applied for ensembles, which can be categorized into two groups: on the one hand are tests that consider the individual ensemble members as discrete samples from a probability distribution, and on the other hand are truly probabilistic skill metrics that require that some form of probability distribution has been derived from the ensemble members prior to verification.

Concerning the first category of tests, a multitude of approaches has been proposed in the literature. The majority of these tests assess ensemble reliability; they ask whether the ensemble members are statistically indistinguishable from truth, and if not, whether systematic deficiencies such as under-dispersion or unconditional bias can be identified. Ensemble reliability is thereby typically interpreted as an average statistical property over the entire verification set; i.e. most of the tests only assess the *average* consistency between ensemble members and observations, but do not provide insight into the flow-dependent *case-to-case* performance and error-characteristics. This shortcoming has not yet been satisfactorily resolved, not least for conceptual reasons. Another shortcoming is that most tests of ensemble quality are designed for point forecasts of one variable, ignoring the fact that often forecasts of more than one variable at more than one location are simultaneously applied. However, some initial promising approaches to multidimensional ensemble verification have been published, which may be the starting point for further developments.

Concerning the second category of tests, essentially all the standard probabilistic verification techniques of Chapter 7 can be applied, with the choice of score being determined by the prediction context (e.g. whether forecasts of binary, categorical or continuous outcomes are issued) and by the skill attribute of interest (e.g. whether reliability or resolution is to be assessed). However, it is important to stress that probabilistic skill metrics here not only measure the quality of the ensemble prediction system itself, but also the appropriateness of the method used to convert sets of discrete ensemble members into probabilistic forecasts. Moreover, many probabilistic scores are sensitive to ensemble size, with skill decreasing as ensemble size is reduced. For certain applications, this may be an unwanted and misleading property, for instance if models of different ensemble sizes are to be compared. Recent research has started to address this issue for specific skill metrics.

## Acknowledgement

This work was supported by the Swiss National Science Foundation through the National Centre for Competence in Research on Climate (NCCR Climate).

# 9

# Economic value and skill

## David S. Richardson
*European Centre for Medium-Range Weather Forecasts*

## 9.1 Introduction

Three types of forecast 'goodness' were identified by Murphy (1993): *consistency, quality* and *value* (see Section 1.4). Consistency and quality have been the main focus of much of this book on forecast verification. However, this chapter will now consider economic value (or utility) and its relationship to quality measures such as forecast skill. Space prevents us from giving a comprehensive review of the economic value of forecast information in a single chapter. Rather, the aim is to introduce the basic concepts of the value of forecast information to users and to explore some of the fundamental implications for forecast verification.

The main aspects of deriving economic benefits from forecasts are incorporated in so-called decision-analytic models (e.g. Murphy, 1977; Katz and Murphy, 1997a). A decision-maker (forecast user) has a number of alternative courses of action to choose from, and the choice is to some extent influenced by the forecast. Each action has an associated cost and leads to an economic benefit or loss depending on the weather that occurs. The task of the decision-maker is to choose the appropriate action that will minimize the expected loss (or maximize the expected benefit). In this chapter, we focus on the simplest of these economic decision models known as the (static) cost-loss model (Ångström, 1922; Thompson, 1952; Murphy, 1977; Liljas and Murphy, 1994).

A simple decision model is introduced in Section 9.2. It is applied first to deterministic forecasts; then it is used to show how probability forecasts can be used in the decision-making process. The benefit of probability forecasts over deterministic forecasts is then assessed.

A particularly useful feature of this simple model is that it provides a link between user value on the one hand and more standard verification measures on the other. In Section 9.3 the relationship between value and the relative operating characteristic (ROC) is explored, and in Section 9.4 a measure of overall value (over all users) and the link with the Brier score are considered. The contrasting effects of ensemble size on the ROC Area and Brier Skill Scores are examined in Section 9.5, and the differences in behaviour of the two scores is interpreted in terms of user value. A variety of applications using the cost-loss model and some general perspectives are considered in Section 9.6.

The chapter will be illustrated with examples taken from the operational Ensemble Prediction System (EPS) of the European Centre for Medium

Range Weather Forecasts (ECMWF) (Palmer *et al.*, 1993; Molteni *et al.*, 1996; Buizza *et al.* 2000). More detailed evaluations of the economic value of these forecasts in specific situations are given by Taylor and Buizza (2003) and Hoffschildt *et al.* (1999). Further examples using the cost-loss model to evaluate ensemble forecasts can be found in Richardson (2000), Zhu *et al.* (2002), and Palmer *et al.* (2000).

## 9.2 The cost/loss ratio decision model

Consider a decision-maker who is sensitive to a specific adverse weather event A. For example, A may be 'the occurrence of ice on the road' or 'more than 20 mm of rain in 24 hours'. We assume that the decision-maker will incur some loss L if this bad weather event occurs and no action is taken. The forecast will be useful only if the decision-maker can take some action to prevent or limit the expected loss due to bad weather. We take the simplest possible situation where the user has just two alternative courses of action – either do nothing (carry on as normal) or take some form of protective action to prevent loss. This action will cost an amount C (additional to the normal expenditure). Examples could be gritting roads to prevent the formation of ice, or arranging to move an outdoor event inside if heavy rain is expected.

There are four possible combinations of action and occurrence with the net cost depending on what happened and what action was taken. If action is taken then the cost is C irrespective of the outcome. However, if action is not taken, the expense depends on the actual weather that occurs: if event A does not occur there is no cost, but if event A does occur then there is a loss L. The situation is summarized in Table 9.1, sometimes known as the *expense matrix (payoff matrix)*. Note that all expenses are taken relative to the 'normal event' of no action and no adverse weather, and that we have (once again) taken the simplest scenario where the protective action completely prevents the potential loss. This may appear to be an oversimplification of the general case of different expenses in each cell of the expense matrix (see Table 3.6), but the expression for economic value developed in the following is essentially the same for both situations (Richard-

**Table 9.1**  The expense matrix: costs (C) and losses (L) for different outcomes in the simple cost-loss decision model

|  |  | Event occurs | |
| --- | --- | --- | --- |
|  |  | Yes | No |
| Action taken | Yes | C | C |
|  | No | L | 0 |

son, 2000). In other words, Table 9.1 captures the salient features of the more general cost-loss model (see Section 9.2.1).

Assume the decision-maker aims to minimize the average long-term loss by taking the appropriate action on each occasion. We assume for simplicity that the weather is the only factor influencing the decision. To set a baseline for economic value, we first consider the reference strategies available in the absence of forecast information. If there is no forecast information available then there are only two possible choices: either always protect or never protect (we will ignore the third possible option of making decisions randomly). If the decision-maker always protects, the cost will be C on every occasion, so the average expense is

$$E_{\text{always}} = C \qquad (9.1)$$

On the other hand, if action is never taken there will be some occasions with no expense and other occasions with loss L. Over a large number of cases, let $s$ be the fraction of occasions when event A occurred (the climatological base rate; see Section 3.2.1). The average expense is then given by

$$E_{\text{never}} = sL \qquad (9.2)$$

In general, $E_{\text{always}}$ and $E_{\text{never}}$ are not equal, and so to minimize losses the decision-maker should choose the strategy with the smallest average expense. The optimal strategy (*decision rule*) is to always take protective action if $E_{\text{always}} < E_{\text{never}}$ and never take protective action if $E_{\text{always}} > E_{\text{never}}$. For this optimal strategy, the mean expense can be written as

$$E_{\text{climate}} = \min(C, sL) \qquad (9.3)$$

This will be referred to as the *climate expense* because the user needs to know the climatological base rate probability ($s$) of the event in order to know whether to always protect or not. For events having base rates greater than the cost/loss ratio C/L the decision-maker should always take protective action (assume the event is going to happen), whereas for rarer events with base rates less than C/L, the decision-maker should never take protective action (assume the event will not happen). For events with a base rate equal to the cost/loss ratio, the expense is the same for both strategies. The climate expense provides a baseline for our definition of forecast value. It has been assumed that the base rate used to determine the choice of default action will stay the same in the future (stationarity assumption). Given perfect knowledge of the future, the decision-maker would need to take action only when the event was going to occur. The mean expense would then be

$$E_{perfect} = sC \qquad (9.4)$$

The fact that this is greater than zero for non-zero base rates reminds us that some expense is unavoidable. The aim of using forecast information in the decision process is to optimally reduce the mean expense from $E_{climate}$ towards $E_{perfect}$. However, the mean expense can never be completely reduced to zero unless either the base rate is zero or there is no cost for taking preventative action.

The value, $V$, of a forecast system can be defined as the reduction in mean expense relative to the reduction that would be obtained by having access to perfect forecasts:

$$V = \frac{E_{climate} - E_{forecast}}{E_{climate} - E_{perfect}} \qquad (9.5)$$

This definition is equivalent to the standard definition of a skill score with climatology as the reference (see Section 2.7). Similar to results for equitable skill scores, zero value will be obtained by constantly forecasting either the event if $s > C/L$ or non-event if $s < C/L$. A maximum value of 1 will be obtained for systems that perfectly forecast future events. When $V$ is greater than zero, the decision-maker will gain some economic benefit by using the forecast information in addition to using the base rate information.

It should be noted that value is sometimes defined in an absolute sense as the saving $S = E_{climate} - E_{forecast}$ (e.g. Thornes and Stephenson, 2001). This definition has some advantage for a specific decision-maker in that it gives a direct measure of the amount of financial benefit (e.g. in units of currency). However, the definition of value used here, sometimes referred to as *relative value*, is more useful in a general context where we wish to compare the forecast value for a range of different users. The perfect saving $S_{perfect} = E_{climate} - E_{perfect}$ gives an absolute upper bound on how much can be saved.

### 9.2.1 Value of a deterministic binary forecast system

A deterministic binary forecast system gives a simple yes/no prediction of whether the weather event A will occur or not. The decision-maker takes protective action when the forecast is for A to occur and does nothing otherwise. The value of such forecasts over a set of previous events can be estimated using the cell counts in a contingency table accumulated over previous events (Table 9.2; see also Chapter 3).

The sample mean expense using the forecasts is easily obtained by multiplying the expenses in Table 9.1 by the corresponding relative frequencies in Table 9.2:

$$E_{forecast} = \frac{a}{n}C + \frac{b}{n}C + \frac{c}{n}L \qquad (9.6)$$

It is convenient to re-express this expense in terms of sample estimates of the likelihood-base rate conditional probabilities: hit rate $H = a/(a+c)$, false alarm rate $F = b/(b+d)$ and base rate $s = (a+c)/n$. This yields

$$E_{forecast} = F(1-s)C - Hs(L-C) + sL \qquad (9.7)$$

Substitution into Equation 9.5 then gives an expression for the relative value of the forecasts:

$$V = \frac{\min(\alpha, s) - F(1-s)\alpha + Hs(1-\alpha) - s}{\min(\alpha, s) - s\alpha} \qquad (9.8)$$

**Table 9.2**  Contingency table showing counts for a single event

| | | Event observed | | |
|---|---|---|---|---|
| | | Yes | No | Marginal totals |
| Event forecast | Yes | $a$ | $b$ | $a+b$ |
| | No | $c$ | $d$ | $c+d$ |
| | Marginal totals | $a+c = ns$ | $b+d = n(1-s)$ | $a+b+c+d = n$ |

where $\alpha = C/L$ is the specific user's *cost/loss ratio*. Equation 9.8 shows that value depends not only on the quality of the system ($H$ and $F$), but also on the observed base rate of the event ($s$) and the user's cost/loss ratio ($\alpha$). The introduction of the forecast user into the verification process brings an extra dimension into the problem: the value of forecast to the user is very much user-dependent. However, it is only the cost/loss ratio that is important rather than the individual values of $C$ and $L$. For simplicity, we have assumed that the cost $C$ provides full protection against the potential loss $L$ (Table 9.1). It should be emphasized that the general case, allowing different costs or losses in each cell of the expense matrix (Table 3.6), still leads to Equation 9.8, with $\alpha$ now representing a more general 'cost-loss' ratio (Richardson, 2000); in the notation of Section 3.5.3, the more general expense situation leads to a 'cost-loss' or 'utility' ratio of

$$\alpha = \frac{U_d - U_b}{U_d - U_b + U_a - U_c}$$

For a given event, the base rate $s$ is fixed; and for a particular forecast system $H$ and $F$ are also fixed, and so $V$ is then only a function of the user's cost/loss ratio. We can show this variation of value with user by plotting $V$ against $C/L$. Since there would be no point in taking action if the protection cost $C$ is greater than the potential loss $L$, we need only consider the range $0 < C/L < 1$.

Figure 9.1 shows the value of ECMWF deterministic forecasts of European precipitation for three different amount thresholds over the whole range of possible cost/loss ratios. The control forecast is a single forecast made using the best estimate of initial conditions. An ensemble of other forecasts is generated by perturbing the initial conditions of the control forecast – the value of the whole ensemble of forecasts will be discussed below. It is clear that the value of the forecasts varies considerably for classes of hypothetical users with different cost/loss ratios. For example, while some users gain up to 50% value for the 1 mm event, users with cost/loss ratios greater than 0.7 or less than 0.2 will find no benefit; for these ranges of cost/loss ratio the forecasts are less useful than climatology (negative values are not shown on the figure). The shape of the curves is similar for the different thresholds, although the maximum value is shifted towards lower cost/loss ratios for the rarer higher precipitation events. In other words, users with small cost/loss ratios (e.g. those who incur especially large losses when the rare event happens and no protection has been taken) benefit the most from forecasts of rare events.

It is straightforward to determine the location and magnitude of the maximum value from the expression for $V$ in Equation 9.8. When $\alpha < s$, Equation 9.8 becomes

$$V = (1 - F) - \left(\frac{s}{1-s}\right)\left(\frac{1-\alpha}{\alpha}\right)(1 - H)$$

(9.9)

and so the value increases for increasing cost/loss ratios. When $\alpha > s$ then

$$V = H - \left(\frac{1-s}{s}\right)\left(\frac{\alpha}{1-\alpha}\right)F$$

(9.10)

and the value decreases for increasing cost/loss ratios. Hence, the maximum value will always occur when $\alpha = s$ (i.e. when the cost/loss ratio equals the base rate). At this point, the mean expense of taking either of the climatological options (always or never protect) is the same: climatology does not help the

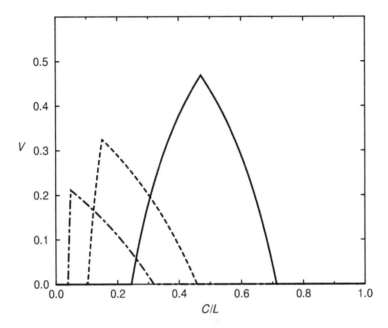

**Figure 9.1**   Value of the European Centre for Medium Range Weather Forecasts (ECMWF) Ensemble Prediction System (EPS) control deterministic forecast of 24-hour total precipitation over Europe at day 5 for winter 1999/00. Curves show value $V$ as a function of user cost-loss ratio $C/L$ for three different events, defined as precipitation exceeding 1 mm (solid line), 5 mm (dashed line) and 10 mm (chain-dashed line)

decision-maker and the forecast offers the greatest benefit. As the cost/loss ratio approaches 0 or 1, the climatological options become harder to beat.

The maximum value obtained by substituting $\alpha = s$ in Equation (9.8) is given by

$$V_{max} = H - F \qquad (9.11)$$

This is identical to the Peirce Skill Score described in Section 3.5.1. Hence, *maximum* economic value is related to forecast skill, and the Peirce Skill Score can be interpreted as a measure of *potential* forecast value as well as forecast quality. Two of the three aspects of goodness identified by Murphy (1993) are therefore related. However, the Peirce Skill Score only gives information about the maximum achievable value, and gives no information about the value for a specific user having a cost/loss ratio different to the base rate. Strictly, the maximum value should be the absolute value of the Peirce Skill Score since forecasts with negative Peirce Skill Scores can always be recalibrated (by relabelling the forecasted event as non-event, and vice versa) to have positive skill.

We can also determine the range of cost/loss ratios over which there is positive value – i.e. the class of users who can derive benefit from using the forecasts. From Equations 9.9 and 9.10, $V$ is positive when

$$\frac{1-H}{1-F} < \left(\frac{\alpha}{1-\alpha}\right)\left(\frac{1-s}{s}\right) < \frac{H}{F} \qquad (9.12)$$

This is more conveniently expressed as a range for $\alpha$ using the cell counts in Table 9.2 instead of $H$ and $F$:

$$\frac{c}{c+d} < \alpha < \frac{a}{a+b} \qquad (9.13)$$

The lower and upper limits for $\alpha$ are sample estimates of calibration-refinement probabilities obtained by conditioning on the forecasts, namely, the upper limit $a/(a+b)$ is one minus the False Alarm Ratio (see Section 3.5.1). Therefore, the range of cost/loss ratios for which the forecasts have positive value is determined by the conditional probabilities of the event occurring given the forecast. There is no value for users with cost/loss ratios sufficiently close to either 0 or 1 unless the forecasts are almost perfect. The condition for the system to have value for at least one user (i.e. non-zero range in Equation 9.13) is that the event is more likely to occur following a 'yes' forecast than following a 'no' forecast.

The difference between the upper and lower limits of cost/loss ratios for which the forecasts have value is equal to $(ad - bc)/(a + b)(c + d)$, which is the Clayton Skill Score – a score similar to the Peirce Skill Score except conditioned on the forecasts rather than the observations (see Section 3.5.1; Wandishin and Brooks, 2002). The numerator $(ad - bc)$ is the same for both Peirce and Clayton skill scores. If $(ad - bc) > 0$ there will at least some range of $\alpha$ for which users will gain value from the forecasts. This condition can be stated equivalently as $(ad/bc) > 1$, so to have positive range of $\alpha$ and hence value for at least one user, the odds ratio $\theta = (ad/bc)$ (see Section 3.5.1) must exceed 1. This condition $(ad - bc) > 0$ sets the absolute minimum standard for a forecast system to have any practical value, and provides another example of how a quality measure (the odds ratio) provides a necessary condition for extracting value from forecasts (Stephenson, 2000). The Peirce Skill Score and the Clayton Skill Score quantify two of the most important features of the value curves (i.e. the maximum value and the range of cost/loss ratios that can gain value from the forecasts), and therefore, are useful measures for quantifying both skill *and* value of binary forecasts.

### 9.2.2 Probability forecasts

A potential stumbling block to the use of probability forecasts, particularly for those more familiar with deterministic forecasts, is the perception that probabilities have no place in the real world where hard yes/no decisions are required. The decision framework of the cost/loss model provides a simple illustration of the importance of the informed use of probabilities in maximizing the value of forecast information.

When presented with forecast information in the form of probabilities, the question facing the decision-maker is how high does the probability need to be before the threat is great enough to warrant protective action being taken. The decision-maker needs to set a *threshold probability*, $p_t$, so that action is taken only when the forecast probability exceeds $p_t$. In this way, the probability forecast is converted to a deterministic binary forecast that can be evaluated using the standard methods presented in Chapter 3. By varying the decision threshold probability $p_t$ over the range 0 to 1, a sequence of hit rates and false alarm rate pairs $(F(p_t), H(p_t))$ can be plotted that trace out the Relative Operating Characteristic (ROC) for the system (Section 3.3.2). Different value curves such as those shown in Figure 9.1 can be plotted for each distinct point on the ROC diagram (i.e. each value of threshold probability).

Figure 9.2 shows a selection of value curves obtained for an ensemble of 50 forecasts, using the simple empirical (frequentist) interpretation of counting the proportion of members predicting the event – see Section 8.4 for details of this and other methods to obtain probability forecasts from an ensemble of forecasts. A different value curve can be produced for each of the 50 possible probability threshold values $p_t = \{1/51, 2/51, \ldots, 50/51\}$. For visual clarity, curves are displayed only for the subset of probability decision thresholds $p_t = \{1/51, 5/51, 10/51, 15/51 \ldots, 50/51\}$ corresponding to 1, 5, 10, 15, ..., 50 members. The envelope curve (heavy solid line) shows the optimum maximum value of the forecasting system, obtained when each user chooses the probability threshold that maximizes the value for their specific cost/loss ratio. The envelope curve is never less than any of the individual curves, which shows that economic value for any particular user is maximized by choosing the optimal probability threshold for their particular cost/loss value. Therefore, no single threshold probability (i.e. a deterministic forecast) will be optimal for a range of users with different cost/loss ratios – this is a strong motivation for providing probability rather than deterministic forecasts.

The importance of choosing the correct probability threshold for each user is shown in Figure 9.3. A user with $C/L = 0.2$ would have relatively large potential losses and would therefore benefit by taking action at a relatively low probability. But a different user, with $C/L = 0.8$, say (relatively high costs), would wait until the event was more certain before committing to expensive protective action. If either user took the decision threshold appropriate to the

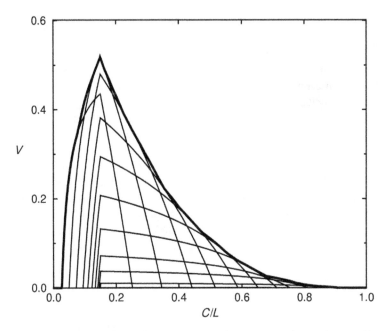

**Figure 9.2**  Value of European Centre for Medium Range Weather Forecasts (ECMWF) Ensemble Prediction System (EPS) probability forecasts of 24-hour total precipitation exceeding 1 mm over Europe at day 5 for winter 1999/2000. Thin curves show value $V$ as a function of cost-loss ratio $C/L$ for different choices of probability threshold ($p_t = 0.02, 0.1, 0.2, \ldots$); heavy solid line shows the envelope curve of optimal value

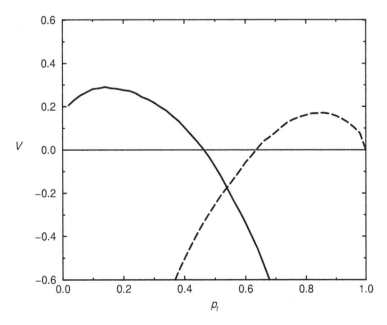

**Figure 9.3**  Variation of value $V$ with probability threshold $p_t$ for the Ensemble Prediction System (EPS) probability forecasts of Figure 9.2 for users with $C/L = 0.2$ (solid line) and $C/L = 0.8$ (dashed line)

other user, value would be reduced and in this case even becomes negative. The naively sensible choice of $p_t = 0.5$ would give no value to either user!

The main advantage of probability forecasts is that different probability thresholds are appropriate for different users. Deducing the appropriate probability at which to act is straightforward and follows similar reasoning to that at the beginning of the chapter for deciding whether or not to act with only climate information. Consider only those occasions where the forecast probability $\hat{p}$ is one particular value, $\hat{p} = q$, say. Should the user act or not? The average expense of taking action is of course

$$E_{\text{yes}}(q) = C \qquad (9.14)$$

The mean expense of not acting, averaged over all cases with $\hat{p} = q$, will be

$$E_{\text{no}}(q) = p'(q)L \qquad (9.15)$$

where $p'(q)$ is the fraction of times the event occurs when the forecast probability is $q$. Hence, users with $\alpha < p'(q)$ should take action, while those users with larger $\alpha$ should not. In general, users should act if $p'(\hat{p})$ is greater than their cost/loss ratio.

For reliable probability forecasts (Chapter 7), $p'(\hat{p}) = \hat{p}$, so the optimal strategy is to act if $\hat{p} > \alpha$ and not act if $\hat{p} < \alpha$. In other words, the appropriate probability threshold for a given user is $p_t = \alpha$, i.e. their own cost/loss ratio. If the forecasts are not reliable then the threshold should be adjusted so that $p'(p_t) = \alpha$. Note, however, the discussion of some of the pitfalls of calibration in Chapter 7.

The value curves shown in the rest of this chapter all plot the envelope curve of optimal value, i.e. they assume the forecasts are correctly calibrated. In this sense they are a measure of potential value rather than actual value, although the difference is small for forecasting systems that are reasonably well calibrated.

### 9.2.3 Comparison of deterministic and probabilistic binary forecasts

Figure 9.4 shows the value obtained for ECMWF probability forecasts at four different precipitation thresholds compared to the value obtained for the control and ensemble mean deterministic forecasts. Since maximum value occurs at the climate frequency, the curves for the more extreme events are concentrated around lower values of cost/loss ratio; to see the differences between the curves more clearly, Figure 9.4c and d are plotted with a logarithmic axis for the cost/loss ratio. The additional value (benefit) of the probability forecasts is clear for each precipitation event. For the 1 mm event (Figure 9.4a), maximum value is similar for all three forecasts, but the probability forecasts have greater value for a wider range of users. For the higher precipitation thresholds, there is increased benefit using the probability forecasts compared to using the deterministic forecasts.

Comparing the ensemble mean and control forecasts, users with low cost/loss ratios will gain more from the control forecast while large cost/loss users will prefer the ensemble mean (although for the 1 mm threshold this is reversed). The ensemble mean is an average field and is therefore less likely to contain extreme values than the individual members. If the ensemble mean forecasts heavy precipitation, it is likely that the majority of members also forecast large amounts. In other words, the ensemble mean forecasts imply relatively high probability – which is more optimal for users with high cost/loss ratios. In contrast the single control forecasts, taken alone, must be treated with more caution – it is a relatively lower probability indication of heavy precipitation and therefore more likely to benefit low cost/loss ratio users. The generally low value for the higher precipitation amounts for the ensemble mean is another indication of the difficulty for an averaged field to produce intense precipitation events.

The ensemble-mean value curve coincides with the probabilistic forecast value curve at some point where users will find the ensemble mean has equal value to the probability forecasts. However, it is apparent that for almost all cost/loss ratios, the probability forecasts have considerably greater value than the ensemble-mean deterministic forecast. The biggest advantage of probability forecasts is their ability to provide this more valuable information. In terms of value, there is no consistent benefit to all users of using the deterministic ensemble mean forecast instead of the control.

While for some users the ensemble mean forecast has the same value as the probability forecasts, the

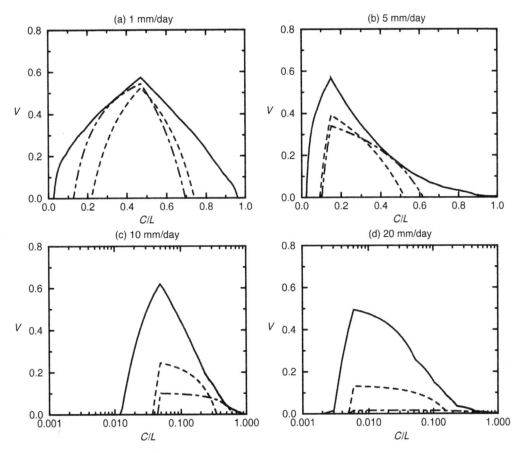

**Figure 9.4**   Value of European Centre for Medium Range Weather Forecasts (ECMWF) Ensemble Prediction System (EPS) forecasts of 24-hour total precipitation exceeding 1, 5, 10 and 20 mm over Europe at day 5 for winter 1999/00. Curves show value $V$ as a function of cost-loss ratio $C/L$ for EPS probability forecasts (solid line) and for the deterministic control (dashed line) and ensemble-mean (chain-dashed line) forecasts

control value curve is always below the probability forecast value curve and therefore all users will gain greater benefit by using the probability forecasts. This is related to the control forecast $(F,H)$ point lying below the probability forecast ROC curve on the ROC diagram (Section 3.3), whereas the ensemble-mean $(F,H)$ point lies on the probability forecast ROC curve. The link between value and the ROC is considered in more detail in the following section.

## 9.3   The relationship between value and the ROC

There are clearly links between the economic value analysis of the cost/loss model and the ROC anal-

ysis discussed in Section 3.3. Value is a function of both $H$ and $F$, and for a probability forecast the appropriate choice of threshold probability $p_t$ is important. The ROC curve is a two-dimensional plot of $(F(p_t), H(p_t))$ for each probability threshold. The value curve can be calculated from the ROC data as long as the base rate $s$ for the event is known. Note that in this case of probability forecasts, the base rate of the event is independent of the choice of threshold. One benefit of the value approach is that it shows how different aspects of the ROC relate to the economic value of the forecasts. This section explores the links between the two approaches.

Figure 9.5 shows the value curves for the 5 mm precipitation event of Figure 9.4b plotted alongside the corresponding ROC curves. One additional

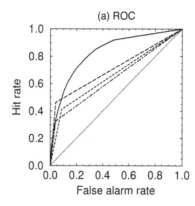
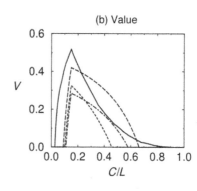

**Figure 9.5** (a) Relative operating characteristic (ROC) and (b) value plots for the 5 mm precipitation event of Figure 8.4b. Solid line, EPS probability forecasts; dashed line, control forecast; chain-dashed line, ensemble mean; long dashed line, ensemble mean day 3 forecast.

curve is shown on the plots of Figure 9.5 – the ensemble mean 3-day ahead forecast, which is an example of a forecast with higher intrinsic quality than the 5-day ahead forecasts used for the other plots. The deterministic control and ensemble-mean forecasts are each represented by single points on the ROC diagram. In Figure 9.5a, these points are shown connected with straight lines to the corners (0, 0) and (1, 1); the interpretation of these straight line segments will be discussed below. The ROC curve for the probability forecasts consists of the set of points $(F_k, H_k)$ determined by taking action when at least $k$ ensemble members predict the event. There are $m$ points on this curve for an $m$-member ensemble. The points are joined by straight-line segments and connected to the corners (0, 0) and (1, 1).

As remarked in the previous section, the value of the control forecast is less than that of the probability forecasts for users with all possible cost/loss ratios. Correspondingly, the control point on the ROC diagram falls below the probability forecasts ROC curve, again indicating that the control forecast performance is worse than that of the probability forecasts. The benefit of the probability forecasts over the deterministic control forecast is unequivocal since all users receive greater value from the probability forecasts than from the control forecast: the probability forecasts are sufficient for the deterministic control forecast in the sense described in Chapter 7. However, it is worth noting that some users will not benefit from either system and that the

differences will be more significant for some users than for others.

The ensemble mean point lies on the probability forecast ROC curve. This suggests that the basic system performance is the same. However, the benefit of the probability forecasts depends on the user: on the value curve it can be seen that the value of the two configurations is the same for some users although for the majority of cost/loss values the probability forecasts are better because of the flexibility allowed by the range of threshold probabilities.

The 3-day-ahead ensemble-mean forecast point lies above the 5-day-ahead probability forecast ROC curve – 3-day-ahead deterministic forecasts have intrinsically higher quality than 5-day-ahead probability forecasts. This translates into higher value for at least some cost/loss ratios. Nevertheless, there may still be a significant proportion of users for whom the 5-day probability forecasts have greater value. This is an example where neither system can be deemed sufficient for the other; the relative benefit again is dependent on the user. It is a useful reminder of the potential benefits of the probabilistic forecasting approach, that some users will gain more value from an 'old' (5-day-ahead) probability forecast than they will from the latest 3-day-ahead deterministic forecast.

The most common summary skill measure for the ROC diagram is the area under the ROC curve (Sections 3.5.1 and 7.5.2). The area $A$ is the fraction of the unit square below the ROC curve, which can be most simply (under-) estimated by summing the area below the points using the trapezium rule. For a deterministic forecast, the single point can be joined directly to the corners $(0, 0)$ and $(1, 1)$ as shown in Figure 9.5 giving the area as the quadrilateral including also the point $(1, 0)$. The area $A$ will generally be smaller than the parametric $A_Z$ described in Section 3.5.1, but $A$ and the connecting straight line segments have useful interpretations. Since the diagonal $H = F$ represents no skill (Section 3.2.4), an area-based skill score can be defined as

$$\text{ROCSS} = 2A - 1 \qquad (9.16)$$

which varies between 0 for points on the diagonal (no skill) to 1 for perfect forecasts. For a deterministic forecast, it is straightforward to show that $\text{ROCSS} = H - F$, in other words, the Peirce Skill Score that gives the maximum attainable value $V_{\max}$ (Equation 9.11). So the skill measure ROCSS based on the 'straight-line' area gives an estimate of the maximum value that can be obtained from the deterministic forecast.

For the probability forecasts, the maximum value over the set of probability thresholds (indexed by $k$) is given by

$$V_{\max} = \max_k(H_k - F_k) = \max_k(\text{ROCSS}_k)$$
$$= \max_k(2A_k - 1) \qquad (9.17)$$

On the ROC diagram this is the maximum distance (horizontal or vertical) between the ROC points and the diagonal line $H = F$. Equivalently, if $K$ is the probability threshold that gives this maximum value, $V_{\max} = 2A_K - 1$ where $A_K$ is the area of the quadrilateral $(0, 0)$ $(F_K, H_K)$ $(1, 1)$ $(1, 0)$. Since the full ROC using all probability thresholds has a larger area than this, ROCSS for the probability forecasts is larger than $V_{\max}$. In terms of usefulness, this reflects the greater benefit provided to a wider range of users by using different probability thresholds for different users. When more points are added to a ROC curve (e.g. by having a greater number of ensemble members and more probability thresholds), the area will increase and so will the value for some users.

Neither measure, though, can be used to determine the value of a forecast system to a general user or group of users. For example, in Figure 9.5

the control forecast has a larger ROCSS or maximum value (0.35) than the ensemble mean (0.28), but for $\alpha > 0.25$, the ensemble mean forecast is the more valuable forecast. Summary measures of forecast value are discussed further in the following two sections.

The straight-line segments joining the ROC points are also informative to the user. The slope of the line joining the deterministic ROC point $(F, H)$ to the lower left corner $(0, 0)$ is simply $H/F$, while the slope of the connecting line to the upper right corner $(1, 1)$ is $(1 - H)/(1 - F)$. The ratio of the two slopes gives a simple geometric method for calculating the odds ratio. In Section 9.2, these two ratios were shown to determine the range of cost/loss ratios for which the forecasts have value (Equation 9.12). The steeper the slope of the line between $(F, H)$ and $(0, 0)$, the more users with higher cost/loss ratios will benefit; and the shallower the slope of the line between $(F, H)$ and $(1, 1)$, the more users with lower cost/loss ratios will benefit. In Figure 9.5a, the line joining the origin to the ensemble mean ROC point is steeper than the corresponding line for the control forecast, and in Figure 9.5b, it can be seen that more users with higher cost/loss ratios benefit from the ensemble mean forecast than from the control forecast. Conversely, the line between the control forecast ROC point and $(1, 1)$ is shallower than the equivalent ensemble-mean forecast line, and low cost/loss ratio users therefore benefit more from the control forecast.

For the probability forecasts, the lines joining the first and last points to the top-right and bottom-left corners again determine the range of users for whom the system will have positive value. Similar reasoning shows that the slopes of the straight lines connecting the intermediate points indicate the range of cost/loss ratios for which each probability threshold is optimal. If $V_k(\alpha)$ is the value associated with the ROC point $(F_k, H_k)$ then from Equations 9.10 and 9.11, $V_k(\alpha) > V_{k+1}(\alpha)$ and $V_k(\alpha) > V_{k-1}(\alpha)$ when

$$\frac{H_{k-1} - H_k}{F_{k-1} - F_k} < \left(\frac{\alpha}{1 - \alpha}\right)\left(\frac{1 - s}{s}\right) < \frac{H_k - H_{k+1}}{F_k - F_{k+1}}$$
$$(9.18)$$

So the range of $\alpha$ is determined by the slopes of the lines joining $(F_k, H_k)$ to the adjacent points. From the definition of the hit rate and false alarm rates, the gradient of the line joining the two ROC points $(F_{k-1}, H_{k-1})$ and $(F_k, H_k)$ can be written as

$$\frac{H_k - H_{k+1}}{F_k - F_{k+1}} = \left(\frac{p'_k}{1 - p'_k}\right)\left(\frac{1 - s}{s}\right) \qquad (9.19)$$

where $p'_k$ is the observed frequency of the event given forecasts are in class $k$ ($k$ out of $m$ members predict the event for the $m$-member ensemble forecasts). By comparing Equations 9.18 and 9.19, it can be seen that $V_k(\alpha)$ will be greater than $V_{k-1}(\alpha)$ and $V_{k+1}(\alpha)$ for values of $\alpha$ between $p'_{k-1}$ and $p'_k$. For positively calibrated forecast systems, we expect $p'_k$ to increase with $p_k$ (the higher the forecast probability, the more likely it should be that the event occurs). On the ROC, this corresponds to the slope of the lines joining the ROC points increasing monotonically as $k$ increases (moving from the upper-rightmost point towards the lower left corner); this is generally the case for a large enough sample of data, although ROCs generated from small data samples may be more variable. So long as the monotonicity holds, the limits for $\alpha$ given above hold not just for $V_k(\alpha)$ compared to $V_{k-1}(\alpha)$ and $V_{k+1}(\alpha)$, but extend to all points: the value $V_k$ associated with forecast probability $p_k$ (and the corresponding $H_k$ and $F_k$) is optimal for users with $p'_k < \alpha < p'_{k-1}$.

In summary, the slope of the first and last line segments joining the ROC points to the endpoints (0,0) and (1,1) give the range of cost/loss ratios for which value is positive. Extending the ROC by adding more points at either end will thus increase the range of users who will receive positive value, while adding additional points between the existing ones will benefit by providing finer resolution of probability thresholds. Both these measures will help to increase the total area under the ROC curve, and hence the overall value of the forecasting system.

The empirical ROCs presented in this section demonstrate the value that can be obtained with the available set of probability thresholds (up to 50 for the 50-member ensemble forecasts). Parameterizing the ROC curves as in Section 3.3 can be used to demonstrate the potential value that would be achieved if all possible probability thresholds could be used (i.e. an infinite ensemble of forecasts). This could provide a useful method for estimating the benefit that could be obtained by using a larger ensemble of forecasts (Richardson, 2000).

## 9.4 Overall value and the Brier Skill Score

The previous section explored the relationship between value and the ROC. Maximum value is easily deduced from the ROC, as is the range of users for whom value will be positive. ROC area increases as more points are added, consistent with the increase in value to a wider range of users. However, forecast value can vary greatly between users and there is no simple relationship between any single overall measure (such as ROCSS or $V_{max}$) and the value to specific users.

Nevertheless it is often desirable to have an overall summary measure of performance. The specification of such a measure is the subject of this section. We will find that the familiar Brier Skill Score can be interpreted as one such measure, given a particular distribution of users. The implication of this will be explored both here and in the next section.

If we knew the costs and losses for every user, we could calculate the total savings over all users and produce a measure of overall value. This would then provide a representative skill measure based on the overall benefit of the forecast system.

Since we do not know the distribution of users appropriate to a given event, and this may vary between events, we use an arbitrary distribution of users. Assume the distribution of cost/loss ratios for users is given by $u(\alpha)$. We can then derive a measure of overall value based on the total saving made by all users. For this general derivation we assume that a user will take the probability forecasts at face value so that each user will take action when the forecast probability $\hat{p}$ is greater than their cost/loss ratio $\alpha$.

Consider the occasions when the forecast probability is equal to a given value $\hat{p} = q$. Users with cost/loss ratio $\alpha < q$ will take action and hence incur cost $\alpha$ (per unit loss). All other users will not act and will incur a loss (assumed without loss of generality to be $L = 1$) when the event occurs. The total mean expense for all users for this particular forecast probability $\hat{p} = q$ is then given by

$$T(q) = \int_0^q u(\alpha)\alpha d\alpha + p'(q) \int_q^1 u(\alpha)d\alpha$$

$$= p'(q) + \int_0^q u(\alpha)(\alpha - p'(q))d\alpha \quad (9.20)$$

where $p'(q)$ is the observed frequency of the observed event for cases when the forecast probability $\hat{p} = q$. The total expense is obtained by taking the expectation over all possible forecast probabilities. For an ensemble of $m$ forecasts, this amounts to a summation over all classes of probability, weighted by the fraction of occasions when the forecast probability is in each class ($g_k$). With some rearrangement this can be written as

$$T_F = T_C + \sum_{k=0}^m g_k \int_{p'_k}^{p_k} u(\alpha)(\alpha - p'_k)d\alpha$$

$$- \sum_{k=0}^m g_k \int_s^{p'_k} u(\alpha)(p'_k - \alpha)d\alpha \quad (9.21)$$

where $T_C$ is the total expense if all users act using the climatological base rate as decision threshold:

$$T_C = \int_0^s u(\alpha)\alpha d\alpha + \int_s^1 u(\alpha)s d\alpha \quad (9.22)$$

as explained by Richardson (2001).

The second and third terms on the right-hand side of Equation 9.21 are generalized forms of the *reliability* and *resolution* components of the Brier score (Chapter 7) weighted by the distribution of cost/loss ratios), as will be seen below. The third term on the right-hand side of Equation 9.21 is the maximum reduction in expense that would be achieved if all users were to act when the forecast probability of the event $p'_k$ exceeds their particular cost/loss ratio $\alpha$. This benefit increases the more that the forecast probabilities differ from the base rate (note $s$ is a limit in the integral). The potential benefit is reduced if users act on forecast probabilities that are not completely reliable. The second term on the right-hand side of Equation 9.21 gives the additional expense incurred. This reliability term depends on the difference between $p_k$ and $p'_k$ (the limits in the integral) and decreases as this difference reduces

since there are then fewer occasions on which the incorrect choice of action is made.

For perfect forecasts, the expense per unit loss for a given user is $s\alpha$ (Equation 9.4) and so the total mean expense for all users is given by

$$T_P = \int_0^1 u(\alpha)s\alpha d\alpha = s\bar{\alpha} \quad (9.23)$$

i.e. the mean cost/loss ratio multiplied by the base rate. A measure of overall value can then be defined as

$$G = \frac{T_C - T_F}{T_C - T_P} \quad (9.24)$$

To see the relationship of this with the Brier score, imagine a set of users with a uniform distribution of cost/loss ratios, i.e. $u(\alpha) = 1$. Equation 9.21 then becomes

$$T_F = \frac{1}{2}\sum_{k=0}^m g_k(p_k - p'_k)^2 - \frac{1}{2}\sum_{k=0}^m g_k(p'_k - s)^2$$

$$+ \frac{s(1-s) + s}{2}$$

$$= \frac{1}{2}(B_{\mathrm{rel}} - B_{\mathrm{res}} + B_{\mathrm{unc}}) + \frac{s}{2} \quad (9.25)$$

where the bracketed term on the right-hand side is the Brier score expressed in the standard decomposition in terms of reliability, resolution and uncertainty (Chapter 7). It is easily seen that for this distribution of users, $G = \mathrm{BSS}$ and so the Brier Skill Score is the overall value in the special case when the users have a uniform distribution of cost/loss ratios.

It is not easy to determine the distribution of cost/loss ratios appropriate for a given forecast system. However, as discussed in Section 9.6, specific examples from case studies and more general economic considerations both indicate that small cost/loss ratios are more likely than large ones. This suggests that the Brier score is unlikely to represent any real-world mean overall value. In the following section, we consider this point further and demonstrate the dependence on the distribution of users of the overall value measure $G$.

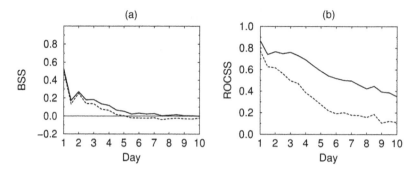

**Figure 9.6**  (a) Brier Skill Score (BSS) and (b) ROC Area Skill Score (ROCSS) for European Centre for Medium Range Weather Forecasts (ECMWF) Ensemble Prediction System (EPS) probability forecasts of 12-hour total precipitation exceeding 10 mm over Europe for winter 1996/97. Curves show skill as a function of forecast lead time for the operational 50-member EPS (solid line) and a 10-member EPS made from a subset of the operational members (dashed line)

## 9.5  Skill, value and ensemble size

The previous two sections have examined the relationships between value and the ROC and Brier scores. In this section, we examine a case study where the ROC and BSS present different indications of the usefulness of the forecasting systems. The cost/loss model allows us to interpret these different conclusions from the perspective of potential users. Our case study aims to demonstrate the effect of varying ensemble size on the value provided to users. We compare the overall value of an ensemble of 50 forecasts of heavy precipitation (more than 10 mm in 12 hours) against that of an ensemble having only 10 ensemble forecasts. The ensemble of 10 forecasts is a random subsample of the original ensemble of 50 forecasts.

Figure 9.6 shows the ROC Area Skill Score (ROCSS) and the Brier Skill Score (BSS) for the two ensemble systems. From the BSS, we can conclude that the impact of ensemble size is generally small and that neither configuration has any useful skill at lead times beyond 5 days. In contrast, the ROCSS shows skill throughout the forecast range with substantial additional benefit from the 50-member ensemble forecasts.

Figure 9.7 shows the value curves for both configurations at day 5 and day 10. Since we are considering an uncommon event ($s = 0.014$), most value is obtained for users with low cost/loss ratios. For larger cost/loss ratios, the value is substantially smaller; by day 10, neither set of ensemble forecasts has any value for users with cost/loss ratios greater than 0.1. BSS, being a measure of the mean overall

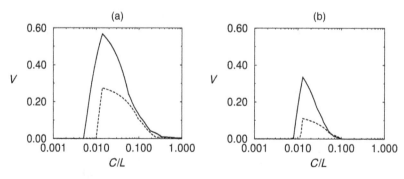

**Figure 9.7**  Value for (a) day 5 and (b) day 10 for the ensemble forecasts of Figure 9.6. Solid line, 50-member ensemble; dashed line, 10-member ensemble.

value for users with uniformly distributed cost/loss ratios, gives little relative weight to the changes in value for low cost/loss ratios. In contrast, ROCSS is greater than $V_{max}$ and so is still positive.

Differences in value between the 10- and 50-member ensemble forecasts are also greatest for low cost/loss ratios. Ideally users should take action at a probability threshold $p_t = \alpha$, so to provide maximum benefit the ensemble must be large enough to resolve these required probability thresholds. The 50-member ensemble gives a better 1/50 resolution in probability thresholds while the small 10-member ensemble has a poorer resolution of 1/10 and so lacks the threshold discrimination needed for small cost/loss ratios. If the probabilities from the 50-member ensemble are restricted to the same 1/10 intervals as the 10-member ensemble, the large differences in value and ROCSS between the two configurations are greatly reduced (not shown).

To summarize the dependence of overall value on the cost/loss distribution of users, Figure 9.8 shows plots of the overall value, $G$, for various sets of users.

As discussed already, BSS represents overall value for a set of users with cost/loss ratios uniformly distributed from 0 to 1. The first three panels of Figure 9.8 illustrate the more probable situation of users with relatively low $C/L$. For cost/loss ratios uniformly distributed in the restricted range (0,0.2) the overall value is substantially increased compared to BSS, especially in the first 5 days, but the differences in $G$ for the 10- and 50-member ensemble forecasts are still generally small (Figure 9.8a). If we concentrate on a narrower band of cost/loss ratios (0.02–0.05), for example, representing users with very large potential loss from an extreme event of heavy precipitation, the overall value again increases and the effect of ensemble size becomes more important (these users have low cost/loss ratios and need comparably low probability thresholds to achieve maximum benefit). Narrowing the cost/loss distribution towards small values increases both overall skill and emphasizes the difference between the different ensemble systems. The overall mean values in Figure 9.8c are approaching the

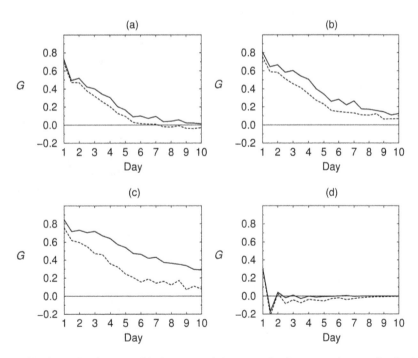

**Figure 9.8**   Overall value $G$ for the ensemble forecasts of Figure 9.6 for four example user distributions. Users are assumed to be distributed uniformly across the $C/L$ interval of (a) 0.0–0.2, (b) 0.02–0.05, (c) 0.01–0.02 and (d) 0.5–0.8. Curves show $G$ as a function of forecast lead time for the operational 50-member EPS (solid line) and a 10-member EPS made from a subset of the operational members (dashed line)

levels of the ROCSS of Figure 9.6b; remember that ROCSS is always greater than $V_{max}$ and can therefore never be exceeded for any distribution of users. Finally, Figure 9.8d shows the contrasting situation for the less likely case of a set of users with relatively high cost/loss ratios (0.5–0.8); for these users there is no economic value in using either forecast system.

Comparing Figures 9.8 and 9.6 shows that in most realistic situations, BSS and ROCSS can be considered lower and upper bounds for overall forecast value. Exactly where the true value lies, between these limits, depends on the distribution of users' cost/loss ratios.

Finally, the above results depend on the event being considered. For more frequent events, greatest value will be achieved at correspondingly larger cost/loss ratios than here, and the differences between BSS and ROCSS will be smaller. The benefit of larger ensembles of forecasts will, however, remain greatest for those users with the lowest cost/loss ratios, for example, users who suffer very large losses when rare extreme events occur. Further discussion of the effects of ensemble size on value can be found in Richardson (2000, 2001).

## 9.6 Applications: value and forecast users

In this chapter we have used a simple decision model to examine the link between traditional skill measures and the potential benefit of forecasts to users. We have seen that different users will benefit differently from the same forecasting system, depending on their sensitivity to the weather event (the users' costs and losses).

For most real-world studies of economic value, the weather sensitivity and decision-making processes are more complicated than in the simple cost/loss model. Even in decision situations where the cost-loss model may seem appropriate, an assessment of the true economic value of a forecasting system must carefully consider a number of aspects that we have not discussed in detail. As noted in Section 9.2.2, a probabilistic forecast system must be well calibrated for the user to be able to take action at the correct probability threshold. It must be remembered that we have presented the results for

the mean expected value: the overall benefit that can be expected by a user making a long sequence of decisions using the forecasting system. This makes the implicit assumption that the user is able and willing to withstand a number of (inevitable) losses in order to make a long-term gain. Therefore care is needed when attempting to draw conclusions about the actual economic benefit that would be gained from using a forecasting system on just a few occasions (e.g. using a seasonal forecast only once a year). See Chapter 11 for a discussion of the particular problem of sample size in seasonal forecasting and of appropriate measures of economic value. Despite these limitations, there are a number of studies that demonstrate the usefulness of the cost/loss model in providing a user-oriented perspective on the quality of weather forecasts. Examples include winter road maintenance (Thornes and Stephenson, 2001; Berrocal et al., 2010), the value of hurricane forecasts to oil and gas producers (Considine et al., 2004), reservoir management (McCollor and Stull, 2008), and air traffic management (Leigh, 1995; Keith, 2003; Roquelaure and Bergot, 2008). It is not easy to determine the distribution of cost/loss ratios appropriate for a given forecast system. There are relatively few investigations that have attached financial costs to the simple model used in this chapter. Where figures are available, they tend towards lower cost/loss ratios: for instance 0.02–0.05 for orchardists (Murphy, 1977), 0.01–0.12 for fuel-loading of aircraft (Leigh, 1995) and 0.125 for winter road-gritting (Thornes and Stephenson, 2001). This agrees with the expectation from general economic considerations (Roebber and Bosart, 1996), while for large potential losses the issue of risk-aversion (not considered here) would also mitigate towards lower cost/loss ratios. Rather than trying to find a definitive user distribution, Murphy (1966) and Roebber and Bosart (1996) have used beta distributions to parameterize various probability distributions of the cost/loss ratio in order to investigate the impact on value.

Recent survey results have shown considerable variation in the threshold probability at which members of the public would take precautionary action against a range of adverse weather events (Morss et al., 2010; Joslyn and Savelli, 2010). These results confirm the general principle that forecast producers need to account for substantially different sensitivity to risk among forecast users. They also indicate that users do not always respond to forecast information in the optimal way expected of an objective user of the cost-loss model. Nevertheless, controlled studies (e.g. Roulston *et al.*, 2006) have shown that non-specialist users are able to exploit uncertainty information to make better decisions. Millner (2009) shows one method to account for an unsophisticated user in the framework of the cost-loss model. Katz and Murphy (1997a) present a large number of applications evaluating the economic value of weather forecasts, including uses of the simple cost-loss model and extensions to more categories and so-called dynamic cost-loss models.

A different way to demonstrate the potential economic value of probabilistic weather forecasts is the game of 'weather roulette' (Hagedorn and Smith, 2009). The quality of a forecasting system is judged by its ability to make a profit by beating the house odds in a weather casino (house odds are set by a suitable reference forecasting system, such as climatology). This is an intuitive concept that can help in communicating the use of probabilities to a non-expert user who is not familiar with such forecasts. Hagedorn and Smith (2009) also show that weather roulette is connected to the Ignorance Score for probabilistic forecasts (see Chapter 7), illustrating another link between forecast value and forecast skill.

## 9.7 Summary

This chapter had two principal aims. Firstly, to introduce the concept of the economic value of forecasts in the context of a simple decision-making process; and secondly, to explore the relationship between economic value and some of the common verification measures used for forecast performance.

The simple cost/loss model gives a straightforward example of how forecast information can be used in decision-making situations. Probabilistic forecasts, when calibrated and used appropriately, are inherently more valuable than deterministic forecasts because they are adaptable to the differing needs of different users. The principal benefit of ensemble forecasting is the ability to produce reasonably reliable probability forecasts, rather than, for example, the deterministic ensemble mean. Forecast value depends not only on the quality of the forecasting system but also on the weather sensitivity of the user. Different users will gain to differing extents from the same forecasts. Indeed, while one user may gain substantial benefit from a forecast system, another user may well derive no additional economic value.

This chapter has shown that several measures of forecast quality provide useful insight into the mean economic value of forecasts for a general set of users having a range of different cost/loss ratios. The Peirce Skill Score gives the maximum possible value that can be obtained from the forecasts, which occurs only for those users who have a cost/loss ratio exactly equal to the base rate. The Clayton Skill Score gives the range of cost/loss ratios over which value can be extracted by using the forecasts. Both scores have the same numerator, and this being positive (equivalently the odds ratio exceeding 1) is a necessary and sufficient condition for at least some user to get value from the forecasts. The Brier Skill Score is an estimate of the mean overall value of the forecasts for a population of users having a uniform distribution of cost/loss ratios. The area under the ROC curve provides a more optimistic estimate of mean overall value for a population of users. The two aspects of forecast 'goodness', quality and utility, are therefore intimately related. In the same way that probability may be defined subjectively in terms of the price of a fair bet, forecast quality may be considered to be the expected utility of forecasts for a population of unknown users.

Commonly used measures of forecast performance such as the Brier Skill Score or ROC Area Skill Score often give differing impressions of the value of forecasts and of the differences between competing systems. Viewing these scores from the perspective of economic value allows these differing results to be interpreted from the user perspective. The Brier Skill Score is equal to the overall mean value for a set of users having a uniform distribution of cost/loss ratios ranging from 0 to 1. If, as studies have shown, most real users generally operate with small cost/loss ratios, the Brier Skill Score will tend to give a pessimistic view of the overall value. It is possible for the BSS to be zero or even negative and yet for the forecasts to have substantial value for a significant range of users (Palmer *et al.*, 2000; Richardson, 2001). Conversely, it can be argued that the ROCSS gives perhaps an overly optimistic view of the value of forecasts since it is always somewhat greater than $V_{max}$.

Given a distribution of cost/loss ratios, the overall mean value can easily be calculated. In the absence of detailed information on the distribution of cost/loss ratios, it is as well to be aware of the assumptions about users implicit in these common skill measures. As an example, while the area under the ROC curve is relatively sensitive to ensemble size, BSS is much less so. BSS would not, then, be an appropriate measure for the evaluation of extreme event forecasts for low cost/loss users; the likely benefit for these particular users from increasing ensemble size would probably not be discernible in the BSS.

Finally, no single summary measure of performance (including the overall value $G$) can be taken as representing the specific benefit to any individual user. Although an increase in $G$ does, by definition, mean that value will increase over the group of users taken as a whole, the benefits to different individual users will vary significantly. It is quite possible that while overall value increases, the value to some users may actually decrease. Such skill-value reversals have been discussed by Murphy and Ehrendorfer (1987), Ehrendorfer and Murphy (1988), and Wilks and Hamill (1995). Only in the exceptional circumstance of one forecasting system being sufficient for all others, will all users be sure to prefer the same system (cf. discussion in Section 3.3).

# 10

# Deterministic forecasts of extreme events and warnings

**Christopher A.T. Ferro and David B. Stephenson**
*Mathematics Research Institute, University of Exeter*

## 10.1 Introduction

A critical gap in verification research arises in relation to the verification of forecasts and warnings of extreme weather events. There is an increasing demand for weather services to provide forecasts of such events, a focus also for current international activities such as the World Meteorological Organization's THORPEX programme. Without appropriate verification methods for extremes, it will be difficult to evaluate and then improve forecasts of high-impact hazards that are required for societal adaptation to climate change (e.g. the increasing intensity and frequency of UK floods). The need for further research in this area was clearly recognized in a report of the second meeting of the European Centre for Medium-Range Weather Forecasts (ECMWF) Technical Advisory Committee Subgroup on Verification Measures (10–11 September 2009), which contains the following paragraph:

> In view of the high priority given by the National Meteorological Services to the accurate prediction of severe weather events, the Subgroup recognised the lack of fundamental research into related verification within the Member States' meteorological services and universities. Without an enhanced research effort, it will continue to be difficult to provide those National Meteorological Services with robust performance measures.

Some of the many challenges facing verification of forecasts of extreme events were reviewed in Casati *et al.* (2008).

Because of recent research progress, this chapter focuses on two types of ubiquitous problems in extreme event forecasting: deterministic forecasts of rare binary events (Section 10.2), and deterministic warnings of severe events (Section 10.3). The discussion highlights the complex issues facing this important area of forecast verification. Rarity of events can be quantified by the base rate probability of the observed event, which will be denoted here by $p$ rather than $s$ as used in Chapter 3.

*Forecast Verification: A Practitioner's Guide in Atmospheric Science*, Second Edition. Edited by Ian T. Jolliffe and David B. Stephenson.
© 2012 John Wiley & Sons, Ltd. Published 2012 by John Wiley & Sons, Ltd.

## 10.2 Forecasts of extreme events

### 10.2.1 Challenges

The tornado forecasts of Finley (1884) that are discussed in Chapters 1 and 3 illustrate some of the challenges that arise when verifying forecasts of extreme weather events. Finley's data in Table 1.1 show that tornados were quite rare: a tornado occurred on only 51 of the 2803 occasions considered, which corresponds to a base rate of 1.8%. We shall take rarity to be a defining feature of extreme events. Two-by-two contingency tables of rare events are characterized by relatively small numbers of hits and misses and, unless the event is highly overforecast, a relatively small number of false alarms too. In practice, some overforecasting is common for severe events like tornados because the consequences of failing to forecast a tornado can be more serious than the consequences of issuing a false alarm (Murphy, 1991b). We see this in Finley's forecasts, where a tornado was forecast to occur on 100 occasions, yielding a frequency bias of almost 200%.

Although Finley achieved a high proportion (96.6%) of correct forecasts, Gilbert (1884) noted that he would have achieved an even higher proportion (98.2%) by forecasting 'no tornado' on every occasion. Furthermore, the proportion correct (PC) achieved by such unskilful forecasts would increase with the rarity of tornados. If only one tornado were observed, for example, then the constant 'no tornado' forecast would be correct on all occasions bar one. It can be shown that the PC always tends to 100% as the base rate decreases to zero unless the event is so strongly overforecast that the false alarm rate does not decrease to zero.

The foregoing analysis illustrates the fact that scoring well on some measures of forecast performance can become easier as the rarity of the predicted event increases. Other measures, by contrast, can become more demanding. The threat score or critical success index (CSI) provides an example. The CSI is the proportion of correct forecasts among those instances when the event was either forecast or observed. As such, the CSI is independent of the number of correct forecasts of the non-event, the entry that tends to dominate contingency tables for rare events. Correctly forecasting that a rare event will not occur is often considered to be a relatively easy task, and consequently the CSI is a popular measure for forecasts of rare events (see Mason, 1989). As event rarity increases, however, the CSI almost always decreases to zero. The only situation in which the CSI does not decay to zero is when the forecasts are so good that the numbers of misses and false alarms both decrease to zero at least as fast as the number of hits.

We have seen that the PC tends to increase and the CSI tends to decrease as event rarity increases, and this is true irrespective of whether the forecasting system is skilful or unskilful. We shall see later that most other measures for two-by-two tables exhibit similarly degenerate behaviour. This degeneracy of measures pertains not only to the two-by-two contingency tables that we shall focus on, but also to multi-category forecasts when some of the categories are rare. Probabilistic forecasts of event occurrence are another case in point as issuing small probabilities for rare events will yield a Brier score close to its optimal value of zero. Brier scores are often found to decrease as rarity increases (e.g. Glahn and Jorgensen, 1970) because rare events can be hard to predict, and probabilistic forecasts tend to be close to the climatological frequency in such circumstances (Murphy, 1991b).

As a result of this degeneracy, comparing forecast performance for events with different rarities is difficult. For example, to say that a forecasting system performs better for rare events than for common events may be true for only some aspects of forecast performance; there may well be other aspects of performance for which the order is reversed. The degeneration of most verification measures to trivial values such as 0 and 1 as event rarity increases also hampers the comparison of different forecasting systems for rare events because differences in performance measures tend to be small. This complication is exacerbated by the following effect too.

The small number of observed events is another source of trouble when verifying forecasts of rare events. Some measures for contingency tables, such as the odds ratio, are undefined if certain elements of the table are zero. More generally, a small number of cases can inflate the sampling variation of verification measures, which increases uncertainty about the quality of the forecasting system. This inflation can arise for at least two reasons. The

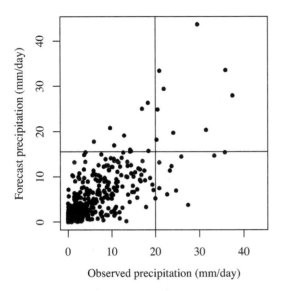

**Figure 10.1** Forecast against observed daily precipitation totals with event thresholds superimposed. Modified from Ferro (2007b)

variance of some measures, such as the mean bias, is inversely proportional to the sample size, and so small samples can inflate uncertainty. Another effect is due to the increasing influence that may be exerted on the value of the verification measure by a small number of forecast-observation pairs. In our setting of two-by-two contingency tables, the measures that depend critically on the small numbers of hits, misses and false alarms are most affected. For example, if only two events are observed then the hit rate jumps from 0 to 0.5 to 1 depending on the success of just two forecasts. Another example of this sensitivity is provided by the mean bias or mean squared error, where one poor forecast of a large value can have a strong, adverse effect on the value of the measure.

We can illustrate these two challenges of degeneracy and uncertainty further with the forecasts of daily precipitation totals that are plotted against radar observations in Figure 10.1. The 649 available forecasts from 1 January 2005 to 11 November 2006 are from a mesoscale version of the UK Met Office Unified Model at a gridpoint in mid-Wales (52.5°N, 3.5°W).

Suppose that we are interested in forecasting whether or not the observed precipitation amount exceeds a particular threshold, and that we forecast the event to occur if the forecast precipitation amount exceeds the same threshold. For example, the event 'precipitation above 10 mm' is forecast to occur if the forecast precipitation exceeds 10 mm. For any particular threshold, we can construct a contingency table containing the numbers of hits, misses, false alarms and correct forecasts of the non-event. From these tables, we can compute verification measures such as the hit rate for each threshold and investigate how the measures change as the threshold increases. Figure 10.2 shows how the hit rate, PC, CSI and odds ratio vary with threshold. The hit rate and CSI decrease towards zero as the threshold increases, indicating worse performance, while the PC and odds ratio increase, indicating better performance. Sampling variation also increases with threshold for the hit rate, CSI and odds ratio, but decreases for the PC.

A third difficulty when verifying forecasts of extreme events concerns measurement error, but this has received little attention in the literature and we shall not discuss it in detail. The measurement errors for extreme events can be relatively large, owing to, for example, instrumental shortcomings when measuring unusually large values or the possibility that observational networks may fail to record the occurrence of events with small space-time scales. A failure to account properly for measurement errors has the potential to penalize good forecasts of extreme events.

### 10.2.2 Previous studies

We noted in the previous subsection that many measures for two-by-two contingency tables degenerate to trivial values as event rarity increases. Stephenson *et al.* (2008a) have obtained the most general results for how verification measures degenerate as the base rate decreases to zero, and they show that traditional measures typically degenerate to trivial values such as 0, 1 or infinity. Earlier work by Mason (1989) and Schaefer (1990) showed that the CSI decreases to zero as the base rate decreases if the hit rate and false alarm rate remain constant. This result does not provide a full characterization of the behaviour of the CSI for rare events, however, because the hit rate and false alarm rate usually decrease to zero with the base rate. Other authors identified

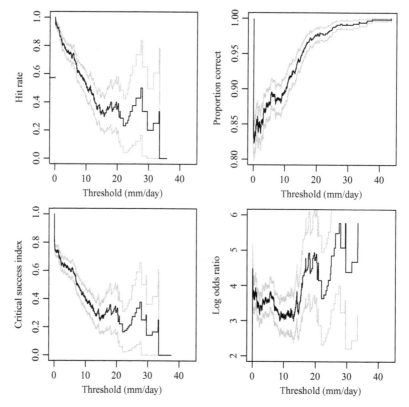

**Figure 10.2** Four verification measures (black lines) calculated for the precipitation forecasts in Figure 10.1 and plotted against threshold with approximate 95% confidence intervals (grey lines) based on a Gaussian approximation to the sampling distribution of the four measures (e.g. Jolliffe, 2007; Stephenson, 2000)

relationships between measures for rare events. For example, Schaefer (1990) claimed that the Gilbert Skill Score approaches the CSI as the base rate decreases to zero. While this is usually true for skilful forecasts, it need not hold in all circumstances: see Stephenson *et al.* (2008a). Doswell *et al.* (1990) considered alternative limiting conditions, equivalent to the false alarm rate decreasing to zero faster than the hit rate, under which the Peirce Skill Score approaches the hit rate. They also claimed that the Heidke Skill Score approaches a transformation of the CSI, but this is true only if the base rate also decreases to zero, which is not implied by their conditions.

The degeneracy of measures has nothing to do with base-rate dependence since even base-rate independent scores such as the Peirce Skill Score and odds ratio typically degenerate (Stephenson *et al.*, 2008a). A base-rate independent measure is one that

remains constant as the base rate changes while the hit rate and false alarm rate of the forecasting system remain unchanged (e.g. Ferro and Stephenson, 2011). Such changes in base rate may be due to natural variations in the observed frequency of events, for example. When the base rate is decreased by changing the *definition* of the event, as in Figure 10.2, there is no reason for the hit rate or false alarm rate to remain constant; degeneracy occurs merely because scoring well on the measured aspects of forecast performance becomes more or less demanding for rare events.

There is also some discussion in the literature about the potential for different measures to be hedged when forecasting rare events. Unfortunately, there is no consensus about the meaning of hedging for deterministic forecasts (Jolliffe, 2008; Ferro and Stephenson, 2011) and this makes it difficult to draw definitive conclusions about the efficacy of different measures. Marzban (1998), for example, identifies hedging with issuing biased forecasts. He finds that many measures encourage hedging in this sense for rare events because they are optimized by biased forecasts. In a similar vein, Doswell *et al.* (1990) claim that the Peirce Skill Score can be hedged by overforecasting because it often behaves like the hit rate for rare events. The effect of overforecasting on the false alarm rate must be controlled in this case, however, and so a sophisticated form of overforecasting is required in which the non-event remains correctly forecast on the vast majority of occasions. As Manzato (2005) reaffirms, the intrinsic equitability of the Peirce Skill Score precludes its improvement by simple hedging strategies. It is important to remember that measures satisfy properties such as equitability and base-rate independence either for all base rates or for none. If a measure enjoys such properties then they should not be discounted for rare events just because another measure without the same properties becomes a good approximation.

Several other topics in the verification of forecasts of rare events have been addressed in the literature, but systematic and comprehensive treatments are wanting. The false alarm rate, $F$, tends to decrease to zero faster than the hit rate as event rarity increases, and therefore ROC curves for rare events tend to have points concentrated near to the lower part of the hit-rate axis, $F = 0$. Wilson (2000) noted that employing the trapezoidal rule in such circumstances can lead to severe underestimation of the area under the ROC curve and recommended fitting the binormal model instead. Atger (2004) used the same approach to improve estimates of the reliability curve for probabilistic forecasts of binary events, in particular for poorly represented probabilities such as large probabilities when forecasting rare events. Lalaurette (2003) proposed an alternative to the ROC curve in the rare event case in which the hit rate is plotted against the false alarm ratio instead of the false alarm rate.

### 10.2.3   Verification measures for extreme events

In the rest of this section we shall restrict our attention to deterministic forecasts of rare, binary events for which some new verification tools have been proposed in recent years. A generic contingency table for such forecasts is given in Table 10.1, and associated verification measures are discussed in Chapter 3. As we have seen, many traditional measures degenerate to trivial values as the base rate $p = (a + c)/n$ decreases to zero. Such measures nevertheless remain important tools for understanding forecast performance, and their degenerate behaviour merely reflects the ease or difficulty of maintaining particular attributes of forecast performance for rare events. However, degenerate behaviour can compromise a measure's usefulness and this provides the motivation for new measures that do not degenerate. To this end, we review in this subsection a set of measures that describe the *rate* at which forecast performance degenerates.

For a particular base rate $p$ and sample size $n$, the contingency table is completed by the forecast rate $q = (a + b)/n$ and the relative frequency of hits, $a/n$. In order to describe how forecast performance degenerates, we need to examine how these two latter terms change as $p$ decreases to zero. Of course, there is no universal reason why these terms should follow any particular pattern as the base rate decreases, and if there is no pattern then attempting to measure a rate of decay will be fruitless: we can only calculate our favourite verification measures for each base rate of interest. However, some theory does suggest that in a wide class of situations we can expect certain patterns of decay to arise, and we discuss these patterns next.

A fairly flexible model for how $a/n$ decays with the base rate $p$ assumes that

$$\frac{a}{n} \sim \alpha(pq)^{\beta/2} \qquad (10.1)$$

**Table 10.1**   A contingency table representing the frequencies with which an event and non-event were forecast and observed

|  | Observed | Not observed |  |
| --- | --- | --- | --- |
| Forecast | $a$ | $b$ | $a + b$ |
| Not forecast | $c$ | $d$ | $c + d$ |
|  | $a + c$ | $b + d$ | $n$ |

as $p \to 0$, where $\alpha > 0$ and $\beta \geq 1$ are constants that control the rate of decay, and the notation $x \sim y$ means $x/y \to 1$. For example, the expected value of $a/n$ is $pq$ when the event is forecast to occur at random with probability $q$, in which case $\alpha = 1$ and $\beta = 2$. In fact, this model (Equation 10.1) is unlikely to be a good description of the rate of decay of $a/n$ unless $q$ decays at the same rate as $p$ (Ramos and Ledford, 2009) so that the frequency bias, $q/p$, converges to a finite, positive constant. If $p$ and $q$ decay at different rates then they can affect the rate of decay of $a/n$ in ways that typically admit no simple description. Furthermore, there is no theory for how $q$ behaves as $p$ decreases. In order to obtain a simple description of the rate of decay of $a/n$ we prefer to recalibrate the forecasts to force $q = p$ and consider the bias separately.

If the event is forecast to occur when a continuous forecast variable $\hat{x}$ exceeds a threshold $\hat{u}$ and is observed to occur when a continuous observation variable $x$ exceeds a threshold $u$ then the forecasts can be recalibrated by choosing $\hat{u}$ and $u$ to be the upper $p$-quantiles of the $n$ values of the forecast and observation variables respectively (Ferro, 2007b; Stephenson et al., 2008a). For example, the upper 0.03-quantiles shown in Figure 10.1 are $u = 20\,\text{mm}$ and $\hat{u} = 15.5\,\text{mm}$, above which fall approximately 3% of the observations and forecasts respectively.

When the forecasts are recalibrated to force $q = p$, the model (Equation 10.1) becomes

$$\frac{a}{n} \sim \alpha p^{\beta} \qquad (10.2)$$

and this holds for a wide class of distribution functions that satisfy a property known as bivariate regular variation (Ledford and Tawn, 1996, 1997; Heffernan, 2000). Deciding in advance whether or not this model (Equation 10.2) is appropriate for describing the performance of a particular forecasting system is probably not feasible and so the adequacy of the model must be checked empirically. Figure 10.3 shows one way to do this. Plotting $a/n$ against $p$ on logarithmic axes will yield an approximately straight line with slope $\beta$ at low base rates if Equation 10.2 holds. Allowing for sampling variation and jumps caused by the relative scarcity of data at small base rates, the plot shows that a model with

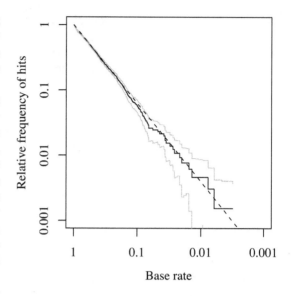

**Figure 10.3** The relative frequency of hits ($a/n$, solid black line) calculated for the recalibrated precipitation forecasts and plotted against base rate ($p$) on logarithmic axes with approximate 95% confidence intervals (grey lines) based on a Gaussian approximation to the sampling distribution of a binomial proportion (e.g. Jolliffe, 2007). The model (Equation 10.2) with $\alpha = 1$ and $\beta = 1.2$ is superimposed (dashed line)

$\alpha \approx 1$ and $\beta \approx 1.2$ is a good approximation to $a/n$ over almost all base rates.

In cases where the model of Equation 10.2 is adequate we can use it to estimate the rate of decay of forecast performance. For recalibrated forecasts with $q = p$, the contingency table is completed by the entry $a/n$ and forecast performance increases with $a/n$ for a fixed base rate. The model (Equation 10.2) therefore describes how forecast performance changes with base rate. Stephenson et al. (2008a) proposed focusing on the parameter $\beta$ that controls the rate at which performance degenerates with the base rate. We know that $\beta \geq 1$ and that random forecasts have $\beta = 2$. As a result, the measure $l = 2/\beta - 1$ of extremal dependence lies in the interval $(-1, 1]$ and has the following interpretation:

- if $l = 0$ then $\beta = 2$ and performance decreases at the same rate as for random forecasts;
- if $l > 0$ then $\beta < 2$ and performance decreases slower than for random forecasts;

- if $l < 0$ then $\beta > 2$ and performance decreases faster than for random forecasts.

One way to estimate $l$ relies on the following results for the hit rate, $H$, and false alarm rate, $F$. Given the model of Equation 10.2,

$$H = \frac{a}{np} \sim \alpha p^{\beta-1}$$

as $p \to 0$ and therefore

$$\log H \sim \log \alpha + (\beta - 1) \log p \sim (\beta - 1) \log p.$$

Similarly,

$$F = \frac{p - a/n}{1 - p} \sim \frac{p - \alpha p^{\beta}}{1 - p}$$

and therefore

$$\log F \sim \log p + \log \left(1 - \alpha p^{\beta-1}\right) - \log(1 - p)$$
$$\sim \log p.$$

Consequently, the extremal dependence index (EDI)

$$EDI = \frac{\log F - \log H}{\log F + \log H}$$

converges to

$$\frac{\log p - (\beta - 1) \log p}{\log p + (\beta - 1) \log p} = l$$

as $p \to 0$. This measure was proposed by Ferro and Stephenson (2011) and is discussed further in Chapter 3. If we calculate the EDI for recalibrated forecasts of rare events then, assuming that the model of Equation 10.2 is appropriate, the EDI will provide an estimate of the extremal dependence measure, $l$.

The extremal dependence index for our recalibrated precipitation forecasts is plotted against the base rate in Figure 10.4. The confidence intervals are computed using the Normal approximation $EDI \pm 2s$, where

$$s = \frac{2 \left| \log F + \frac{H}{1-H} \log H \right|}{H(\log F + \log H)^2} \sqrt{\frac{H(1-H)}{np}}$$

is the estimated standard error (Ferro and Stephenson, 2011). As the base rate decreases in Figure 10.4, the EDI stabilizes at approximately $l = 2/3$, which corresponds to the value $\beta = 1.2$ illustrated in Figure 10.3. The performance of these forecasts, were they perfectly calibrated, therefore decreases significantly slower than does the performance of random forecasts as the event threshold is increased. This picture of forecast performance is complemented by the plot of the bias in Figure 10.4. Precipitation occurrence is overforecast by up to 20% at lower thresholds but is underforecast by up to 50% at higher thresholds.

The EDI has several attractive properties: it is base-rate independent, difficult to hedge, asymptotically equitable, regular, and can take values in the whole interval $[-1, 1]$ regardless of the base rate, where zero demarcates better-than-random forecasts and worse-than-random forecasts. For details refer to Ferro and Stephenson (2011), who also propose an alternative version of the EDI that is complement symmetric and known as the symmetric extremal dependence index, or SEDI. Both of these measures are considered preferable to the extreme dependency score, EDS, proposed by Stephenson *et al.* (2008a) and its transpose symmetric version, SEDS, proposed by Hogan *et al.* (2009).

### 10.2.4 Modelling performance for extreme events

We saw at the start of this section that verification measures for extreme events can be subject to large sampling uncertainty. In this subsection we discuss a way to reduce this uncertainty. The idea is to fit a statistical model to the joint distribution of forecasts and observations, and derive verification measures from the fitted model instead of estimating them directly from the sample. This modelling approach often yields more precise estimates of the desired measures, particularly when samples are small, and it has been applied many times in the literature: see Bradley *et al.* (2003) and references therein, for example.

Instead of recalculating the contingency table for each pair of thresholds $u$ and $\hat{u}$, as we did in the previous subsection, we can reduce

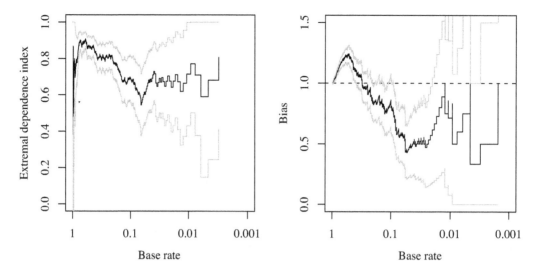

**Figure 10.4**  Left: extremal dependence index (EDI, black line) for the recalibrated precipitation forecasts plotted against base rate ($p$) on a logarithmic axis with approximate 95% confidence intervals (grey lines) based on a Gaussian approximation. Right: frequency bias ($q/p$, black line) for the uncalibrated precipitation forecasts plotted against base rate ($p$) on a logarithmic axis with approximate 95% confidence intervals (grey lines) based on a Gaussian approximation

uncertainty by imposing a parametric model that constrains how the table entries will change as the base rate decreases. We consider again the case of recalibrated forecasts for which we introduced a model (Equation 10.2) in the previous subsection: the relative frequency $a/n$ is assumed to follow $\alpha p^{\beta}$ when $p$ is small. This model implies that the normalized contingency table (where all entries are divided by the sample size $n$) for recalibrated forecasts will look like Table 10.2 for all small base rates.

If such a model is a good description of reality then we no longer need to recalculate entries $a$, $b$, $c$ and $d$ for each base rate of interest. Instead, we estimate the model parameters $\alpha$ and $\beta$ once only from all available forecast-observation pairs,

**Table 10.2**  The modelled version of the normalized contingency table for recalibrated forecasts

|  | Observed | Not observed |  |
| --- | --- | --- | --- |
| Forecast | $\alpha p^{\beta}$ | $p - \alpha p^{\beta}$ | $p$ |
| Not forecast | $p - \alpha p^{\beta}$ | $1 - 2p + \alpha p^{\beta}$ | $1 - p$ |
|  | $p$ | $1 - p$ | $1$ |

insert these estimates into Table 10.2, and then derive whichever verification measures we want for selected base rates $p$. For example, if we estimated $\alpha = 1$ and $\beta = 1.2$ for the precipitation forecasts discussed earlier then we would substitute these values into Table 10.2 and derive measures such as the hit rate,

$$H \sim p^{1.2}/p = p^{0.2},$$

and proportion correct,

$$PC \sim 1 - 2p + 2p^{1.2},$$

and evaluate them for any base rate, $p$, of interest.

Remember that our model (Equation 10.2) is designed to hold only for small base rates. Therefore, a threshold base rate, $p_0$, must be identified below which the model holds, and the model should be used for base rates $p \leq p_0$ only. We shall discuss how to choose $p_0$ later, after we have seen how to estimate $\alpha$ and $\beta$.

Suppose that we have a sample of $n$ pairs $(\hat{x}_t, x_t)$ for $t = 1, \ldots, n$ of the forecast and observation variables introduced earlier. Define the transformed values $z_t = \min\{\hat{x}'_t, x'_t\}$ for each $t$, where

$$\hat{x}'_t = -\log\left[1 - \frac{1}{n+1}\sum_{s=1}^{n} I\left(\hat{x}_s \le \hat{x}_t\right)\right],$$

$$x'_t = -\log\left[1 - \frac{1}{n+1}\sum_{s=1}^{n} I\left(x_s \le x_t\right)\right],$$

$I(E) = 1$ if $E$ is true and $I(E) = 0$ if $E$ is false. If the model (Equation 10.2) holds for base rates $p \le p_0$ for some small $p_0$ then Ferro (2007b) shows that the $z_t$ form a sample from the probability distribution of a random variable $Z$ for which

$$\Pr(Z > z) = \alpha e^{-\beta z} \qquad (10.3)$$

when $z \ge w_0 = -\log p_0$. Maximum-likelihood estimates for the parameters $\alpha$ and $\beta$ can then be obtained via the formulae

$$\hat{\beta} = \max\left\{1, \left[\frac{1}{m}\sum_{t=1}^{n} \max\{0, z_t - w_0\}\right]^{-1}\right\}$$

and

$$\hat{\alpha} = \frac{m\exp\left(\hat{\beta}w_0\right)}{n}$$

where $m$ is the number of the $z_t$ that exceed $w_0$.

There is a trade-off involved in choosing $p_0$: we need $p_0$ to be small enough for the model to fit, but we would also like $p_0$ to be large so that we use as many of the $z_t$ as possible to estimate $\alpha$ and $\beta$. Therefore, we should choose $p_0$ to be the largest base rate below which the model holds. One way to choose $p_0$ is to select the largest base rate below which the graph shown in Figure 10.3 of $a/n$ against $p$ on logarithmic axes is linear. Two other graphical tools are helpful too. If the model (Equation 10.3) holds at threshold $w_0$ then it must also hold for all $w > w_0$. Therefore, we can estimate $\alpha$ and $\beta$ for a range of values for $w_0$ and select the lowest threshold above which the estimates remain approximately constant. An example is shown in Figure 10.5, where the estimates of $\beta$ are found to increase as $w_0$ increases from 0 to 2 but then stabilize, relative to the sampling variation, at around $\beta = 1.25$. The parameter estimates of $\alpha$ are not shown because they are highly correlated with those of $\beta$ and also stabilize at $w_0 \approx 2$. The percentile bootstrap confidence intervals in Figure 10.5 are obtained by repeatedly resampling the original forecast-observation pairs and recalculating the estimates for each resample (Davison and Hinkley, 1997, pp. 202–203). Another graphical tool uses the fact that the expected excess of the $z_t$ above a threshold $w$ is

$$E(Z - w | Z > w) = \frac{1}{\beta}$$

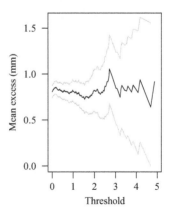

**Figure 10.5**   Left: maximum-likelihood estimates (black line) of the parameter $\beta$ calculated for the precipitation forecasts and plotted against threshold ($w_0$) with non-parametric percentile bootstrap 95% confidence intervals (grey lines). Right: mean excess (black line) calculated for the precipitation forecasts and plotted against threshold ($w_0$) with approximate 95% confidence intervals (grey lines) based on a Gaussian approximation. Modified from Ferro (2007b)

when the model (Equation 10.3) holds. A graph of the mean excess against threshold should therefore be flat once a suitable $w_0$ is reached. Figure 10.5 provides an example, where again the mean excess stabilizes at about $w = 2$. As is the case here, the exact choice of threshold is rarely clear cut and so it is often appropriate to investigate the sensitivity of results to $w_0$.

Given our choice of $w_0 = 2$, which corresponds to the base rate $p_0 = \exp(-w_0) = 0.135$, we obtain parameter estimates $\alpha = 0.91$ (0.65, 1.49) and $\beta = 1.23$ (1.07, 1.44) where non-parametric percentile bootstrap 95% confidence intervals are given in brackets. Since $\beta < 2$, performance decays slower than for random forecasts. Furthermore, the recalibrated forecasts perform better than random forecasts for all base rates that satisfy $\alpha p^\beta > p^2$, and for our parameter estimates this inequality is satisfied for all $p < p_0$. Apart from the underforecasting of rare events evident in Figure 10.4, therefore, these are very good rare-event forecasts. The fit of the model (Equation 10.3) can be assessed with quantile-quantile and probability-probability plots for the $z_t$, such as those shown in Figure 10.6. These show no significant departures of the model from the data.

Now that we have a model for the contingency table for base rates smaller than $p_0 = 0.135$, we can derive estimates of whichever verification measures we want and plot them against the base rate. Figure 10.7 compares these model-based estimates of the hit rate, PC, CSI and log odds ratio with the direct estimates that we obtained in Section 10.2.1. Notice how the model estimates smooth the direct estimates, have narrower confidence intervals, and extrapolate verification measures to base rates lower than those previously observed. This confirms that our model is a good description of how the performance of the recalibrated forecasts degenerates with base rate.

### 10.2.5 Extreme events: summary

We have focused on two-by-two contingency tables and have seen that traditional verification measures tend to degenerate to trivial values as we consider increasingly rare events. Nonetheless, we should still use these measures to verify forecasts of rare events if the measures are pertinent to the aims of the verification exercise; we just need to be aware of their tendency to either increase or decrease with event rarity. If sampling uncertainty also increases then identifying statistically significant differences between the performances of forecasting systems for rare events can be difficult. Fitting an appropriate statistical model to the forecasts and observations can mitigate this problem to some extent.

We have focused on deterministic forecasts of binary events in this section because most published material for rare events has addressed that case. We hope that future research will expand the suite of techniques designed to help in the verification of forecasts of extreme events.

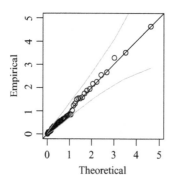
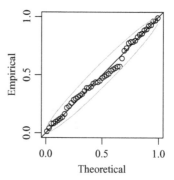

**Figure 10.6** Quantile-quantile and probability-probability plots with pointwise non-parametric percentile bootstrap 95% confidence intervals for the model fitted to the precipitation forecasts at base rate $p_0 = 0.135$. Modified from Ferro (2007b)

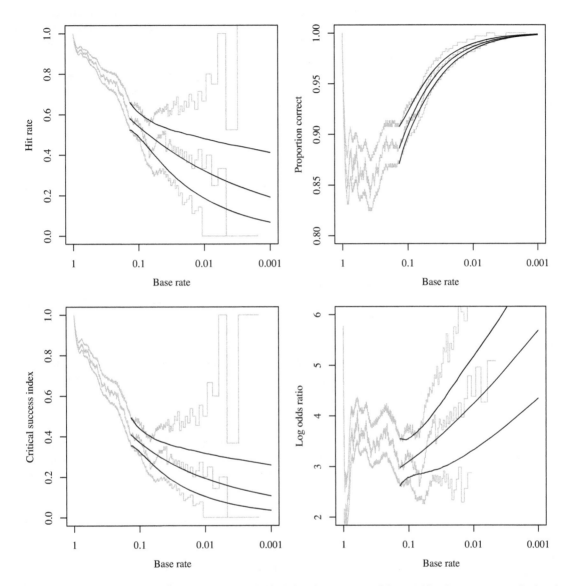

**Figure 10.7**   Direct (grey lines) and model-based (black lines) estimates of four verification measures calculated for the recalibrated precipitation forecasts and plotted against base rate on a logarithmic axis with non-parametric percentile bootstrap 95% confidence intervals. Modified from Ferro (2007b)

## 10.3  Warnings

### 10.3.1  Background

Weather-related warnings are deterministic forecasts of future time periods when severe weather events are expected to be observed. Warnings are simple to communicate and they are routinely is-sued by national weather services. For example, the UK Met Office currently issues many types of warning such as severe weather warnings, ex-treme rainfall alerts, marine wind warnings, heat health warnings, aviation and defence warnings, etc. Stephenson *et al.* (2010) present a comprehensive review of verification of Met Office warnings, which forms the basis of the summary presented here.

Warnings are also routinely issued for other types of environmental hazard such as floods. The warnings are often categorical forecasts such as the UK Environment Agency's four categories consisting of *severe flood warning, flood warning, flood watch* and *all clear*. Such forecasts are difficult to verify for a number of reasons. Firstly, the events at a particular location are often extremely rare; for example, a severe flood might have occurred only once or not at all since historical records began. The number of joint events of severe flood warning and a severe flood happening are therefore likely to be extremely small and so contingency tables are full of very small numbers, which makes statistical analysis difficult. The problem is sometimes alleviated by spatial pooling (regional analysis) of data from similar yet independent catchments. Another major problem with flood verification is in defining the actual event – flooding is often extremely sensitive to small local elevation differences that are not easily measured. Krzysztofowicz (2001) presents a compendium of reasons for moving away from deterministic to probabilistic forecasting of hydrological variates. In some sense, flood-warning categories should perhaps be interpreted as crude probability statements about a single event rather than as deterministic categorical forecasts.

The simple warning-in-effect/warning-not-in-effect binary format of warnings allows the public to take quick action to minimize potential losses. If the warnings have been obtained by thresholding probabilistic forecasts, then ideally the probabilities should be issued as this provides decision-makers with necessary uncertainty information to allow them to make optimal decisions based on their specific losses. However, many users prefer the weather service to implicitly make the decision for them by issuing simpler deterministic forecasts, so such forecasts are likely to continue for the foreseeable future. Despite the ubiquity of such warnings, there are surprisingly few published research articles dealing explicitly with how best to verify them.

### 10.3.2 Format of warnings and observations for verification

A simple warning consists of three times: a time $t_0$ when the warning is issued; a start time $t_1 \geq t_0$ when the warning comes into effect (i.e. when the severe weather is expected to start occurring); and an end time $t_2 \geq t_1$ when the risk is expected to decrease. The lead time $t_1 - t_0 \geq 0$ is a crucial quantity for forecast users who require time to take action; it can be zero if severe weather is considered imminent (nowcasting). Multiple warnings can overlap; for example, UK Met Office early (large lead-time) storm warnings are frequently in effect when imminent flash warnings are issued. Warning times are usually recorded at regular discrete times; for example, UK Met Office marine and severe weather warnings are given as clock times to the nearest hour. In addition, warnings often include crude measures of the anticipated severity of the event – for example, the *amber* and *red* levels used by the UK Met Office to signify severe weather, and extremely severe weather, respectively.

Warnings are generally issued when a meteorological variable, either within a geographical region (e.g. severe weather and marine warnings) or at a specific site (e.g. defence warnings), is expected to exceed a predefined threshold. A problem for warnings over geographical regions is that observations may only be available at fairly arbitrary, unevenly spaced points in space. A related matter, which is also relevant to discrete sampling in time, is that an event not being observed does not necessarily mean that it did not happen (e.g. small-scale features such as tornados can be missed). Hence, hit rates based on observation networks can easily underestimate the true hit rate for small spatial systems. It is not obvious how best to choose geographical areas for warnings, and it should be recognized that the size of the chosen region has a large impact on the base rate and other related verification measures (Stephenson *et al.*, 2010). Observations of severe weather are prone to several other sources of sampling, measurement and representation error that can confound the verification. Barnes *et al.* (2007) suggest that when there is reliance on volunteers or unofficial observations (e.g. in North America), there may be tendency to look harder for an event when it has been forecast than when it has not; this is not relevant when observations are automatic.

Figure 10.8a,b shows an illustrative typical example of 100 days of warnings and corresponding observations of severe weather. The warnings and verifying observations were generated artificially by

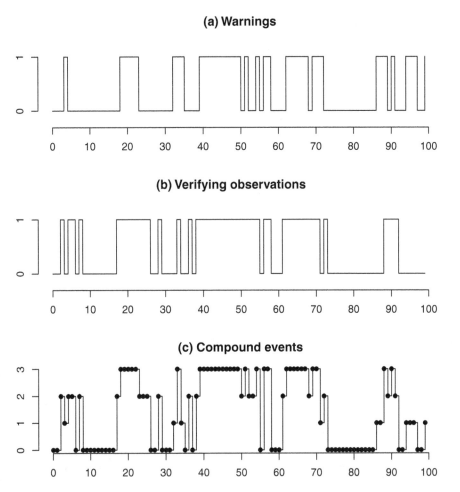

**Figure 10.8** Illustration of 100 days of (a) warnings, (b) verifying observations and (c) the resulting compound events. Occurrence of warning or observed event is signified by 1 in panels (a) and (b) and the lines join up the 100 discrete values. Data were generated by thresholding simulations of red noise generated by a first-order autoregressive time series model

using exceedances above a threshold of 100 values from a simulation of a serially correlated first-order autoregressive time series process, $x_{t+1} = \rho x_t + \varepsilon_t$, where $\varepsilon_t$ are independent Gaussian variables. Values of 1 indicate that warnings were in effect or that severe weather has been observed. It can be seen that there is some overlap between periods when warnings were in effect and when severe weather had occurred. The purpose of verification is to summarize the essential characteristics of this overlap. Note, however, that important information about when the warnings were issued (and hence invaluable lead-time information) has been lost in this binary representation.

### 10.3.3 Verification of warnings

*Counts of hits, misses and false alarms*
Following in the footsteps of early pioneers such as Finley (1884), verification of warnings has traditionally relied upon categorical scores based on counts of the following types of overlapping observation-forecast *compound event*:

- Hit – severe weather observed while a warning is in effect.
- Miss – severe weather observed while no warning is in effect.

- False alarm – severe weather not observed while a warning in effect.
- Correct rejection – severe weather not observed while a warning is not in effect.

If defined correctly, these counts represent a $2 \times 2$ cross-classification that can then be summarized using standard categorical measures (see Chapter 3).

### Methods for counting compound events

Compound events of simultaneous warnings and observations such as hits, misses and false alarms can be counted easily if one considers warning and observation events to be the binary events that occur simultaneously at fixed sampling times. For example, for hourly observations of wind speed above a predefined threshold, one can count the number of exceedances and non-exceedances when warnings are and are not in effect, and hence compile the contingency table. However, it should be noted that the counts here represent the numbers of hours in each category rather than the number of distinct meteorological events.

Meteorological events such as tornados and other storms occur sporadically in time and persist for varying durations. The mean rate of such events can be estimated by dividing the number of events in a given time interval by the length of the time interval. Rates can also be calculated conditional upon when warnings are and are not in effect. However, it is problematic to calculate the rate of observed non-events since the duration of a non-event is undefined. Similarly, for irregular duration warnings, it is impossible to count the number of 'no warning in effect' events since it is not known how long such non-events last. It may be easy to see that there are five tornados in a month for a given area, but much more difficult to say how many 'no tornado' events there were. There is no easy answer to this, though one ad hoc method is to count the non-events by dividing the length of the non-event periods by the average length of the event periods.

The problem with counting non-observed events and no-warnings, has led meteorologists to believe that the number of correct rejections, $d$, cannot be calculated reliably for weather warnings. If $d$ is not available, then it is not possible to calculate many of the usual binary verification scores such

as the base rate $p = (a + c)/n$, which requires $d$ to find $n$, the Peirce Skill Score, the Equitable Threat Score, etc. Brooks (2004) discusses tornado forecasting and notes that given $a, b, c$ then $d$ can be found if the base rate $p$ is known, so he attempts to estimate the base rate in order to infer $d$. His base rate is not the overall occurrence of tornados, which is extremely small, but their occurrence in circumstances where a tornado might conceivably have been forecast (the 'difficult' cases). In calculating the base rate, it seems reasonable to consider the period when warnings are actually issued rather than include long periods when there is little risk of severe weather. Mason (1989) also suggests restricting attention to only 'difficult' cases in order to infer $d$, which he calls the 'no-no frequency'. However, Glahn (2005) points out two difficulties with the suggestion of Brooks (2004). There is a question of how to estimate the base rate for the 'difficult' cases and also how to decide the threshold between easy and difficult cases. Verification measures will depend on the choice of two thresholds: where to draw the line between easy and difficult situations as well as how large the probability of the event needs to be before a warning is issued. It should also be noted that the problem of counting correct rejections, non-observed events when warnings are not in effect, equally applies to counting the number of non-observed events when warnings are in effect i.e. false alarm events.

Counting of compound events is illustrated here for two different strategies applied to the synthetic data. Figure 10.8c shows the sequence of compound events by plotting a compound event variable equal to twice the warning variable plus the observation variable. The compound event variable $Z$ takes values of 0, 1, 2, 3 signifying correct rejection, miss, false alarm and hit events, respectively. Strategy A simply counts the compound events over all the possible sampling times. For example, the number of false alarms is the number of sampling times (e.g. days in our artificial example) when $Z = 2$. Faster sampling rates will lead to larger counts. Rather than consider distinct sampling times, Strategy B counts the number of distinct time periods when $Z$ takes the values (e.g. the number of steps rather than daily points in Figure 10.8c). For example, the third warning period, which lasted from days 33 to 35, had the sequence of observations {010} and

**Table 10.3** Summary of counts and resulting measures obtained for the synthetic data using two different counting strategies. Confidence is defined as $C = 1 - \text{FAR} = a/(a + b)$ where $\text{FAR} = b/(a + b)$ is the False Alarm Ratio (not to be confused with the False Alarm Rate $F = b/(b + d)$ – see Chapter 3)

| Measure | Symbolism | Strategy A: day counts | Strategy B: event counts |
|---|---|---|---|
| Hits | $a$ | 31 | 10 |
| False alarms | $b$ | 10 | 7 |
| Misses | $c$ | 19 | 15 |
| Correct rejections | $d$ | 40 | 12 |
| Total counts | $n = a + b + c + d$ | 100 | 44 |
| Base rate | $p = (a + c)/n$ | 0.5 | 0.57 |
| Bias | $B = (a + b)/(a + c)$ | 0.82 | 0.68 |
| Hit rate | $H = a/(a + c)$ | 0.62 | 0.40 |
| Confidence | $C = a/(a + b)$ | 0.76 | 0.59 |
| Threat score | $\text{TS} = a/(a + b + c)$ | 0.52 | 0.31 |

compound events {232} and so contained two false alarm events separated in day 34 by one hit event. Since events are separated from each other by non-events, this strategy tends to give similar counts for hits and false alarms, and similar counts for misses and correct rejections. Table 10.3 shows that the different strategies yield substantial differences in counts and statistics derived from them. Large differences can be noted in the sample size, which is greater for strategy A that counts each individual day, and in traditional measures of skill such as hit rate, confidence and threat score. Different conclusions about the warning system would be inferred using these different counting approaches.

*Fuzzy definitions of compound events*

The basic definitions of hits, misses and false alarms are often modified to take account of slight mistiming in warnings and the consequent near hits and near misses. One might argue that the traditional definitions used to verify warnings overly penalize slight errors in timings of the warnings or observed variables being either just slightly above or below the fixed threshold used to define an event. Similar to the double penalty problem in spatial verification (see Chapter 6), slightly delayed warnings can easily lead to an early miss *and* a later false alarm. Generalized definitions of hits, false alarms and misses have been developed to try to help alle-

viate these problems. For example, for an imminent UK Met Office gale warning the period for judging a hit is from $t_1$ to $t_1 + 12$ hours, whereas the period for judging a false alarm is $t_1 - 6$ hours to $t_1 + 24$ hours (Stephenson *et al.*, 2010). The UK Met Office has also tried several other ideas such as subdividing missed events according to temporal distance from the warning, introducing a near-hits category for which the event occurred but in a smaller proportion of the area than required for a hit, and using different wind gust thresholds for hits and false alarms. Barnes *et al.* (2007) refer to false alarms, unwarned events (misses) and perfect warning (hits) and propose additional categories, namely 'underwarned events' and 'overwarned events', or even a continuum to replace the small number of categories.

The extended categories are verified by collapsing counts of these events into a $2 \times 2$ contingency table and then using categorical measures such as those presented in Chapter 3. However, it should be noted that these generalized events do not necessarily form a consistent two-way cross-classification. For example, a *near miss* event cannot easily be described as the simultaneous occurrence of a warning-not-in-effect event and an observed severe weather event. Furthermore, the set of compound events may not describe all possible outcomes (e.g. they might not cover late false alarms). If the compound events do not form an exhaustive and exclusive description of all possible events then conditional probabilities such as hit rate or false alarm rate will be incorrectly estimated using a $2 \times 2$ classification. New statistical approaches need to be developed for such events rather than using an inappropriate $2 \times 2$ contingency table.

### Measures used for verification of events

Because counts of correct rejections are considered to be either unreliable or unavailable, verification measures for warnings are often restricted to measures based on counts of hits, misses and false alarms. For example, warnings at the UK Met Office are routinely verified using hit rate $H = a/(a + c)$, false alarm ratio FAR $= b/(a + b)$, and threat score TS $= a/(a + b + c)$.

A good warning system is expected to have a high hit rate, a low false alarm ratio and a high threat score. Credible warning systems should ideally have a high probability of a rare severe weather event occurring given a warning:

$$C = 1 - \text{FAR} = a/(a + b),$$

known by the UK Met Office as *confidence*. A useful illustration of loss of credibility for low-confidence warnings is provided by Aesop's classic fable of a shepherd boy who lost credibility with villagers when no wolf appeared after he repeatedly cried wolf. The UK Met Office typically sets targets as high as $C > 0.8$ for various operational warning systems (Stephenson *et al.*, 2010). However, Barnes *et al.* (2007) suggest that there is little evidence that a high value of FAR causes users to disregard warnings of severe events. The FAR values quoted by Barnes *et al.* (2007) for a number of different types of event, are all well above 0.2 and so have confidences less than 0.8.

For increasingly rare events and finite frequency bias, hit rates generally tend to zero (see Section 10.1). Since $1 - \text{FAR} = H/B$, high confidence can only be maintained for vanishing hit rates $H$ by having a frequency bias that also vanishes at the same or faster rate. However, such underforecasting can be incompatible with overforecasting rare severe events in order to reduce the risk of large losses incurred by missed events. For many warnings (e.g.

aviation), overforecasting is considered preferable if it increases safety, even if it degrades the apparent performance of the forecasts according to the chosen verification measures. For example, there is intentional overforecasting of storm and violent storm force winds at the German weather service. There is an important trade-off between false alarms and misses that ultimately depends on the losses incurred by the users for these two types of warning error.

In summary, high confidence may be achieved by issuing fewer warnings (i.e. not crying wolf), but then underforecasting can compromise safety by failing to warn about events. Hence, overly high confidence targets for warnings of rare severe events might not be in the public interest. In general for most users, the loss due to a missed event exceeds the loss incurred by a false alarm.

### 10.3.4  Warnings: summary

To summarize, warnings are widely issued by national weather services and so need to be properly verified. Existing verification methods represent the warnings and observed severe weather as binary events and then consider binary categorical measures such as the threat score. There is no unique way to count the events, and different counting strategies can give very different counts and values for verification measures. Counting binary events fails to explicitly include useful information about the lead time of the warnings, which is important for effective response to warnings. Various modified compound events have been proposed such as near-hits, which can help to account for slight errors in timings of warnings. However, it should be noted that such modifications can lead to an inconsistent cross-classification, which invalidates the use of $2 \times 2$ categorical measures. For increasingly rare events, the false alarm ratio will generally increase unless the forecaster issues fewer warnings than observed events, and so high confidence targets will be difficult (if not impossible) to achieve. Warnings generally tend to be issued more often than the observed event in order to avoid missing events that can lead to catastrophic losses (and criticism of the forecasters!). It is clear from this review that the verification of warnings faces several important challenges that could benefit from further research.

## Acknowledgements

The authors of Chapter 10 wish to thank the following people for useful and stimulating discussions on warnings: Harold Brooks, Bob Glahn, Brian Golding, Martin Göber, Tim Hewson, Ian Jolliffe, Marion Mittermaier, Michael Sharpe, Dan Wilks and Clive Wilson.

# 11

# Seasonal and longer-range forecasts

**Simon J. Mason**

*International Research Institute for Climate and Society, Palisades, NY, USA*

## 11.1  Introduction

As discussed in Chapter 1, verification procedures should be tailored to the specific questions that are being asked, and to the nature of the forecasts as they are presented. Procedures for verifying seasonal and longer-range forecasts need only differ from those for shorter-range forecasts to the extent that different questions about the quality and value of the forecasts are asked, that data are available to answer those questions, and that the forecasts at the different timescales are presented in dissimilar formats. In this regard, two key issues have to be considered when performing verification analyses on forecasts at these longer timescales: limited sample size and low levels of predictability.

In most parts of the world seasonal forecasts were initiated only in the 1990s or later, and are rarely issued more frequently than once per month, and so there are currently very few examples of operational forecasts with more than about two decades of history. [Forecasts for the Indian monsoon, which were started in the 1880s (Blanford, 1884), are a notable exception.] Thus, sample sizes of seasonal forecasts typically are highly limited, while at the longest timescales there may not be any verifiable earlier predictions at all. Although it may be possible to generate hindcasts, it is often difficult to do this in a way that does not introduce an element of artificial skill (as discussed in Sections 1.4.2 and 11.3.1), and so there is a danger of overestimating the quality of the forecast system. In addition, generating the hindcasts may not even be viable: in decadal forecasting, for example, potential predictability is believed to come largely from subsurface ocean conditions, but observational data for initializing the models are severely lacking prior to the 1990s. Even when there is a history of forecasts and corresponding observations available, the quality of the forecasts over this period is unlikely to have remained constant because of model revisions and changes in observation accuracy. Consequently, verification results will give an indication of the average performance of the forecasts over the period of analysis, but will not necessarily give an accurate indication of the expected quality of subsequent predictions. The net effect of this sample size problem is that uncertainty estimates on measurements of the quality (or value) of seasonal and longer-range forecasts are typically large, and so assessing 'skill', whether against a baseline or against a competing forecast system, can be difficult (Tippett *et al.*, 2005).

The second overriding consideration in the verification of seasonal and  longer-range forecasts is

*Forecast Verification: A Practitioner's Guide in Atmospheric Science*, Second Edition. Edited by Ian T. Jolliffe and David B. Stephenson.
© 2012 John Wiley & Sons, Ltd. Published 2012 by John Wiley & Sons, Ltd.

that levels of predictability are almost invariably much lower than those of weather forecasts. This relatively poor predictability is an inherent part of the climate system itself, but is compounded by the fact that computational demands and poor availability of observational data mean that the models used to make predictions, and the initialization of such models, are of weaker quality than for the weather forecasting problem. By far the majority of verification analyses conducted to date have sought to address the simple question of whether the forecasts have any 'skill'. As discussed in Section 11.3.1, this question is often poorly posed: careful construction of the verification question, and interpretation of the results, may be required to avoid unnecessarily pessimistic conclusions about the potential usefulness of some forecasts.

In this chapter the primary focus is on verification of seasonal forecasts for the simple reason that longer-range predictions almost invariably do not have sufficient sample sizes to perform a verification analysis. However, that is not to say that longer-range predictions cannot be evaluated at all, and some guiding principles are provided in Section 11.5 at the end of this chapter. Before discussing

verification procedures for seasonal forecasts, the most common forecast formats are briefly described (Section 11.2) because the appropriate verification options are constrained by the type of information that is being communicated in the forecasts. What constitutes 'skill' and ways of measuring it are reviewed in Section 11.3. In Section 11.4 some issues regarding the verification of individual forecasts are discussed.

## 11.2    Forecast formats

### 11.2.1    Deterministic and probabilistic formats

Most seasonal forecasts fall into one of two broad categories (Section 2.2): firstly, one or more 'deterministic' predictions of a seasonally averaged or integrated meteorological variable (e.g., mean temperature or total rainfall; Figure 11.1); and, secondly, a set of probabilities for the verification to fall within each of two or more predefined ranges (e.g. Figure 11.2). The deterministic forecasts are most often statistical or dynamical model outputs,

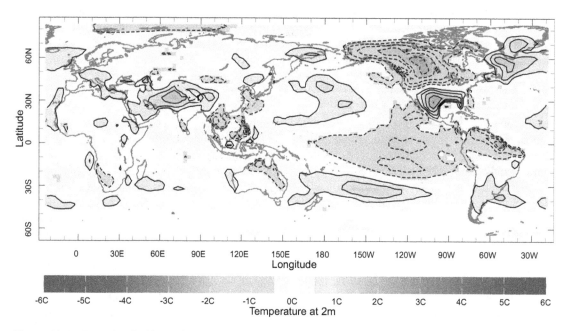

**Figure 11.1**  Example of a 'deterministic' seasonal prediction made in February 2011 using the ECHAM 4.5 model (Roeckner *et al.* 1996). The prediction is expressed as the ensemble mean March–May 2011 temperature anomaly with respect to the model's 1971–2000 climatology. A full colour version of this figure can be found in the colour plate section

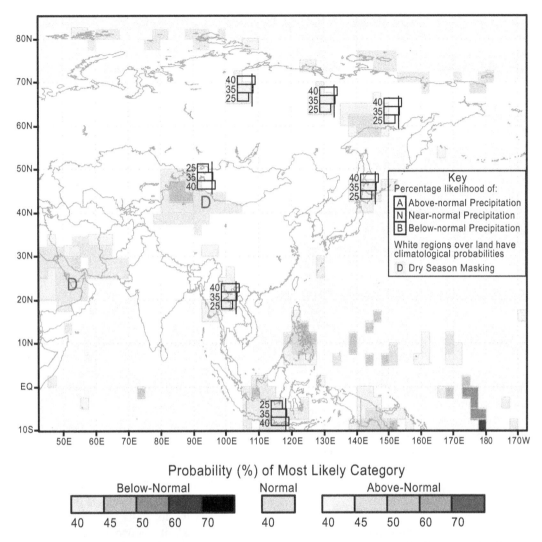

**Figure 11.2**   Example of a 'probabilistic' seasonal prediction issued in February 2011 by the International Research Institute for Climate and Society. The prediction shows the probabilities that the March–May 2011 precipitation total will be in one of the three categories 'above-normal', 'normal' and 'below-normal', with these categories defined using the terciles of March–May totals between 1971 and 2000. The probabilities of the most likely category are shaded, but the probability for the normal category is constrained to a maximum of 40% based on prior verification analyses. Probabilities for all three categories are only shown for large areas and for areas with relatively sharp probabilities. The three horizontal bars are scaled by the corresponding forecast probability, and the thin vertical line indicates the climatological probability of 33%. A full colour version of this figure can be found in the colour plate section

and the predictand is usually expressed as a 'best-guess' value on a continuous scale (van Oldenborgh *et al.*, 2005). This value may represent an area-average, which is typically the case for dynamical model outputs, or the forecast may be for a specific location, such as a meteorological station. Other seasonal forecast formats include counts, such as

hurricane frequencies (Owens and Landsea, 2003; Vitart, 2006; Wang *et al.*, 2009) or rain-day frequencies (Moron *et al.*, 2006; Robertson *et al.*, 2009), and dates, such as the onset of a rainfall season (Moron *et al.*, 2009). When a set of deterministic forecasts is available, the ensemble mean is often represented as the 'best-guess' forecast, and the uncertainty in the prediction is represented by some measure of the ensemble spread, or the model outputs may be presented in the probabilistic format described below. For statistical models, prediction intervals can be used to represent the uncertainty in the prediction.

The most common probabilistic format is for three climatologically equiprobable categories to be defined using a reference ('climatological') period, and the probability that the verification will fall within each of these categories is indicated (Livezey and Timofeyeva, 2008; Barnston *et al.*, 2010). Forecasts from the Regional Climate Outlook Forums (RCOFs) are typical examples of such forecasts (Ogallo *et al.*, 2008), and this format is followed closely by many national meteorological and hydrological services (NMHSs). However, examples of climatologically unequal categories, and of the use of more than three categories, are not uncommon (Ward and Folland, 1991; Tippett and Barnston, 2008). The Australian Bureau of Meteorology (BoM) uses an equiprobable two-category system, providing the probability that the forecast parameter will be above- or below-median (Fawcett, 2008). This two-category format is followed frequently in climate change projections, where the proportion of models indicating a change in one direction is indicated; usually care is taken to communicate that this proportion should not be taken as a forecast probability.

## 11.2.2  Defining the predictand

For some forecasts it is not always clear whether the forecast is for an area-average or is valid for all specific locations. This problem is most common where some subjective input has been introduced into a probabilistic forecast. Since different results can be realized depending upon how the forecast is interpreted, this ambiguity is an undesirable property. In such cases it may not be possible to verify the predictand precisely as it has been defined, and a new interpretation may need to be imposed. Reinterpretation of a forecast prior to verification may be quite intentional even when there is no ambiguity. For example, seasonal forecasts are often presented as coarse spatial and temporal averages, which tend to limit their usefulness because the predictand is

of little relevance in most practical settings (Vogel and O'Brien, 2006; Hansen *et al.*, 2011), but it is perfectly valid to transform or reinterpret the forecast to a variable of more direct interest, and then verify the reinterpreted forecasts. While the verification results would no longer necessarily indicate whether or not the forecasts themselves are 'good' in Murphy's (1993) 'quality' sense, they would indicate whether or not the forecasts can be successfully reinterpreted to be more directly useful for some specific purpose. See Chapter 9 for a more detailed discussion of measuring the potential usefulness of forecasts.

## 11.2.3  Inclusion of climatological forecasts

Climatological forecasts are frequently issued in seasonal and longer-range forecasts either because of no skill or because of no signal for the current target period. Climatological probabilities are an explicit indication that each of the possible outcomes is as likely to occur as it has done over the climatological period, and they should be seen as distinct from areas of no-forecast where no statement is made about changed or unchanged probabilities. Climatological forecasts are usually included in verification analyses, but no-forecasts excluded. However, if there are a large number of climatological forecasts, these can dominate the verification analyses, and the forecasts may score poorly because of the lack of sharpness (e.g., Wilks, 2000; Wilks and Godfrey, 2002; Livezey and Timofeyeva, 2008; Barnston and Mason, 2011). While this poor scoring is appropriate because the forecasts do not contain much useful information, it can give the impression that the occasional non-climatological forecasts are not particularly useful. When comparing forecasts (perhaps for another region or season, or from another forecast system), the climatological probabilities should be included in the analysis because credit should be given for issuing sharper forecasts if those forecasts contain potentially useful information, while if they do not the forecasts should score badly. However, if the objective is to determine whether the forecasts are believable there may be justification in omitting the climatological forecasts.

## 11.3 Measuring attributes of forecast quality

As discussed in Section 1.1.2, the WMO's Commission for Basic Systems (CBS) established a Standardized Verification System for Long-Range Forecasts (SVSLRF; World Meteorological Organization, 1992) as part of a set of minimum requirements for qualification as a Global Producing Centre (GPC) for long-range forecasts. The SVSLRF addresses verification requirements for deterministic and probabilistic forecasts. The recommended verification scores for deterministic forecasts are based on the mean-squared error and its decomposition into terms measuring conditional and unconditional biases and Pearson's correlation (Chapters 2 and 5). The probabilistic procedures include reliability and relative operating characteristics (ROC) diagrams. All these procedures are discussed extensively in Chapters 3 to 5, 7 and 8, and so only issues related to their specific application to seasonal forecasts are discussed in this Chapter. The WMO's Commission for Climatology (CCl) guidelines for the verification of seasonal forecasts are targeted exclusively at probabilistic forecasts (Mason, 2011). The CCl guidelines include considerable overlap with the CBS guidelines for probabilistic forecasts, and so again details of the procedures are provided in Chapters 7 and 8. The aim in this section and in Section 11.4 is to highlight some of the peculiar issues in applying such verification procedures to seasonal and longer-range forecasts.

### 11.3.1 Skill

As discussed in Chapter 1, there are many possible reasons for verifying seasonal and longer-range forecasts. However, by far the most dominant question in the verification of seasonal and longer-range forecasts has been whether the forecasts have any 'skill'. There are particular difficulties in addressing this question with forecasts at these timescales, and so the measurement of this attribute is considered in detail here. Of course, other attributes are of interest, and modellers in particular are often interested in more detailed analyses that can reveal systematic errors in their models. Methods for identifying conditional and unconditional errors are therefore required, but one problem that often arises in verification of seasonal forecasts is that procedures are often selected that measure multiple attributes of forecast quality making the interpretation of the results difficult. It is argued in this section that procedures that measure individual attributes are to be preferred.

The underlying objective in measuring the skill of seasonal and longer-range forecasts against a naive forecast strategy such as guessing or perpetual forecasts (always forecasting the same thing) is almost invariably to answer the question of whether the forecasts are worth considering. Unfortunately, this question has often been poorly formulated, which has resulted in frequent misinterpretation. Much of the problem is that 'skill' is a vague attribute: as discussed in Sections 1.4 and 2.7, skill is a relative concept – a forecast has skill if it is better than another set of forecasts. But better in what respect? Skill requires reference to another attribute of forecast quality, and this attribute is frequently left undefined. Instead, skill has often been imprecisely interpreted as whether the forecasts have outscored climatological forecasts or some other naive forecast strategy without considering what attributes the chosen score might be measuring. In seasonal forecasting, because of the weak levels of predictability and suboptimal quality of prediction models, the diagnosis of skill can generally be reduced to the search for some potentially useful information. For deterministic forecasts this interpretation translates to a requirement that observed values should increase and decrease at least to some degree with the forecasts, while for probabilistic forecasts categories should verify more and less frequently as the probability increases and decreases. Of course, the forecasts are potentially useful if the observations vary in the opposite direction to that implied by the forecasts, and this possibility can be measured by negative skill.

Given the limited sample sizes of operational long-range forecasts, skill is often estimated using hindcasts to obtain larger samples. There are a number of problems in trying to generate a set of hindcasts that will give accurate indications of the expected skill of operational forecasts, and these problems are discussed separately below.

When assessing the skill of seasonal forecasts regardless of their format, trends in the data have to be considered. Correlations, for example, between two series that both contain trends are likely to be non-zero, even if the year-to-year variability is not successfully predicted. Similarly, probabilistic forecasts are likely to score well if the probabilities for the category in the direction of trend are consistently inflated. If trends are ignored, spurious forecasts may easily be falsely identified as being skilful, while the quality of low-skill forecasts may be exaggerated. It is often recommended that trends be removed before any skill calculations, although an argument could be made that the successful prediction of a trend should at least be acknowledged. One solution is to measure the skill of trend and interannual components separately, and to quote both. Another, related, solution is to consider presenting the forecast with reference to a shorter and more recent climatological period, which is likely to be of more relevance to many user communities anyway since it will focus the forecast on comparisons with more readily remembered climate variability.

### Skill of deterministic forecasts

By far the most commonly used skill measure for deterministic forecasts is Pearson's product moment correlation coefficient. This coefficient is discussed extensively in Section 5.4.4, and so is considered only briefly here. Pearson's correlation is implicitly scaled as a skill score, with reference strategies of perpetual forecasts and of random forecasts both having expected scores of zero. Its distributional properties for random forecasts are well known and so it is possible to calculate statistical significance analytically, and to estimate its sampling uncertainty on condition that assumptions about independence of the forecasts and the observations, and of their respective distributions, are met. Pearson's correlation is a preferred measure of choice also because of its wide use and hence familiarity, and its relationship to the percentage of explained (or predicted) variance, which provides it with a reasonably intuitive interpretation.

A further feature of Pearson's correlation is often considered an advantage: it ignores conditional and unconditional biases, which can be quite large in seasonal forecasts derived from global dynamical models, but which, in principle, should be easily correctable given a sufficient sample of forecasts to estimate the biases accurately. In practice, the correlation and the biases are not typically independent (DelSole and Shukla, 2010; Lee et al., 2010), but the common variability that the correlation measures seems a reasonable minimum requirement for forecasts to have some potentially useful information: if observed values do not increase and decrease with the forecasts at least to some extent then there seems little reason to consider them.

As discussed in Chapter 5, there are a number of problems with using Pearson's correlation coefficient as a skill measure. The interpretation of the coefficient's value is complicated by the fact that it is a function not only of the potential skill of the forecasts, but also of the precise distribution of the data. An example is shown in Figure 5.5, which illustrates that large Pearson correlations can result from the influence of only a few extreme values. Although this problem can be addressed to some extent by calculating bootstrapped estimates of uncertainty in the correlation, the problem remains that the results can be misleading. Consider the example shown in Figure 11.3, which compares a set of forecasts and observations of January–March seasonal rainfall totals for 1971–2000 for Kalbarri, in Western Australia. The forecasts were calculated as the mean of 85 ensemble members, each with different initial conditions, using the ECHAM4.5 model (Roeckner et al., 1996), and have a correlation with the observed rainfall of about 0.39 (90% bootstrap confidence limits of 0.08 and 0.64). If the three wettest years, which are not known a priori, are omitted from the analysis the correlation drops to 0.08 (−0.23 to 0.40). Is it to be concluded that virtually all the skill is provided by only 10% of the cases? What is clear is that the large bootstrap confidence intervals need to be taken seriously, especially when distributional assumptions are not strictly met, and sample sizes are small. Much of the underlying difficulty is that in small samples the most extreme values can contribute much of the total variance. In Figure 11.3, for example, the three wettest years represent over 50% of the total variance (and the wettest two years over 45% of the total), and so regardless of the quality of the forecasts the score will be heavily weighted by the forecasts on only a very few of the observations.

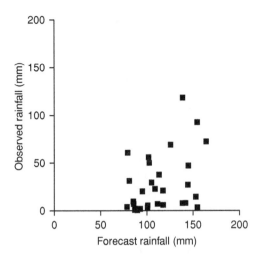

**Figure 11.3** Forecasts and observations of January–March seasonal rainfall totals for 1971–2000 for Kalbarri in Western Australia (27°42'43''S, 114°09'54''E). The forecasts were calculated as the mean of 85 ensemble members, each with different initial conditions, using the ECHAM4.5 model (Roeckner *et al.*, 1996)

A difficulty with Pearson's correlation is that it imposes a stricter definition of 'skill' than the requirement defined above that observed values should increase and decrease with the forecasts. Pearson's correlation imposes the additional criterion that the observed values should increase and decrease by precisely defined amounts as the forecasts vary. Although a better set of forecasts will predict more precise increases and decreases in the observations than will a poorer set, if the objective is to identify whether forecasts are potentially useful, and if the predictability is inherently weak, the weaker skill definition is likely to be more appropriate.

As argued in Section 5.4., it would be better still not to use Pearson's correlation at all in cases where its assumptions are violated: the intuitive sense of what constitutes a 'good' correlation that experienced practitioners may have is largely irrelevant, and even misleading, when the data are not normally distributed. Instead, alternative measures of skill should be considered, specifically Spearman's and Kendall's correlations. These measures are better suited to verification of variables with non-Gaussian distributions, such as precipitation and the counts and onset dates that were mentioned in Section

11.1. While Spearman's is the more widely used of the two correlations because of its close association with Pearson's correlation, the advantages of Kendall's $\tau$ are worthy of consideration. In addition to the advantages listed in Section 5.4.6, Kendall's $\tau$ has an intuitive interpretation: Kendall's $\tau$ (depending on how ties are handled, as discussed in the following sub-section) can be transformed simply to a scale ranging from 0 to 1 to represent the probability that the forecasts successfully discriminate the larger value of any two observations (Mason and Weigel, 2009). A second, related advantage is that Kendall's $\tau$ has close affinities to other widely used verification measures such as the area beneath the ROC curve (Chapters 2 and 7), as discussed in the following subsection.

### Skill of probabilistic forecasts

The most commonly used skill measure for probabilistic forecasts is the ranked probability score (RPS), and its skill score (RPSS; Sections 7.3.2 and 8.4.2), although the ignorance score (Sections 7.3.2 and 8.4.2) and similar information theory-based scores are becoming increasingly popular. In the context of objective forecasts, in which the forecast probabilities are estimated by counting the proportion of ensemble members predicting a value in each of the categories, the RPSS is biased because of reliability errors that result in turn from sampling errors in estimating the forecast probabilities given limited ensemble sizes (Section 8.4.3). Adjustments can be made to the score to remove this source of bias, but such an option is not available for subjectively derived probabilities. The need for the correction points to one of the difficulties in interpreting the RPS and its skill score: they measure multiple attributes, and so forecasts can score imperfectly if the forecasts are good in some respects, but poor in others. The RPS, Brier score, and ignorance score can each be decomposed into reliability, resolution and uncertainty terms (see Chapter 7 for further details), and in each case skill is achieved if the resolution term is larger than the reliability term. While this requirement for skill may be meaningful in some contexts, it is unnecessarily strict when trying to identify whether forecasts might be potentially useful (Mason, 2004). When verifying seasonal forecasts, which generally

**Table 11.1** Three idealized sets of seasonal forecasts for above-median rainfall, and corresponding observations (1 indicates above-median and 0 indicates below-median), with Brier score calculations for the base rate (50%) and for the three sets of forecasts

| Year | Observed | Forecasts | | | Brier score | | | |
|---|---|---|---|---|---|---|---|---|
| | | I | II | III | Base rate | I | II | III |
| 1 | 1 | 80% | 60% | 60% | 0.25 | 0.04 | 0.16 | 0.16 |
| 2 | 1 | 80% | 60% | 60% | 0.25 | 0.04 | 0.16 | 0.16 |
| 3 | 1 | 80% | 60% | 60% | 0.25 | 0.04 | 0.16 | 0.16 |
| 4 | 0 | 80% | 60% | 40% | 0.25 | 0.64 | 0.36 | 0.16 |
| 5 | 0 | 80% | 60% | 40% | 0.25 | 0.64 | 0.36 | 0.16 |
| 6 | 0 | 20% | 40% | 40% | 0.25 | 0.04 | 0.16 | 0.16 |
| 7 | 0 | 20% | 40% | 40% | 0.25 | 0.04 | 0.16 | 0.16 |
| 8 | 0 | 20% | 40% | 40% | 0.25 | 0.04 | 0.16 | 0.16 |
| 9 | 1 | 20% | 40% | 60% | 0.25 | 0.64 | 0.36 | 0.16 |
| 10 | 1 | 20% | 40% | 60% | 0.25 | 0.64 | 0.36 | 0.16 |
| | | | | Average | 0.25 | 0.28 | 0.24 | 0.16 |

suffer from overconfidence (i.e. poor reliability) and weak resolution, skill scores can often be negative, and there is then a danger of rejecting potentially valuable forecasts as useless.

Consider an idealized example in which ten seasonal rainfall forecasts are to be evaluated against corresponding observations. For the sake of simplicity it will be assumed that there are only two equiprobable categories. In one set of ten forecasts (marked I) five of the ten forecasts indicate an 80% probability of above-median rainfall, while the remaining five indicate a 20% probability (Table 11.1). Above-median rainfall occurs on 60% (i.e. three out of five) of the occasions that the forecast indicated an 80% probability, and on 40% (i.e. two out of five) of the occasions that the forecast indicated a 20% probability. The 80% forecasts correctly indicated an increase in the probability of above-median rainfall, and the 20% forecasts correctly indicated a decrease, but did so (somewhat typically for seasonal forecasts) overconfidently. Brier score (Section 7.3.2) calculations are shown in the Table (similar results are obtained using the ignorance score), and the skill score is −0.12, which suggests that the forecasts are worse than climatological forecasts. If the forecasts had been perfectly reliable (marked II), the score would naturally improve despite there being no gain in resolution, and the skill becomes marginally positive (0.04). Simi-

larly the skill can be raised by improving the resolution at the cost of reliability: for forecast set III the forecasts are under-confident, but have maximum resolution, which more than offsets the loss in reliability (the Brier skill score is 0.36). The progression in skill from set I to set III indicates that skill can increase on this measure, but with no indication of whether that is because reliability or resolution has improved, and there is no guarantee that either of these attributes has not deteriorated.

From a Brier and ignorance score perspective forecast set I is worse than information only about the base-rate, but to conclude that it would therefore be better not to have the forecasts at all is surely incorrect: the forecasts successfully indicate increased and decreased chances of above-median rainfall. The problem with set I is that the reliability errors are larger than the gain in resolution, but because both components are being measured together the resolution may be missed unless the skill is diagnosed carefully. Such difficulties in interpretation result from the skill scores imposing an arbitrarily high maximum acceptable level of reliability error for a given level of resolution. The situation is somewhat analogous to that of Pearson's correlation, which requires the observations to increase and decrease by precise amounts along with the forecasts, rather than just to increase or decrease; so also the reliability terms in the Brier and ranked probability skill scores require the predicted events to be more and less frequent by precise amounts as the forecast probability increases and decreases.

Although the previous examples illustrate that when reliability is measured with resolution there are difficulties in interpreting the result, the reliability term cannot be completely ignored for now. On its own the resolution term is not generally considered a satisfactory indication of skill: as long as the observed relative frequency is much higher for some forecast probabilities than for others the resolution term of the Brier, RPS and ignorance scores is large. In effect, the resolution term is measuring whether the expected observation differs given different forecasts, regardless of whether or not the observations vary arbitrarily with the forecasts. Imposing the requirement that the observed relative frequency should increase as the forecast probability increases therefore seems quite reasonable.

The interpretation problems that can affect scores that measure multiple attributes, or scores such as the resolution score that have an unsatisfactorily weak definition of skill arise only if such scores are calculated in isolation: when accompanied by analyses of reliability diagrams (Section 7.6.1), for example, the scores can be valuable summaries, and their decompositions can be informative. A primary difficulty in measuring resolution and constructing reliability diagrams for seasonal and longer-range forecasts is the severe sample-size restriction. The sampling errors in constructing the graph are likely to be prohibitively large, at least for some of the points (Section 7.6.1; Bröcker and Smith, 2007a). One possible solution is to bin the forecasts into only a few bins, although there is then likely to be a deterioration in skill (Stephenson et al., 2008b). Typically reliability diagrams can only be constructed meaningfully by pooling forecasts over large areas. When pooling forecasts, corrections need to be made for the effects of decreasing grid areas towards the poles, either by sampling fewer grids at higher latitudes (Wilks, 2000; Wilks and Godfrey, 2002) or by weighting each grid by its area (Barnston et al., 2010; Barnston and Mason, 2011).

A more widely adopted approach is to calculate the frequency of hits only for the category with the highest probability (e.g. Livezey and Timofeyeva, 2008). If one of two categories with tied highest probabilities verifies, a half-hit is usually scored, or a third-hit if one of three categories with tied highest probabilities verifies. In some of the RCOFs a half-hit is scored if the middle category has the highest probability, and one of the outer categories verifies, but the probability for that category is higher than for the other extreme. Instead of redefining the score, and thus complicating its interpretation, a more detailed perspective of the resolution of the forecasts could be obtained by calculating the frequency of hits for the highest probability category, but also calculating how often the category with the second highest probability verifies, etc., through to how often the category with the lowest probability verifies.

A widespread criticism of seasonal forecasts is that the sharpness of the forecasts is low (or overconfident when sharpness is strong). A major problem with the frequency of hits is that it does not consider sharpness at all. However, once the probabilities themselves are considered it is difficult to avoid mixing measurement of resolution and reliability. As an alternative, measures of discrimination could be considered. Whereas resolution (in its more strict sense defined earlier) indicates whether the frequency of a category occurring increases or decreases with its forecast probability, so discrimination indicates whether the forecast probability increases and decreases as the category increases or decreases in frequency (see Chapter 2). As discussed below, although measures of discrimination are insensitive to sharpness, they do at least consider the rankings in the probabilities, but the analysis is not complicated by consideration of reliability. One other advantage of measuring discrimination instead of resolution for typical seasonal forecast formats is that it is easier to measure the conditional distribution of the forecasts on the observations than vice versa because there are usually only three possible outcomes (or very few) whereas there are many possible forecast probabilities. Given small sample sizes the sampling errors in the conditional distribution of the forecasts are therefore likely to be smaller than in the conditional distribution of the observations.

The relative operating characteristics (ROC) graph (Chapter 3), and the area beneath its curve, are widely used measures of discrimination in seasonal forecast verification, and these procedures are explicitly recommended in the SVSLRF and in the

CCl verification guidelines (Mason, 2011). As a measure only of discrimination, the area beneath the ROC curve is insensitive to some reliability errors (Kharin and Zwiers, 2003; Glahn, 2004), which may render it an inadequate summary measure of forecast quality, but is a distinct advantage when combined with measures of other attributes. Its insensitivity to the overconfidence that is commonly observed in seasonal forecasts makes the score useful for identifying skill, and the graphs can be helpful in more detailed diagnoses of forecasts at these timescales (Mason and Graham, 1999; Kharin and Zwiers, 2003).

The ROC area is calculated under the assumption of a two-category forecast system, and separate ROC areas can be calculated for each of the categories. Since most probabilistic seasonal forecasts have three or more categories, a generalized version of the ROC area may be a useful summary of the discriminatory power of the forecasts. This generalized discrimination score (Mason and Weigel, 2009) calculates the probability that given two observations the forecasts can successfully discriminate the observation in the higher category. For example, assuming that the predictand is rainfall, what is the probability of successfully discriminating the wetter of two observations? In the classical ROC, the test can be applied in a three-category system, for example, to calculate the probability that an above-normal observation could be successfully distinguished from an observation that was not above-normal, but a separate test would have to be conducted to distinguish normal and below-normal observations. A normal and below-normal observation would therefore be treated as indistinguishable when the ROC test is applied to above-normal events. In the generalized version of the test all the categories can be distinguished.

The generalized discrimination score, $D$, can be calculated as follows. Assume a forecast system with $m$ mutually exclusive and exhaustive categories (i.e., each observation has to be in one and only one of the categories). As mentioned, $m$ typically is 3, but regardless of how many, the categories are ranked from lowest values (category 1) to highest (category $m$). Each forecast is a vector of probabilities, $\mathbf{p}$, which consists of $m$ probabilities, one for each category, which must total to 1.0. Next assume that category $k$ verified $n_k$ times, and that the $i^{th}$ of

the $n_k$ vector of probabilities for when this category verified is given by $\mathbf{p}_{k,i}$. The generalized discrimination score can be defined as

$$D = \frac{\sum_{k=1}^{m-1} \sum_{l=k+1}^{m} \sum_{i=1}^{n_k} \sum_{j=1}^{n_l} I\left(\mathbf{p}_{k,i}, \mathbf{p}_{l,j}\right)}{\sum_{k=1}^{m-1} \sum_{l=k+1}^{m} n_k n_l} \qquad (11.1)$$

where

$$I\left(\mathbf{p}_{k,i}, \mathbf{p}_{l,j}\right) = \begin{cases} 0.0 & \text{if } F\left(\mathbf{p}_{k,i}, \mathbf{p}_{l,j}\right) < 0.5 \\ 0.5 & \text{if } F\left(\mathbf{p}_{k,i}, \mathbf{p}_{l,j}\right) = 0.5 \\ 1.0 & \text{if } F\left(\mathbf{p}_{k,i}, \mathbf{p}_{l,j}\right) > 0.5 \end{cases} \qquad (11.2)$$

and where

$$F\left(\mathbf{p}_{k,i}, \mathbf{p}_{l,j}\right) = \frac{\sum_{r=1}^{m-1} \sum_{s=r+1}^{m} p_{k,i}(r) p_{l,j}(s)}{1 - \sum_{r=1}^{m} p_{k,i}(r) p_{l,j}(r)} \qquad (11.3)$$

In Equation 11.3 $p_{k,i}(r)$ is the forecast probability for the $r^{th}$ category, and for the $i^{th}$ observation in category $k$.

It can be shown that when $m = 2$, Equation 11.1 reduces to the area beneath the ROC curve (Mason and Weigel, 2009). Similarly, if $m$ is set in Equations 11.1–11.3 to the number of observations (assuming there are no ties), and if the forecasts are deterministic (represented by $x$ with a subscript) then $D$ becomes

$$D = \frac{2 \sum_{k=1}^{n-1} \sum_{l=k+1}^{n} I(x_k, x_l)}{n(n-1)} \qquad (11.4)$$

which is related to Kendall's $\tau$ by

$$\tau = 2D - 1 \qquad (11.5)$$

(Mason and Weigel, 2009). These relationships are examples of how the generalized discrimination score is essentially equivalent to various so-called non-parametric statistical tests. For example, Kendall's $\tau$ was compared with Pearson's correlation in the previous section, where it was explained

**Table 11.2** Example of seasonal forecasts and corresponding observations (A indicates above-normal, N indicates normal, and B indicates below-normal), with cumulative profits and losses based on an initial investment of $100. The interest earned or lost each year is shown in the last column, together with the effective rate of interest over the ten years

| Year | Observed | Forecasts | | | Profit | Interest |
|------|----------|-----|-----|-----|--------|----------|
| | | B | N | A | | |
| | | | | | $100.00 | |
| 2001 | A | 20% | 50% | 30% | $90.00 | −10.0% |
| 2002 | A | 20% | 55% | 25% | $67.50 | −25.0% |
| 2003 | A | 25% | 35% | 40% | $81.00 | 20.0% |
| 2004 | B | 15% | 30% | 55% | $36.45 | −55.0% |
| 2005 | N | 45% | 35% | 20% | $38.27 | 5.0% |
| 2006 | A | 20% | 50% | 30% | $34.45 | −10.0% |
| 2007 | N | 35% | 40% | 25% | $41.33 | 20.0% |
| 2008 | A | 20% | 50% | 30% | $37.20 | −10.0% |
| 2009 | A | 25% | 35% | 40% | $44.64 | 20.0% |
| 2010 | B | 40% | 35% | 25% | $53.57 | 20.0% |
| | | | | | Effective | −6.1% |

that Kendall's $\tau$ is a correlation based on the ranked values. Similarly, the area beneath the ROC curve is equivalent to the Mann–Whitney $U$-statistic (Mason and Graham, 2002), which is a non-parametric version of the more widely used Student's $t$-test for comparing central tendencies. When applied to forecasts, the $U$-test assesses whether there is any difference in the forecasts when an event occurs compared to when the event does not occur (and, thus, whether the forecasts can discriminate between events and non-events). More specifically, it indicates whether the forecast (whether probability or value) was higher, on average, when an event occurred compared to when not.

Thus, by using the generalized discrimination score, a consistent test can be applied to measure whether forecasts of virtually any format have skill in the sense defined earlier: do the observations increase, whether in value or in frequency, as the forecast value or probability increases, without specifying by how much the increases and decreases should be? This measure is useful in low-skill settings, where it may be acknowledged upfront that the forecasts themselves may be poorly calibrated, whether overconfident or biased. It is helpful, and more informative, to measure the quality of the calibration separately.

To illustrate the importance of considering calibration separately when testing whether forecasts may be potentially useful, consider a fictional set of ten probabilistic forecasts and corresponding observed categories as shown in Table 11.2. The ranked probability skill score (RPSS) for these forecasts is marginally negative (approximately −0.005); similarly the Brier skill scores for all three categories are negative. These results suggest the forecasts are effectively useless. However, given that the category with the highest probability occurs four times, while that with the lowest probability occurs only once, it seems reasonable to assume that the forecasts may have some useful information. The RPSS and Brier scores are negatively impacted by what appears to be hedging on the normal category. Acknowledging that the probabilities are poorly calibrated, but that increases and decreases in probabilities may be meaningful, the generalized discrimination score can be used to indicate whether the forecasts may be potentially useful.

Instead of comparing each forecast with its corresponding observation, as is typical of most verification scores, Equation 11.1 is calculated by comparing each year with all other years that have different observations. For example, starting with years with below-normal rainfall ($k = 1$), the first year available ($i = 1$) is 2004. This year is compared to all the years with normal rainfall ($l = 2$), the first of which is 2005. Given that the observations differ, Equation 11.2 indicates whether the forecast for 2005 successfully indicated that 2005 was likely to be the wetter of the two years. The answer to this question is based on the probability that a value randomly drawn from the distribution represented by the forecast for 2005 will exceed one randomly drawn from that represented by the forecast for 2004, conditioned upon the two values differing (Equation 11.3). For 2004 and 2005, Equation 11.3 gives approximately 0.20. Because this value is less than 0.5, the forecasts fail to discriminate the year with the higher rainfall category (Equation 11.2). Proceeding to the next year with normal rainfall ($l = 2$), 2004 is compared with 2007. For these two years Equation 11.3 gives 0.25, and so again the discrimination is incorrect. Since there are no more years with normal rainfall ($n_2 = 2$), 2004 is then compared with all the years with above-normal rainfall ($l = 3$). This procedure is then repeated for 2010 ($i = 2$), and then the years with normal rainfall ($k = 2$) are compared with the years with above-normal rainfall ($l = 3$). For the example, $D \approx 0.68$, indicating that the forecasts discriminated the observed categories with a success rate of about 68%, and suggesting that the forecasts may be potentially useful.

### Skill of hindcasts

It has long been recognized that in-sample estimates of performance provide overestimates of operational performance (Allen, 1974; Davis, 1976; Rencher and Pun, 1980; Wilkinson and Dallal, 1981). The need for out-of-sample estimates of skill is more of an issue with statistical models than it is with dynamical models (except in the context of skill-weighted multi-model combinations, which is essentially a statistical procedure anyway) because statistical models generally can be more easily tuned to compare favourably with the verification data. However, because dynamical model parameterizations are generally tuned to optimize performance over a verification period, independent verification is still required.

Cross-validation (Section 1.4.2) is the most commonly used method of trying to obtain independent estimates of predictive skill (Michaelsen, 1987). In the atmospheric sciences the most common approach to cross-validation is to predict each observation once, omitting that observation, and possibly some adjacent observations, to reconstruct the model. The re-specification of the model at each cross-validation step should involve not only recalculating the model parameters, but also reselecting the predictors to be included (Elsner and Schmertmann, 1994). In any predictor selection procedure there is a danger of selecting additional spurious predictors or the wrong predictors entirely. As the candidate pool of predictors is enlarged, the danger of choosing spurious predictors increases, and thus the probability of the hindcast skill estimates overestimating those of the operational skill also increases (Barnett et al., 1981; Katz, 1988; Brown and Katz, 1991). The same is true of cross-validated skill estimates if the procedure is not implemented carefully. The problem can be reduced by leaving more than one observation out at each step, but there are few guidelines as to how many observations should be omitted. Xu and Liang (2001) recommend omitting as much as 40–60% of the observations, and even more if the candidate pool of predictors is large. This proportion may be impractical given the small sample sizes available for seasonal forecasting, but the clear message is that, given the vast pool of candidate predictors many modellers consider, the risk of overestimating operational performance is high.

An alternative to cross-validation is the verification of retroactive forecasts (Mason and Baddour, 2008). Retroactive forecasts are generated by withholding the later part of a data set, selecting and parameterizing the model on the first part of the data, and then predicting the subsequent values, possibly repeating the model construction process as observations from the second part are predicted. This process attempts to reproduce the forecasts that would have been made operationally given access to current data sets and models (Mason and Mimmack, 2002; van den Dool et al., 2003). There are, however, two sources of bias. Firstly, even if implemented properly, the procedure is likely to underestimate operational performance because the model should improve gradually over time as more data become available. A more serious source of bias, however, occurs because it is virtually impossible to avoid including predictors based on knowledge of their association with the predictand over the full sample period. Since some of these predictors may be spurious, it is essential that there is a strong theoretical base to their selection prior to producing any hindcasts.

In conclusion, all hindcasting procedures will unavoidably have some biases in their estimates of operational forecast skill. While there are sources of both positive and negative bias, the positive biases are likely to outweigh the negative given how hindcasts are most frequently calculated. One specific recommendation is that leave-one-out cross-validation should almost always be avoided even if there are no problems with temporal autocorrelation. Further research is required to make more specific recommendations about how many years to omit in a cross-validation procedure, but considerably larger numbers than those most frequently used almost certainly need to be considered, especially when the candidate pool of predictors is large. Retroactive skill estimates are normally to be preferred to cross-validated estimates because they have fewer of the problems outlined above (Jonathan *et al.*, 2000). They are not calculated as often as cross-validated skill estimates because of limited sample sizes, but retroactive verification is worth attempting even if only 5 or 10 years are predicted (Landman *et al.*, 2001; Landman and Goddard, 2002; Shongwe *et al.*, 2006), and even very wide uncertainty estimates on verification scores can be useful information.

## 11.3.2 Other attributes

If skill is to be defined in terms of a single attribute (discrimination, or possibly resolution), as proposed in Section 11.3.1, it is essential to measure additional attributes subsequent to concluding that the forecasts may be at least worth considering. The measurement of accuracy and reliability associated with the central tendency of the ensemble and of its distribution can be addressed using procedures detailed in Chapters 7 and 9. For probabilistic procedures, Chapter 8 provides extensive coverage of options for diagnosing over- and under-confidence,

and unconditional biases. Attributes or reliability diagrams are particularly useful in this regard, although forecasts will inevitably have to be pooled over large areas and possibly different seasons to allow for sufficiently large sample sizes.

One aspect of seasonal forecasts that is of interest and is partly a reflection of limited sample size, and partly of longer-term variability, is the degree to which the seasonal forecasts over a limited period of perhaps a few years have indicated the extent to which the observed climate over this period has differed from that of the reference climatological period. For example, if the forecasts have successfully indicated that the verification period would be generally dry, some skill should be acknowledged, but this may not be identified using some of the procedures described above. In areas of significant decadal variability or with long-term trends, for example, the discrimination skill may have been poor because of an inability to distinguish which years are drier than others when all or most of the years are dry. Measurement of the unconditional bias in the forecasts is appropriate in this regard, and procedures for measuring the bias of deterministic forecasts are described in Chapter 5. For probabilistic forecasts, tendency diagrams provide a simple visual indication of any unconditional bias (Mason, 2011). These diagrams compare the average forecast probabilities for each category with their observed relative frequencies; if the forecasts had been reliable, one would expect the observed relative frequencies to be approximately equal to the average probabilities. In the example provided in Figure 11.4 based on the data in Table 11.2, it is evident that above-normal rainfall occurred much more frequently than the other categories, but the forecasts implied that the normal category would be most frequent (perhaps a reflection of a tendency to hedge).

One attribute that is of common interest subsequent to the demonstration of at least some skill, is the potential economic value of forecasts. Appropriate measures are discussed in Chapter 9, but one that is particularly well suited to the standard probabilistic format of seasonal forecasts is the effective interest rate (Hagedorn and Smith, 2009; Tippett and Barnston, 2008), which in turn is based on the ignorance score (Roulston and Smith, 2002; Benedetti, 2010). The ignorance score, *Ign*, can be

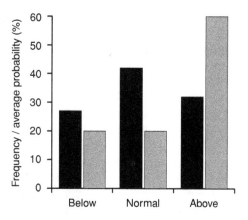

**Figure 11.4** Example tendency diagram for the data from Table 11.2. The black bars show the average forecast probabilities for each category, and the grey bars show the observed relative frequencies for each category

transformed to the effective interest rate using

$$\text{effective interest rate} = \left(2^{Ign(ref)-Ign} - 1\right) \quad (11.6)$$

where $Ign(ref)$ is the ignorance score for the reference (base-rate forecasts). The effective interest rate provides an indication of the average returns an investor would make if (s)he invested on the forecasts, and received fair odds against the climatological probabilities. For example, given three equiprobable categories, the returns on the verifying category would be three times the investment. The investor will then make a profit whenever the forecast probability on the verifying outcome exceeds the base rate. For the data from Table 11.2, the effective interest rate is about −6% per year, suggesting that the forecasts are not useful.

Equation 11.6 is only valid if the forecasts are for a single location and if all the forecasts are for discrete periods (e.g. a specific 3-month season over a number of years) since it assumes that earnings (and losses) are carried over from forecast to forecast. If some of the forecasts are for different locations or for overlapping periods (or, more specifically, if any of the target periods expire after any of the release dates for subsequent forecasts), then the initial investment has to be divided between each of the $s$ locations and periods, and the effective interest rate has to be averaged using the ignorance score for

each instance:

$$\text{average effective interest rate}$$
$$= \frac{1}{s} \sum_{k=1}^{s} \left(2^{Ign(ref)-Ign_k} - 1\right) \quad (11.7)$$

where $Ign_k$ is the ignorance score for the $k^{\text{th}}$ location/season. For the data in Table 11.2, the average interest would have been −2.5% if independent investments had been made on each forecast. However, as discussed in Section 11.3.1, the forecasts do have good discrimination and so could potentially be useful if they could be calibrated reliably.

Even with very good forecasts, the investor could occasionally make a loss because categories with probabilities lower than the base-rate should verify sometimes (otherwise they would be unreliable). However, in the long run, if the forecasts are good, the gains will exceed the losses, and the effective interest rate will be greater than zero. Given that the returns on the investments each time are a direct function of the forecast probability, in order for the effective interest rate to be positive the reliability of the forecasts is important, and the forecasts therefore must have skill higher than the minimum requirement as defined in Section 11.3.1. A plot of gains and losses over time provides a useful graphical illustration of potential forecast value. Such a graph can be constructed by plotting

$$\left(\prod_i \left(\frac{1}{s} \sum_{k=1}^{s} \frac{p_{k,i}}{c_{k,i}}\right)\right) - 1 \quad (11.8)$$

on the $y$-axis against time, $i$, on the $x$-axis, where $s$ is the number of locations/seasons, $p_{k,i}$ is the forecast probability for the verifying category at location/in season $k$, and $c_i$ is the corresponding base rate. An example is provided in Figure 11.5, with corresponding data in Table 11.2, using the same forecasts and observations as for the generalized discrimination score example.

### 11.3.3 Statistical significance and uncertainty estimates

Regardless of whether it is the skill of operational forecasts or of hindcasts that is being estimated,

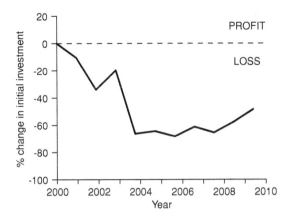

**Figure 11.5** Cumulative profits and losses diagram based on data from Table 11.2

whether a model revision improves the forecast skill compared to an earlier version it is generally impractical to demonstrate a significant improvement because the uncertainty estimates on the skill levels are so large due to limited sample sizes, and so any improvement in skill may be acceptable. However, when considering how to weight different models in some form of skill-based multi-model average the need to demonstrate robust differences in model skill levels becomes more important, otherwise the unequally weighted model average will reflect sampling errors in differences in model skill, and will therefore likely perform less well than an equally weighted average (Kang and Yoo, 2006; Weigel *et al.*, 2010).

some indication of whether the measured skill provides a basis for concluding that the forecasts are good or bad is required (Section 1.4.3). The preferred approach has been to calculate statistical significance, or *p*-values, which indicate the probability that a result at least as good as that measured could have been achieved by chance. Some of the problems in interpreting *p*-values when sample sizes are large (Mason, 2008) are rarely an issue for seasonal and longer-range forecasts, but other problems of interpretation remain (Nicholls, 2001; Jolliffe, 2004, 2007). Confidence intervals remain underutilized. Statistical procedures for calculating *p*-values and confidence intervals for the measures discussed above are described in other chapters, and so only a few comments are included here that pertain specifically to seasonal forecasts.

When skill is calculated for specific seasons and locations, temporal autocorrelation in the data is often not a major problem except in the presence of trends, and so confidence intervals and *p*-values can often be calculated using distributional assumptions (Jolliffe, 2007) when available. However, when data are pooled from different locations, or if field significance is being assessed (Livezey and Chen, 1983; Wilks, 2006a), then spatial correlation has to be accounted for. Similarly, if data from overlapping or adjacent seasons are being pooled, temporal correlation can affect the results, and block bootstrapping may be required (Barnston and Mason, 2011).

Statistical significance for differences in model skill levels is rarely calculated. When considering

## 11.4 Measuring the quality of individual forecasts

It is a perfectly reasonable question to ask whether the forecast for a specific season was good or bad even if the forecasts are probabilistic. Mathematically, most of the probabilistic verification scores discussed in this book could be calculated using forecasts representing different locations rather than different times. However, many such calculations would involve incorrect interpretations of the forecasts. For example, consider a set of forecasts for ten locations all for the same season, and all of which indicate an 80% probability of above-median rainfall. If above-median rainfall occurred at 60% of the stations rather than at 80%, one cannot necessarily conclude that the forecasts were overconfident: the forecasts were not stating that 80% of the area would be wet, only that at each station on 80% of the occasions on which an 80% probability of above-median rainfall is issued can above-median rainfall be expected to occur. Any attempt to measure the reliability of the forecast probabilities by considering the forecasts at different locations for an individual season represents an incorrect interpretation of the forecast.

The primary reason why the reliability of the forecasts cannot be measured by verifying spatially rather than temporally is that in most practical settings forecasts for different locations will not be independent. In effect, there is a sample size problem: because of strong spatial correlation there are very few independent realizations in a forecast for any individual season.

Unfortunately, although the generalized discrimination score may be useful for a relatively short series of forecasts, there may be problems of interpretation when it is used for verifying individual forecasts, especially when the spatial domain is small. When a discrimination score is applied spatially, it indicates whether the forecast correctly discriminated wet from dry areas, not whether the forecast gave a good indication of whether the specific season would be unusually wet or dry. The forecast may successfully have indicated a high likelihood of unusually wet or dry conditions over the entire domain, but this information is ignored because of the score's insensitivity to calibration, To answer this second question, the specific contribution that the forecast in question would make to the generalized discrimination score could be calculated: the probability that the specific season of interest would have been correctly identified as wetter or drier (or warmer or colder, or whatever) than each other season for which forecasts are available, could be calculated. [To calculate this probability, set $k$ or $l$ in Equation 11.1 to the verifying category, and then $n_k$ or $n_l$ to 1.]

The frequency of hits is widely calculated as a verification measure for individual forecasts. The score indicates the proportion of the area in which the verifying categories had the highest probability, but, as discussed in Section 11.3.1, the score is more informative when scores for the second highest probability category, etc., are calculated. The frequency of hits for the various probability rankings still ignore much of the information in the probabilities, and are unable to credit sharp probabilities. The linear probability score (Wilson *et al.*, 1999) and the average interest rate (Equation 11.7) are worth considering, despite the fact that both scores lack propriety (Bröcker and Smith, 2007b). This lack of propriety is not necessarily a problem if no attempt is made to optimize these values or conclude naively that one forecast is better than another simply because the score is higher. Scores for individual years are generally calculated to tell us something about the temporal variability of predictability (Livezey and Timofeyeva, 2008), just as in Figure 11.4, for example.

Instead of calculating scores for individual seasons, much more can be discerned from a detailed diagnostic of dynamical model outputs (Jakob, 2010). By diagnosing the model's atmospheric structure, useful insights into its strengths and weaknesses can be derived (e.g., Lyon and Mason, 2009).

## 11.5 Decadal and longer-range forecast verification

For forecasts at decadal and longer timescales there are at best too few realizations to perform any meaningful significance testing on the kind of scores described above. Hindcasting is not a realistic option to expand the sample size of decadal forecasts because of the lack of subsurface ocean observations required to initialize the models (Smith *et al.*, 2008), and unpredictable events such as major volcanic eruptions add an important noise component to the observed climate that exacerbates the sampling problem. Traditional verification analyses that compare a set of forecasts with the corresponding observations may therefore not be a viable option. However, there are some evaluations that can be usefully performed that may not directly answer the most immediate questions of interest regarding forecast quality at these long timescales, but do at least provide some information that may help in deciding whether the forecasts are worth considering (Fildes and Kourentzes, 2011).

A common starting point in place of rigorous verification analyses is some measurement of consistency in predictions. Perhaps the simplest such measure that is widely used for climate change projections is the proportion of models agreeing upon the sign of the anomaly in the target variable (Whetton *et al.*, 2007; Hawkins and Sutton, 2009). It is usually assumed that if this proportion is close to 50% then there is little agreement between the models, and so confidence in any prediction should be low. However, this procedure rests upon the rather unreasonable assumption that the models are independent, and upon the only sometimes reasonable assumption that the underlying data are symmetrically distributed. Further, in the situation that models are closely agreed upon minimal change, the level of agreement in the sign of the anomaly may be very low, and some measure of spread in the predictions would be more informative.

Following a similar principle of consistency in predictions, but involving more sophisticated diagnostics, so-called 'perfect model' experiments test how well the model is able to predict one of its own ensemble members (typically as measured by the root mean squared error) as additional data are assimilated into the model (Dunstone and Smith, 2010; Meehl *et al.*, 2010). If the model is able to predict its own behaviour more successfully as more data thought to be relevant to predictability are assimilated then there is some basis for suspecting that the assimilated data indicate a process of variability in the real world that the model may be able to predict. This assumption is, of course, problematic (Stainforth *et al.*, 2007), but there seems little point in verifying a model that cannot even predict itself, and in the absence of any verification results against real-world observations, such improvements in the signal-to-noise ratio provide some grounds for credibility. However, ensemble-member and inter-model consistency should, at best, be considered a very weak form of validation, and at least some attempt at comparison between model outputs and observed data should be made (Fildes and Kourentzes, 2011).

A starting point of any verification procedure should be to evaluate the accuracy with which the model's climatology matches that of the observations (Caminade and Terray, 2010; Gent *et al.*, 2010). If possible, this assessment should be performed over a number of climatological periods to test for robustness of results, especially if the skill levels of models are being compared (Macadam *et al.*, 2010). The mean squared error and its decomposition into conditional and unconditional biases (Chapter 5) can be used, although the interpretation of results given non-normally distributed data can be complicated, and the calculation of absolute errors may be more appropriate (Section 5.3). A variety of other statistics have been proposed, all generally based on mean squared or absolute errors, and differences are largely a matter of scaling, and sensitivity to extremes (Watterson, 1996). Regardless of the measure, if a model is not reproducing the observed climate realistically, there is no compelling reason to assume that simulated variability and change in its climate will match that of the

real world. This assertion is certainly borne out at the seasonal timescale, where skill is a function of models' unconditional biases (DelSole and Shukla, 2010; Lee *et al.*, 2010), but it is not necessarily the case that an accurate model climatology implies forecast skill (Knutti *et al.*, 2010).

Even though the number of realizations may be trivially small, it is still worth calculating verification scores with whatever data are available. While it may be impossible to demonstrate statistically significant skill, the extent to which the models improve their simulation of the observed large-scale climate variability as improved data sets are assimilated, for example, reinforces the belief that the models may be able to make useful predictions (Doblas-Reyes *et al.*, 2006; Keenlyside *et al.*, 2008; Mochizuki *et al.*, 2010). Some account may need to be taken for the loss of skill resulting from the unpredictability of major volcanic eruptions and their effects on climate. Selecting start dates that avoid periods with major eruptions (Troccoli and Palmer, 2007) may be useful for model validation, but gives a biased estimate of operational forecast skill.

While there is no guarantee that models that produce skilful forecasts at one timescale will be skilful at other timescales, verification information for timescales for which more data are available can be informative. For example, predictability at decadal timescales is premised partly on the ability to predict sea-surface temperatures from subsurface conditions (Meehl *et al.*, 2009), and so skill at the seasonal scale, which depends largely on ocean-atmosphere coupling, may provide some indication of skill at decadal scales (Palmer *et al.*, 2008; Caminade and Terray, 2010). Of course, there may be other sources of predictability at decadal scales such as 'committed climate change' and future greenhouse gas emissions (Meehl *et al.*, 2009), the effects of recent volcanic eruptions (Troccoli and Palmer, 2007), land-surface feedbacks and cryospheric effects, and so more detailed diagnostics are therefore usually to be recommended (Scaife *et al.*, 2009), including investigations into the processes of climate variability (Giannini, 2010). Even where there is no discrimination or resolution skill at seasonal timescales, information about the reliability of the ensemble spread provides some basis for assessing the reliability of the spread at longer timescales (Palmer *et al.*, 2009).

## 11.6 Summary

Although many of the procedures used in seasonal forecast verification are similar to those used at shorter timescales, problems of limited sample size and low levels of predictability are invariably major factors in verification analyses at these and longer timescales. Both limitations contribute to a strong focus on measuring 'skill', although conclusions can be misleading if skill is not precisely defined. It has been argued in this chapter that widely used definitions of 'skill' for seasonal forecasts are unduly strict, and that some commonly used verification measures may therefore not be the most appropriate ones to use. For deterministic forecasts, for example, skill can be defined as increases and decreases in the observed values as the forecasts increase and decrease. This definition points to a measure of association based on the ranks of the forecasts. Similarly, for probabilistic forecasts, skill can be defined as increases and decreases in the frequency of events or a verifying category as the probability increases and decreases. This definition can be measured either by resolution or by discrimination, although the latter is usually easier to measure when sample sizes are small. In either case, it is helpful, and more informative, to consider the measurement of reliability as a separate verification question. The generalized discrimination score is proposed since it can be applied to an extensive range of forecast and verification data formats, and provides a useful indication of whether there is any potentially useful information in the forecasts. Separate tests for conditional and unconditional biases, and other reliability checks should be applied subsequently.

Partly because of the infrequency with which seasonal and longer-ranger forecasts verify, there is widespread interest in whether a specific forecast was good or bad. When forecasts are expressed probabilistically this question becomes complicated because attributes such as resolution, discrimination and reliability can change their meaning, and may become inappropriate. Much of the difficulty arises from the fact that the number of spatially independent forecasts is likely to be very low, and so these attributes cannot be measured meaningfully. However, some measures can be informative, including ones that are not strictly proper, as long as they are interpreted appropriately and their limitations recognized.

At longer timescales the sample sizes can become so small that no meaningful verification results can be realized. However, even in these cases measures of adequacy of model climate, and of forecast and/or model consistency can be helpful. It is also possible to use verification results for shorter timescales to provide some indication of credibility.

# 12

# Epilogue: new directions in forecast verification

**Ian T. Jolliffe and David B. Stephenson**

*Mathematics Research Institute, University of Exeter*

## 12.1 Introduction

The preceding chapters present a contemporary review of the wide range of forecast verification methods that have been developed and are currently being employed in weather and climate forecasting centres around the world. Since the burst of activity caused by Finley's forecasts in the 1880s, forecast verification has been undergoing exciting new developments with ever more measures/scores and techniques continually being invented (and reinvented!). The increasing amounts of weather and climate forecast products and verification data will inevitably continue to drive the demand for more verification. Forecasters will continue to ask 'How can the forecasting system be improved?'; users will continue to ask 'How useful are these forecast products?'; and administrators will surely continue to ask 'Has the forecasting performance of our institution improved?' In addition to these driving forces, the abundance of and increased reliance on forecasts in other disciplines provides exciting opportunities for innovative cross-disciplinary future work in forecast verification.

This final chapter will highlight some of the most important key concepts that have been introduced in previous chapters, will briefly discuss some of the different methods developed in other disciplines, and will finally outline which topics within verification are currently of greatest interest.

## 12.2 Review of key concepts

What is forecast verification? A suitably general definition might be that *forecast verification is the exploration and assessment of the quality of a forecasting system based on a sample or samples of previous forecasts and corresponding observations*. But what is meant by 'quality'? Murphy (1997) explained that forecast quality has many different attributes that can be estimated using a wide variety of different sample statistics. Despite the obvious appeal, it is clear that no unique score such as mean squared error (MSE) can fully summarize the joint probability distribution between pairs of previous matched observations and the respective forecasts. Forecast verification is therefore a multidimensional

---

*Forecast Verification: A Practitioner's Guide in Atmospheric Science*, Second Edition. Edited by Ian T. Jolliffe and David B. Stephenson.
© 2012 John Wiley & Sons, Ltd. Published 2012 by John Wiley & Sons, Ltd.

problem with many possible scores/measures. It is this richness (or curse of dimensionality!) that makes the subject so perplexing yet so fascinating. At risk of going into an infinite regress, metaverification screening measures have even been invented, such as *propriety*, *equitability* and *consistency*, for scoring the quality of verification scores (Murphy, 1997).

At first sight, it is easy to become completely bewildered by the multitude of possible verification scores and the associated philosophical issues in forecast verification. It is therefore helpful to go back to basics and reconsider the fundamental reason why we make forecasts: *to reduce our uncertainty about the (unknown) future state of a system.* Since the dawn of civilization, humankind has attempted to cope with the uncertain knowledge about what will happen in the future by searching for clues in the past and present that help to reduce the range of possible outcomes in the future. By doing so, it is hoped that the uncertainty about the future will be reduced. This concept of reducing uncertainty about the future by *conditioning* on existing clues (forecasts) is the key to understanding the quality of forecasts. It is the dependency between observations $x$ and forecasts $\hat{x}$ that makes forecasts useful – forecasts that are completely independent of observations are of absolutely no use in predicting the future. The dependency is best quantified by considering the *conditional probability* of the observations given the forecasts $p(x|\hat{x}) = p(x, \hat{x})/p(\hat{x})$. This conditional probability is the *calibration* factor in the *calibration-refinement factorization* of the joint probability introduced by Murphy and Winkler (1987). The *conditional uncertainty* can be measured most easily by the variance, $\text{var}(X|\hat{X})$, of observations conditioned/stratified on a particular given value of forecast – the variance of the observations in the subset/class of previous cases in which the issued forecast was exactly equal to $\hat{X}$. A simple and revealing identity can be derived (see DeGroot, 1986, section 4.7) relating the mean conditional variance to the total unconditional variance of the observations:

$$E_{\hat{X}}[\text{var}_X(X|\hat{X})] = \text{var}(X) - \text{var}_{\hat{X}}[E_X(X|\hat{X})]$$
(12.1)

The mean of the variance of the observations given forecast information, $E_{\hat{X}}[\text{var}_X(X|\hat{X})]$, is equal to the unconditional variance var$(X)$ of the observations (total uncertainty) minus the variance of the conditional means $\text{var}_{\hat{X}}[E_X(X|\hat{X})]$ of the observations given the forecasts. In other words, the mean uncertainty given forecast information is equal to the *uncertainty* of the observations minus the *resolution* of the forecasting system, as defined by Murphy and Winkler (1987). So for a forecasting system to reduce uncertainty it must have non-zero resolution (see Chapters 2 and 7 for more discussion about the importance of resolution). This is why resolution is one of the most important aspects in forecast verification. For the linear regression model of observations on forecasts (see Section 2.9), the resolution divided by the variance of the observations is $r^2$, the fraction of variance explained, and so the variance given the forecasts is reduced by a factor of $1 - r^2$ compared to the original variance of the observations. Similar conditioning arguments apply to the case of probability forecasts with the sole difference being that the conditioning variable (the forecast probability) is now a real number from 0 to 1 (or a vector of $K - 1$ probabilities for $K > 2$ categories) rather than being the same type of variable as the observation.

The identity in Equation 12.1 also helps to explain why the MSE score (or Brier score for probability forecasts) and its partitioning are useful quantities to consider. By conditioning on the forecasts, the MSE can be decomposed into the sum of a mean variance component $E_{\hat{X}}[\text{var}_X(X|\hat{X})]$ and a conditional bias (reliability) component $E_{\hat{X}}[(E_X(X|\hat{X}) - \hat{X})^2]$ as follows:

$$E[(X - \hat{X})^2] = E_{\hat{X}}[E_X\{(X - \hat{X})^2|\hat{X}\}]$$
$$= E_{\hat{X}}[\text{var}(X|\hat{X})] + E_{\hat{X}}[(E_X(X|\hat{X}) - \hat{X})^2]$$
(12.2)

Substituting for the mean variance component using the identity in Equation 12.1 then gives the calibration-refinement decomposition of MSE:

$$E[(X - \hat{X})^2] = \text{var}(X) - \text{var}_{\hat{X}}[E_X(X|\hat{X})]$$
$$+ E_{\hat{X}}[(E_X(X|\hat{X}) - \hat{X})^2]$$
(12.3)

In other words, the MSE is the sum of an *uncertainty* term, a negated *resolution* term and a *reliability* term. The first two terms are exactly the same as those that appear in identity Equation 12.1 and so can be used to describe how much variance is explained by conditioning on the forecasts. The third term, reliability, is of less importance since it can be reduced to very small values by recalibrating (bias correcting) the forecasts, $\hat{X}' = E_{\hat{X}}(X|\hat{X})$, using a sufficiently large sample of previous forecasts and the assumptions of stationarity in the observations and the forecasting system. It should be noted that simplified presentation of probability forecasts using fewer bins can degrade the resolution and overall skill of the forecasts (Doblas-Reyes *et al.*, 2008).

By conditioning instead on the observations rather than the forecasts, the MSE can also be expressed as a sum of a sharpness term var($\hat{X}$), a discrimination term $\text{var}_X[E_{\hat{X}}(\hat{X}|X)]$, and a reliability (of the second kind) term $E_X[(E_{\hat{X}}(\hat{X}|X) - X)^2]$. However, the calibration-refinement decomposition given in Equation 12.3 is more natural for interpreting the reduction of uncertainty obtained by conditioning on forecast information. These additive properties of the quadratic MSE score make it particularly appealing for interpreting the reduction in uncertainty. However, it should be noted that this discussion has focused on only first and second moment quantities (means and variances) and so does not necessarily capture all the possible features in the joint distribution of forecasts and observations (Murphy, 1997). The relevance of first- and second-order quantities is especially debatable for the verification of forecasts of non-normally distributed meteorological quantities such as precipitation and wind speed. Furthermore, care also needs to be exercised in practice since sample estimates of the MSE are unduly sensitive to outlier forecast errors (non-resistant statistic) as is discussed in more detail in Sections 12.3.2 and 12.4.

## 12.3 Forecast evaluation in other disciplines

It is perhaps not surprising that meteorologists faced with the job of making regular daily forecasts over the last century have invented and applied many different verification approaches. However, meteorologists are not the only people who have to make forecasts or who have considered the forecast ver-

ification problem. This section will briefly review some developments in other fields and provide some references to literature in these areas. It is hoped that this will stimulate a more progressive cross-disciplinary approach to forecast verification. The disciplines discussed in this section are certainly not the only ones in which forecasts are made and evaluated. For example, Krzanowski and Hand (2009) note the use of ROC curves in machine learning, biosciences, experimental psychology and sociology in addition to those disciplines discussed here. Many of these different disciplines have their own terminology. For example, in social sciences a common measure of association for $K \times K$ contingency tables is Cohen's kappa (Cohen, 1960). For $2 \times 2$ tables this is identical to the Heidke skill score. Less specifically, Campbell and Diebold (2005) note that what meteorologists call skill scores are referred to as $U$-statistics in the econometric literature.

### 12.3.1 Statistics

Forecast verification is inherently a statistical problem – it involves exploring, describing and making inferences about data sets containing matched pairs of previous forecasts and observations. Nevertheless, much of the verification work in atmospheric sciences has been done by meteorologists without the collaboration of statisticians. Notable exceptions have produced major breakthroughs in the subject – for example, the papers mentioned throughout this book by Allan Murphy and the decision-theorist Robert Winkler.

Although statisticians have been heavily involved in time series forecasting, verification in these studies has often not gone beyond applying simple measures such as MSE – see Chatfield (2001) for a clearly written review. Chatfield (2001) classifies forecasts of continuous real valued variables into *point forecasts* (deterministic forecasts), *interval forecasts* (prediction intervals/limits in which the observation is likely to occur with a high specified probability), and *density forecasts* (the whole probability density function). Interval forecasts are to be preferred to point forecasts since they provide an estimate of the likely uncertainty in the prediction whereas point forecasts do not supply this information. Rather than consider deterministic forecasts as perfectly sharp probability forecasts in which the uncertainty is zero, it is more realistic to consider point forecasts as forecasts in which the uncertainty is unknown. In other words, deterministic point forecasts are not perfectly sharp forecasts with probabilities equal to 1 and 0, but instead they should be assigned unknown sharpness. As reflected in the contents of the previous chapters, the intermediate case of interval forecasts has been largely overlooked by meteorologists and would benefit from some attention in the future. One exception is the article by Murphy and Winkler (1974) that discussed the performance of *credible interval* (Bayesian prediction interval) temperature forecasts. Interval forecasts for normally distributed variables can be most easily constructed from point forecasts by using the MSE of previous forecasts to estimate the interval width (Chatfield 2001). A problem with many interval and probability forecasts is that they often assume that the identified forecast model is perfect, and therefore only take into account *parametric uncertainty* (sampling errors on estimated model parameters) rather than also including estimates of *structural model uncertainty* (i.e. the model formulation may be incorrect). A clear discussion of how to take account of model uncertainty in forecasts can be found in Draper (1995) and the ensuing discussion. The main emphasis in the statistical literature when dealing with forecast uncertainty is generally not to attempt to verify the forecasts, but to provide an envelope of uncertainty around the forecast, which incorporates parameter uncertainty, random error and possibly model uncertainty.

Several interesting articles have been published in the statistical literature that address the verification/evaluation of (weather) forecasts. Gringorten (1951) presents the principles and methods behind the evaluation of skill and value in weather forecasts, and concludes by explaining his motivation for publishing in a statistical journal (universality of the verification problem). Slonim (1958) describes the trentile deviation scoring method for deterministic forecasts of continuous variables – a crude forerunner of the LEPS approach (Potts *et al.*, 1996). Goodman and Kruskal (1959) review measures of association for cross-classification and provide a fascinating history and perspective on the $2 \times 2$ canonical verification problem raised by the Finley affair.

Not surprisingly, probability forecasts have also received attention from statisticians. Inspired by questions raised by the meteorologist Edward S. Epstein, Winkler (1969) wrote an article on likelihood-based scoring rules for probability assessors (subjective probability forecasts). He points out that under certain conditions the log-likelihood score becomes equivalent to Shannon's measure of entropy/information. This important fact was also mentioned in Stephenson (2000) for the case of binary forecasts. Winkler published several other single author articles around this time on how best to evaluate subjective probability assessors. More recently, Jose *et al.* (2008) have shown further links between measures of entropy and weighted forms of pseudospherical (Section 7.3.2) and power scoring rules.

In a pivotal article published in *Applied Statistics*, Murphy and Winkler (1977) demonstrated that subjective probability forecasts of US precipitation and temperature were well calibrated. This meteorological example of the calibration of sequential probability forecasts stimulated much future work on Bayesian calibration by Dawid and others. By considering sequential probability forecasts, Dawid (1982) demonstrated that recalibration of sequential probability forecasts leads to incoherent (not the price of a fair bet) probability statements. In a profound, yet difficult to read, paper DeGroot and Fienberg (1983) address this problem and discuss proper scoring rules for comparing and evaluating pairs of probability forecasters. They explain how the concepts of calibration and refinement relate to the concept of *sufficiency* that underpins much of statistical inference. In their review of probability forecasting in meteorology, Murphy and Winkler (1984) present an interesting and complete historical account of the subject. Probability forecasting and associated scoring rules were later succinctly reviewed from a statistical perspective by Dawid (1986). In a clearly written article, Schervish (1989) presents a general decision theory method for comparing two probability assessors and relates it to the calibration problem. Testing the validity of sequential probability (*prequential*) forecasts was revisited in Seillier-Moiseiwitsch and Dawid (1993). A paper by Winkler (1996) with published contributions from nine discussants illustrates the interest among statisticians in scoring rules.

The ideas of scoring rules and their properties such as propriety and consistency continue to be a subject for statistical research – see, for example, Gneiting and Raftery (2007) and Gneiting (2011). Both of these papers include discussion of quantile forecasts, as does Jose and Winkler (2009). Jose *et al.* (2009) construct families of scoring rules that allow for distance between events to be taken into account (they do not have the property of locality – Section 7.3.2).

Another area of statistics that can be viewed as forecasting is discriminant analysis, where measurements on a number of variables are used to assign observations to one of a number of classes (McLachlan, 1992). It is interesting to note that a common objective in discriminant analysis is to formulate a rule that minimizes the probability of misclassification. For a two-group problem, this is equivalent to minimizing proportion correct, a measure that has been shown in Chapter 3 to have rather poor properties.

This short review of verification articles in the statistical literature is by no means complete, yet hopefully provides some useful starting points for further studies in the subject.

### 12.3.2  Finance and economics

Financial analysts and economists are most likely the next biggest producers of forecasts to be issued on a regular or frequent basis. The verification (or *ex post evaluation* as it is sometimes referred to in this discipline) of (*ex ante*) point forecasts has been reviewed by Wallis (1995), Diebold and Lopez (1996), and Armstrong (2001). Loss functions that are commonly used to score point forecasts include:

- Mean Squared (Forecast) Error (MS(F)E)
- Mean Absolute Error (MAE)
- Mean Absolute Percentage Error (MAPE)
- Relative Absolute Error (RAE)

Diebold and Mariano (1995) describe significance tests for MSE and MAE.

Several studies have shown that MSE is not a reliable or resistant sample statistic for evaluating samples of *ex ante* forecasts (Armstrong and Fildes, 1995; and references therein). For example, the sensitivity of MSE to outlier errors (non-resistance)

makes it unsuitable for reliably ranking forecasts taking part in forecast comparisons/competitions (Armstrong and Collopy, 1992). Mean Absolute Error is a more resistant score, as discussed in Chapter 5 of this book. Mean Absolute Percentage Error is obtained by taking the sample mean of $|\hat{x}_t - x_t|/|x_t|$ and helps to account for variations in variance related to changes in mean (*heteroskedasticity*). For this reason, MAPE may be a useful measure to adopt in the verification of precipitation forecasts. Relative Absolute Error is the sample mean of the ratio of absolute errors of two different sets of forecasts and is therefore a comparative mean skill measure. Unlike skill scores based on mean scores, RAE takes the ratio of errors before averaging over the whole sample and so accounts for changes in predictability of events that take place throughout the sample period. Hyndman and Koehler (2006) give a critical review of the measures listed above, together with a number of variants. Their recommendation is to use mean absolute scaled error. This is MAE scaled by the in-sample MAE from 'the naive (random walk) forecast', in other words a persistence forecast.

Many forecasts in economics and finance take the form of a time series, as does the corresponding set of verifying observations. Typically verification measures look at 'vertical' distances between the two series, but Granger and Jeon (2003) note that 'horizontal' distance, which they call time-distance, is also of interest when the possibility of one series leading or lagging the other is considered. This occurs if the 'shape' of the forecast time series is well forecast but shifted in time. A similar phenomenon occurs in forecasting spatial features (see Chapter 6), where the shape of a feature may be well forecast, but its speed of movement is faster or slower than expected. Granger and Leon (2003) formalize the concept of 'time-distance' in terms of various metrics, which they show to be useful in some circumstances.

Various methods have been developed and employed for evaluating financial and economic interval forecasts (Christoffersen, 1998; Taylor, 1999). Typical methods are based on comparing the number of times the observations fall in the prediction intervals with the number of times expected given the stated probability of the interval. Formal tests of skill have been developed based on the binomial distribution. Motivated by the growth in risk management industries, there has been much interest recently in extreme one-sided intervals defined by values falling outside specified rare quantiles (e.g. below the 5% or above the 95% quantile). The forecasted rare quantile is known as *Value-at-Risk* (VaR) and gives information about possible future large losses in portfolios, etc. – see Dowd (1998). Various methods are used to verify VaR systems but many have low power and so fail to reject the no-skill hypothesis even for systems that do have real skill. Unlike point forecasts, these types of interval forecast need good predictions of future volatility (standard deviation) in order to be accurate, and economists have developed many methods for forecasting variance that could be of potential use to the meteorological community.

The most complete description of the future value of a real variable is provided by forecasting its future probability density function. In a special edition of the *Journal of Forecasting* completely devoted to density forecasting in economics and finance, Tay and Wallis (2000) present a selective survey of density forecasting in macroeconomics and finance and discuss some of the issues concerning evaluation of such forecasts. They describe the *fan chart* method for displaying density forecasts obtained by grey shading time series plots of a set of regularly spaced quantiles. Tay and Wallis (2000) point out that no suitable loss functions have been defined for assessing the whole density and that in general the loss function would depend on the specific forecast user. Recent attempts to evaluate density forecasts quantitatively have used the *probability integral transform*

$$\hat{p}_t = \hat{F}(x_t) = \int_{-\infty}^{x_t} \hat{p}(x_t')dx_t' \qquad (12.4)$$

to transform the observed value $x_t$ at time $t$ into predicted cumulative probabilities $\hat{p}_t$ that should be uniformly distributed from 0 to 1 given perfect density forecasts $p(\hat{x}_t)$. Either histograms or cumulative empirical distributions of $\hat{p}_t$ can be plotted and compared to that expected for uniformly distributed probabilities. Formal tests such as the Kolmogorov–Smirnov and likelihood-ratio tests are used to test uniformity. This approach is the basis for the rank histogram diagram discussed in Chapters 7

and 8. A review of the literature on density forecast evaluation is given in Chapter 15 of Dowd (2005), and Dowd (2007) extends some of the ideas to the case of multi-period density forecasting. Methods for multivariate density evaluation and calibration are also reviewed in Diebold *et al.* (1999).

A topic of much recent interest is sports forecasting and betting. Issue 3 of the *International Journal of Forecasting* (2010) is a special issue on this topic. Within it Grant and Johnstone (2010) argue that although probability forecasts are in theory superior to deterministic forecasts, there can be circumstances in which assessing categorical forecasts based on the underlying probabilities can be preferable to assessing the probability forecasts themselves.

### 12.3.3  Medical and clinical studies

Shortly after the publication of the first edition of this book, a book by Pepe (2003) appeared. Despite its very different title (*The Statistical Evaluation of Medical Tests for Classification and Prediction*) it is a book on 'forecast verification' in a medical context. The most common form of 'forecast' in medical studies is a diagnostic test whose outcome is used to infer the likely presence of a disease. This corresponds to the classic $2 \times 2$ contingency table, which is discussed in detail in Chapter 3. In disease diagnosis the 'forecast' is 'disease present' if the diagnostic test is positive and 'disease absent' if the test is negative. This set-up is given in Table 12.1, using the same notation ($a$, $b$, $c$, $d$) for the table as elsewhere in the book.

A number of properties derived from such a table are in common use in medical diagnostic testing (Bland, 1995, section 15.4; Pepe, 2003, section 2.1):

- $a/(a + c)$ = sensitivity (Bland) = true positive fraction (Pepe)
- $d/(b + d)$ = specificity (Bland) = 1 − false positive fraction (Pepe)
- $a/(a + b)$ = positive predictive value (Bland, Pepe)
- $d/(c + d)$ = negative predictive value (Bland, Pepe)

It can be seen that three of these properties are related to quantities defined in Chapter 3, with

**Table 12.1**  Diagnostic medical tests – possible outcomes

| Test (forecast) | Observed | | |
| --- | --- | --- | --- |
| | Disease present | Disease absent | Total |
| Positive test | $a$ | $b$ | $a+b$ |
| Negative test | $c$ | $d$ | $c+d$ |
| Total | $a+c$ | $b+d$ | $a+b+c+d=n$ |

different names. Thus sensitivity is simply the hit rate, specificity $= 1 -$ false alarm rate, and positive predictive value is $1 -$ false alarm ratio. Negative predictive value, sometimes known as detection failure ratio (Section 3.2.3) is less used in atmospheric science, perhaps because for most circumstances it is more important to correctly forecast the occurrence of a meteorological event than it is to forecast its absence. In the case of medical diagnosis, however, it is crucial that negative diagnoses are correct most of the time. Otherwise, an opportunity for medical intervention at an early stage of a disease may be missed.

Two further measures used by Pepe (2003) are the positive and negative diagnostic likelihood ratios (DLRs). The positive DLR, $a(b + d)/b(a + c)$, is the ratio of the odds of having the disease, given a positive test result, namely $a/b$, to the odds of having the disease given no knowledge of the test result $(a + c)/(b + d)$. The negative DLR is defined similarly, but for the odds of not having the disease.

Altman and Royston (2000) discuss an *index of separation* (PSEP), which in the $2 \times 2$ case equals (Positive Predictive Value + Negative Predictive Value $- 1$) and is the same as the Clayton skill score (Section 3.5.1). More generally, with $K$ diagnostic categories, but only two outcomes (disease presence or absence), PSEP is equal to the difference between the proportion of diseased individuals for the least and most favourable diagnostic categories. The idea is similar to that of *resolution* (see Section 2.10).

For many diagnostic tests, the values of sensitivity and specificity can be varied by adjusting some threshold that determines whether the test result is declared positive or negative. Plotting sensitivity against $(1 -$ specificity) as the threshold varies gives a ROC curve (Section 3.3). A number of variations

on the basic ROC curve have been developed in a medical context. For example, Venkatraman (2000) constructs a permutation test for comparing ROC curves, Rodenberg and Zhou (2000) show how to estimate ROC curves when covariates are present, and Baker (2000) extends the ROC paradigm to multiple tests. Two whole chapters, and some other sections, of Pepe (2003) are devoted to ROC curves. The ubiquity of the ROC curve in medical diagnostic testing is illustrated by the fact that a search in the Web of Knowledge using the key words 'ROC curve' AND 'medical' in March 2011 produced 2492 articles from the previous 5 years.

Two differences stand out between verification in atmospheric science on the one hand and evaluation of medical diagnostic testing on the other. Atmospheric science has more measures in common use than medical testing, but what is done with the measures is less sophisticated. There is a stronger emphasis in medical testing on modelling the measures in terms of covariates, on study design, and on using statistical inference to quantify uncertainty associated with the measures and associated models. The importance of these topics reflects the greater involvement of statisticians in medical diagnostic testing than in atmospheric science forecast verification.

The topic of modelling the measures continues to attract attention. For example, Gu and Pepe (2009) look at how adding extra covariates to a set of baseline covariates affects diagnostic likelihood ratios, while Saha and Heagerty (2010) describe time-dependent versions of sensitivity and specificity. The data structures in these and other papers are more complex than is likely to be encountered in atmospheric science applications, but it is nevertheless valuable for researchers in one discipline to be aware of related work in other literatures.

## 12.4  Current research and future directions

This final section will attempt to highlight some (but not all) of the areas of forecast verification that are active areas of research and that we believe could benefit from more attention in future studies. The identification of these areas is partly based on our insight gained in reviewing and editing all the chapters in this book and in our writing of Chapters 1, 12 and the Glossary. The first issue of *Meteorological Applications* in 2008 was devoted to forecast verification and one of its papers was a review by Casati *et al.* (2008). This article has also helped us formulate our thoughts for this final section.

Forecast verification is based on sample statistics calculated from samples of previous matched pairs of forecasts and observations. Since the sample size is always *finite*, the resulting verification statistics are always prone to sampling errors/uncertainties. Some scores such as MSE are more sensitive (less resistant) to outlier errors than are other scores and are therefore prone to larger sampling errors. For small samples such as those obtained for seasonal forecasts, it is important that we are aware of these issues and use more resistant measures (e.g. MAE). The resistance of scores to outlier errors needs to be considered when screening scores for suitability. Sampling uncertainty can also sometimes be reduced by taking advantage of the large number of spatial degrees of freedom to pool forecasts over suitable regions. Regional pooling could be a useful (essential) approach for increasing the effective sample size for verification in cases when there are few samples in time. Note, however, that pooling will not provide much benefit in cases where there are large spatial correlations (e.g. teleconnection patterns). Also, pooling is only justified if the region over which it is done can safely be assumed to be almost homogeneous with respect to the features of interest. Pooling over heterogeneous regions can easily produce misleading results (Hamill and Juras, 2006).

There is a need to move beyond purely descriptive sample statistics in forecast verification. Verification aims to make *inferences* about the *true* skill of the forecasting system based on the sample data available from previous sets of forecasts and observations. Therefore, sample scores should only be considered as finite sample *estimates* of the true scores of the system (i.e. the scores given an infinite number of previous forecasts). The sample estimates differ from the true values because of sampling uncertainties and so it is important not only to quote scores but also to provide estimates of sampling error. Few studies in the verification literature actually provide any form of error estimate (or confidence intervals) on their quoted scores – the scores are implicitly taken at face value as being perfectly correct, which amounts to incorrectly assuming that the verification data set is infinite. Approximate sampling errors can be estimated either analytically by finding the sampling distribution of the score from the null distribution of forecasts and observations, or by computational resampling methods such as bootstrap (Efron and Tibshirani, 1993). Often simple insightful analytic expressions can be found without having to resort to less transparent computational methods. For example, it is easy to derive analytic expressions for approximate sampling errors on ratios of counts of independent events such as hit rates, etc. by using the binomial distribution (Stephenson, 2000). In addition, the neglected topic of statistical power would benefit from further research, as would the verification of complete probability density functions. Jolliffe (2008) and Mason (2008) provide discussion of how inference may be incorporated when calculating verification measures.

Another area that could benefit from more research is in statistically modelling the joint distribution of forecasts and observations. The high $K^2 - 1$ dimensionality of the verification problem estimated by considering the counts in a $K \times K$ contingency table is severely overestimated since not all the counts in the $K^2$ cells are independent of one another. In fact, when the forecasts and observations are normally distributed (as is sometimes a reasonable approximation – e.g. temperatures) then the problem has only five dimensions related to the five parameters of the joint bivariate normal distribution (e.g. two means, two variances and one correlation). So in this case the binning approach (a crude non-parametric estimate of the joint probability) would overestimate the dimensionality whenever more than two categories were used. It should also be noted that five dimensions is less than the total number of distinct aspects of quality identified by Murphy (1997) and so not all aspects of quality are independent in these cases. It therefore makes sense to develop parametric models of the joint distribution, yet surprisingly few studies have attempted to do this (Katz *et al.*, 1982). A parametric modelling approach would also have the advantage of providing estimates of sampling uncertainties on scores. Another promising approach is to model the likelihood of observations given forecasts using suitable regressions (Krzysztofowicz and Long, 1991). Likelihood modelling would allow more Bayesian approaches to be used that are good at merging information provided by different forecasting systems (Berliner *et al.*, 2000). However, it should be noted that such modelling approaches depend on the validity of the underlying model assumptions and structural uncertainty therefore also needs to be taken into account (Draper, 1995).

In the final chapter of the first edition we speculated on what were likely to be the most important developments in forecast verification over the next few years. The areas we identified overlap considerably with those discussed by Casati *et al.* (2008) as important recent developments. The list compiled by Casati *et al.* was based on those topics that played an important role in the third of a series of workshops on verification, organized by a Joint Working Group on Verification (JWGV) under the World Meteorological Organization (WMO)/World Weather Research Program (WWRP) and the WMO Working Group on Numerical Experimentation (WGNE). This workshop took place in Reading in 2007, and around the time the present book is published the fifth workshop will take place in Melbourne in December 2011. The initial list of topics for this workshop again overlaps considerably with that of Casati *et al.* (2008). The sections in Casati *et al.* (2008) cover the following:

1. spatial verification methods;
2. probabilistic forecasts and ensemble verification;
3. verification of extreme events;
4. operational verification;
5. verification packages and open sources;
6. user-oriented verification strategies;
7. forecast value.

The first three of these, with slightly amended wording, appear in the initial list for the 2011 workshop. The suggested topic for the workshop 'User Issues Including Communicating Verification to Decision Makers' has elements of items 4, 6 and 7, while 'Verification Tools' is similar to item 5. New topics for 2011 relate to seasonal and longer forecasts, and to 'propagation of uncertainty'. The importance of most of the topics is reflected in the second edition of our book.

Spatial verification has developed considerably in recent years and continues to do so, as evidenced by the considerably expanded version of Chapter 6 in our new edition. There has been less new research in verification of probabilistic and ensemble forecasts, but such forecasts are increasingly common and further work is needed on their verification. Two chapters are devoted to these topics in this new edition. Forecasts of extreme events have become increasingly relevant in the context of climate change. Verification of such forecasts is thus an important topic and although some progress has been made, as shown in our new Chapter 10, the underlying problem of non-stationarity associated with climate change remains to be tackled thoroughly. Related to this is the verification of more general climate projections. Because of the difficulty posed by the lack of verifying observations, this topic is in its infancy, but is discussed, along with seasonal forecasting, in our new Chapter 11. The new Appendix describes the current position relating to 'verification tools'.

On a final note, it is worth reiterating that the subject of forecast verification could benefit enormously from a wider and more open access to previous forecasts and observations. The majority of operational centres provide forecast products without also providing past forecast data or clear documentation on previous forecast performance (Thornes and Stephenson, 2001). It should be a rule of good practice that forecast providers make such information about their forecasting systems publicly available and should also provide easy access to samples of their past forecasts and verifying observations (e.g. via the internet). Such practice would then enable third parties to examine for themselves the past performance of the forecasting system and ultimately would provide much useful feedback for future forecasting improvements.

A related problem is that of selective publication. For example, there are many potential precursors for earthquake prediction. The International Association for Seismology and Physics of the Earth's Interior (IASPEI) panel for the evaluation of precursors found that a false impression of skill had been created in the literature by the practice of publishing only successful forecasts out of the huge number of precursors tested at locations all around the world. By selecting only the successful cases, such a reporting bias can easily lead to misleading confidence in the true skill of forecasts. The moral of this story is that it is very important also to publish results when forecasting systems do not perform well – this then gives a more balanced view of the whole subject. Seasonal climate forecasting is in danger of falling into a similar trap by isolating *posterior only specific events* in certain years that could have been forecast with skill (e.g. Dong *et al.*, 2000). It is therefore important for forecasters to provide not only forecasts of future events but also clear information on past forecast performance and ideally also an uncensored sample of previous forecasts and observations.

In conclusion, this book on forecast verification has covered many aspects in this amazingly rich and important subject. We hope that this book will enable a wider community of people to understand the complexities and issues involved in forecast verification, and thereby encourage the development and application of improved verification methods. We also hope that this book has opened up some exciting new areas for future research in the subject of forecast verification – only time will really tell!

## Acknowledgements

We are grateful to Robert Fildes, Margaret Pepe and Robert Winkler for useful suggestions regarding important recent literature on forecast verification/evaluation in disciplines other than atmospheric science.

# Appendix
## Verification software

**Matthew Pocernich**

*National Center for Atmospheric Research, Boulder, Colorado, USA*

Nearly every method discussed in this book will require use of a computer. This appendix studies the different aspects of software programs useful for forecast verification and discusses some options available for the practitioner. Just as the utility of verification statistics is a function of the needs of the forecast user, the value of any type of verification software is dependent on the intended use. Ultimately a program should be chosen to best meet a user's needs. This appendix discusses some general aspects of software that might be considered as well as discussing some specific options. With regard to specific software, it is telling that the authors of the different chapters of this book use a wide variety of software programs and packages. While much software has been developed for different aspects of verification, there are few universally used products. What a person ultimately chooses to use for forecast verification in part seems to depend on personal background, experience with software, what software is available, the type of software most commonly used by their peers, as well as the type of forecast they are verifying. Software programs and computing capabilities are constantly evolving and improving. The information contained in this appendix was current at the time of writing, but undoubtedly at some point it will be out of date

while the concepts will retain meaning. To supplement the appendix, software and data sets used in the book will be provided via the book's website: http://emps.exeter.ac.uk/fvb.

This appendix is organized as follows. The first section discusses qualities of good software programs. Examples of these properties include accuracy, documentation, ease of use and computational efficiency. The second section discusses general categories of users, which can range from the individual researcher, to a project team, to an institution. Explicitly identifying a user's needs will often help identify the appropriate type of software. The third section discusses specific software and programming options. There are many programming and statistical programming languages used in the atmospheric science community, each with strengths and weaknesses, advocates and critics.

## A.1 What is good software?

Before discussing the virtues and limitations of verification software, a moment will be spent discussing the attributes of good software in more general terms. While there are no strict definitions about what is good and bad software, the following

*Forecast Verification: A Practitioner's Guide in Atmospheric Science*, Second Edition. Edited by Ian T. Jolliffe and David B. Stephenson.
© 2012 John Wiley & Sons, Ltd. Published 2012 by John Wiley & Sons, Ltd.

attributes should be considered when evaluating the available tools.

### A.1.1 Correctness

First and foremost, a software program or function must correctly implement the desired method. The program needs to exactly translate the intent of the method. Obviously, no one intends to create faulty software, but the more complex the software program, the more opportunities there are for errors. So how do you know if the software you are using is well written? Software engineering contains many practices designed to reduce the risk of errors, most notably testing the uses of small subroutines and documentation. Smaller, simpler functions are easier to test and verify. By building complex programs out of more simple pieces, one may reduce the chance of errors. Cleanly written code allows functions to be more easily checked. Depending on the software, this level of scrutiny may not be possible.

### A.1.2 Documentation

Clear, well-written documentation ensures that a function is correctly used. This reduces the chance of user-caused errors. Furthermore, good documentation and examples can make learning a new type of software easier. The format and structure of documentation should also be considered, with a focus on cross-referenced, hyperlinked and searchable documentation.

### A.1.3 Open source/closed source/commercial

Open source software allows people to use, share and modify code. There are many positive and negative attributes to open source software. With open source software, it is possible to examine the uncompiled code used to create the software. This allows you to check the application and to understand exactly how values are calculated. Even with the most thorough documentation, some details are omitted and seeing how functions are written is often the best option. In the research context, this also allows you to extend or modify the implementation of a program. Open source software is typically distributed free of charge, another significant benefit. Open source licensing protects the creator from liability due to errors. While good for the creator, the user assumes responsibility for the program being correct.

When programs are distributed as closed source, or *compiled*, it may mean that a third party has assumed some responsibility for the accuracy of the software. Depending on the context, having a company or other party assume this responsibility justifies the expense. Commercial software may also have more options for support and training, such as training courses and phone support. These features may be useful and valuable assets worth the costs, and may dictate that a user chooses a commercial product over an open source product. A downside for commercial software is the cost. Aside from the actual cost, high costs can reduce the number of users and therefore reduce the size of the community.

### A.1.4 Large user base

Depending on the importance and resources of the verification project or task, it may be necessary for the analysis to be checked by an outside group or by others internally. While verification statistics can be calculated using a variety of programs with a variety of languages, it is desirable that code can be checked and reviewed. This is most easily accomplished when others understand the software and the language.

Having more users increases the chances that errors will be discovered. This occurs when users apply an algorithm to different and diverse data. Archived and searchable help lists with questions and answers are a useful resource, which benefits from a larger number of users. However, taking the number of users as a criterion for choosing a software or language has limited utility in that it is typically impossible to accurately estimate the size of a community of software users.

## A.2 Types of verification users

Analogous to the idea that the value of a forecast is dependent on the users, the value of verification software is dependent on the type of user. As a user, one should explicitly consider one's needs when evaluating software options. Some general groups of people who use verification software include the following.

### A.2.1 Students

For individuals new to forecast verification, simplicity and transparency take precedence over computational efficiency and sophistication. Operations performed by the code should be easily viewed and interim values produced. This allows the users to understand the methods by slowly stepping through them.

### A.2.2 Researchers

Graduate students and researchers in atmospheric science may be working with new types of data, new types of forecasts and a new group of users. For this reason, it is desirable that the language be flexible enough to accept new types of forecasts, and allow the development of new verification methods.

### A.2.3 Operational forecasters

Operational forecasters may have an interest in evaluating a fixed set of forecasts using a fixed suite of measures over an extended period of time. This requires that some consideration be given to forecast data archiving. For this type of user, a fixed set of methods may be sufficient and it may not be necessary to examine the code in detail.

### A.2.4 Institutional use

Operational forecast centres and research institutions may have different priorities with respect to forecast verification. Institutions may select or create verification software that will exist for an extended period of time, evolve with the development of new methods, and serve multiple users. Institutions may also have more resources to devote to forecast verification. These resources may include

access to software engineers who can greatly expand the capabilities of verification systems by creating customized displays and implementing new methods.

## A.3 Types of software and programming languages

Programming languages describe the tools used to create programs. Low-level programming languages include Fortran, C, C+, Java and others. These languages serve as the basic building blocks for many other programs, and a skilled programmer can certainly use any of these languages to implement any of the methods discussed in this book. However, the initial effort required for the inexperienced non-programmer to get started with a low-level programming language is high. For this reason, they are not typically viewed as verification tools. Higher levels of programming languages are interpreted languages – languages that allow the user to directly enter instructions to the program by using scripts or typing at a prompt. Languages such as SAS, MATLAB, Minitab, R, IDL and NCL fall under this classification. For some of these languages such as SAS, MATLAB and Minitab, graphical user interfaces (GUIs) have been created to allow the user to operate the programs with a mouse, but processes can be automated, and functions developed and modified by creating scripts and writing code. The software and some of their properties are presented in Table A.1.

Spreadsheets such as Microsoft's Excel are operated interactively with a GUI as opposed to scripts. Scientists and statisticians may dismiss them as viable verification tools and there are a lot of reasons not to use them for research or in operations. However, Excel is probably the most commonly used program for data analysis and there is no reason to think that this is any different in the atmospheric sciences. There are certainly arguments to be made for the pedagogical advantages for using spreadsheets to teach methods to a broad audience. Finally, there are web-based technologies that could allow a user to provide forecast and observation data that would utilize code and capacity hosted on another system. However, none were found when preparing this

**Table A.1** Summary of software available for forecast verification

| Software/program | Key user communities | Open source[a] | Comments | Verification capabilities |
|---|---|---|---|---|
| Spreadsheets | Ubiquitous | Varies[b] | Arguably the most commonly used type of software for statistics | No explicit verification functions |
| SAS | Health sciences, business, industry | No | SAS is a dominant language used in biostatistics and medical research | Verification-related functions described in health science terminology |
| Minitab | Statistical process control (Six Sigma), academic | No | Frequently used to teach and implement statistical process control | Limited user-contributed functions |
| MATLAB | Statistics, atmospheric sciences, academic | No | Commonly used in the atmospheric science research community | No publicly available verification functions. Many members of the research community have developed personal verification libraries |
| R | Statistical research | Yes | Basic functionality is equivalent to other statistical programming languages. Contributed packages address a very large variety of techniques | Contain several verification explicit packages including ones that perform basic verification functions, ROC curves as well as ensemble verification |
| IDL | Atmospheric sciences or science related | No | | Some contributed verification functions available |
| NCL (NCAR Command Language) | Weather and climate research | Yes | Addresses many practical issues related to weather and climate research | Designed to process climate model data Few explicit verification tools |
| MET (Model Evaluation Tool) | Weather forecasting and research | Yes | Well supported through online support, workshops and tutorials | Designed as a forecast verification program. While MET will work with most types and expressions of forecasts, MET is designed to work most seamlessly with WRF forecasts |
| EUMETCAL | Weather forecasting and research | Online | | Provides tutorial on forecast verification |

[a] Or available without cost.

[b] While Microsoft Excel is the most commonly used spreadsheet and not free, others open source spreadsheets are available such as Open Office.

appendix. Results of verification processes are commonly posted online and often the user may define regions and time periods, but still these tools do not yet permit the user the ability to process new data.

For the purposes of discussion, software will be classified into the following groupings: (i) spreadsheets; (ii) statistical and mathematical programming languages; (iii) institutionally supported software created by weather forecasting and research institutions; and (iv) displays of verification results.

## A.3.1  Spreadsheets

Prominent statistician Brian Ripley remarked in 2002, 'Let's not kid ourselves: the most widely used piece of software for statistics is Excel'. Whether or not this is true in the atmospheric sciences can be debated, but there are certainly scientists who use spreadsheets for research. Many verification methods for point forecasts and observations can be easily implemented and possibly most easily understood using a spreadsheet. These topics include theoretical aspects of verification discussed in Chapters 2 and 3, contingency table statistics discussed in Chapter 4, and the basics of verifying continuous forecasts, considered in Chapter 5 (Sections 5.1–5.4). While the use of spreadsheets for statistics might be dismissed by scientists and statisticians as not professional or appropriate, spreadsheets allow a new user the ability to immediately see all steps in basic verification procedures. Individual forecasts can be listed by row, with forecasts, observations and errors listed in separate columns. Most spreadsheet programs such as Microsoft Excel and Open Office Calculate include basic statistical functions to help in calculating statistics such as mean square error for continuous forecasts. With the observations or events correctly coded, table or tabulate functions will provide values needed to calculate most contingency table statistics. Moderately sophisticated graphics can be created using conditional formatting, which allows a plotting symbol to be formatted based on specified conditions. Graphs also have a brush-over feature that allows a limited degree of interaction with the figures. Serious limitations for using a spreadsheet include a lack of documentation, which scripts inherently provide, data size limitations, and the additional effort required

to create finished verification graphics. There are no add-ons or macros developed to directly calculate verification statistics on spreadsheet programs – though for someone with some macro programming (like Visual Basic) skills, such an implementation would not be difficult.

## A.3.2  Statistical programming languages

The next level up in complexity and ability from spreadsheets consist of statistical programming languages such as SPSS, SAS, Minitab, MATLAB and R (this list of programs is by no means complete). These programs differ from spreadsheets in that they can handle much larger data sets, have more advanced statistical functions, take longer to learn and are most typically used by writing scripts. Scripts allow procedures to be documented and replicated – both important practices in good verification protocols. Working with scripts is a distinctly different environment for a user familiar with a GUI environment. Functions written in these languages can be readily shared with others who use the same language. A consideration is that the choice of language or software to some degree defines the community of people who can share and collaborate in depth when analysing data. For that reason, your choice of software may in part be dictated by the culture of the company or institution where you work. Again, most of the basic verification concepts discussed in Chapters 2, 3, 4 and 5 can be readily implemented using basic statistics functions provided in all of these packages.

### Minitab
Minitab has been established as a tool used to implement the 'Six Sigma Quality Control' programs at many companies. Statistical process control (SPC) has been adopted by many companies around the world to improve quality and reduce defects. While obvious parallels may exist between problems phrased in the context of quality control and the verification of weather and climate forecasts, there is little interaction between the fields. Some verification-related functions included in Minitab are the ROC plot and basic classification statistics similar to the contingency table type

statistics. Most commonly, Minitab is operated using a GUI, which some people may find easier to use than scripting languages. Minitab is sometimes used in teaching introductory statistics classes. Consequently textbooks exist, which can serve as valuable learning tools.

### SAS

Within the field of biostatistics and pharmacology, SAS is the dominant statistical programming language. In part this is due to the fact that SAS is one of the languages accepted by the US Food and Drug Agency for use in the drug approval process. Because of the strong similarities between forecast verification and the evaluation of treatment effectiveness or diagnosis accuracy, SAS offers many tools and resources for forecast verification. Not only are plots such as the ROC curve (Chapter 3) included as built-in functions, there are books published by SAS Press devoted to topics such as ROC curves, sampling error and built-in functions. The documentation presents some verification techniques using a different vocabulary. For example, in health science-related fields hit rates and $1 -$ false alarm rates are referred to as sensitivities and specificities, respectively.

SAS is a very well-documented programming language. Procedures and methods are outlined in manuals, and SAS users often have local users' groups as well as national conferences. For these reasons, SAS should be considered as a potentially useful resource for the forecast verification community. Over the years, the similarities between medical and atmospheric science research have been pointed out by many (Gabriel, 2000), but the communities remain quite separate with limited interaction.

SAS is a commercial product. Comparatively, the SAS licence can be expensive, with the price varying depending upon the capabilities included.

### MATLAB

MATLAB describes itself as a technical programming language and is widely used within the atmospheric research community. A search for the term MATLAB in the American Meteorological Society (AMS) journals returns hundreds of records, confirming its popularity within the atmospheric research community. Commonly, MATLAB is used to explore and prototype new methods and procedures. Certainly among many research groups and at many institutions there are groups of users to provide advice and support. MathWorks – the parent company of MATLAB – maintains a repository of contributed MATLAB functions. However, with the exception of a ROC plot, few contributed functions exist specifically for forecast verification. Most figures presented in Chapters 3 and 7 were produced by MATLAB using code developed by the authors.

As a commercial product, a licence is required to use MATLAB. Furthermore, basic statistical functions are not included in the base MATLAB product and require the purchase of a statistical toolbox.

### IDL

IDL is another mathematical programming language that is moderately popular in the atmospheric science community. Some active members of the verification community who use IDL have contributed common code. This includes code to calculate fractional skill scores and other neighbourhood methods (Section 6.5) as well as creating Taylor diagrams (Section 5.4).

Beth Ebert of the Centre for Australian Weather and Climate Research (CAWCR) has placed links to IDL code on the CAWCR website. These include the following:

Methods for verifying Probability of Precipitation (POP) (http://www.cawcr.gov.au/projects/verific ation/POP3/POP3.html)

Spatial Methods and Fractional Skill Scores (http://www.cawcr.gov.au/staff/eee/#Interests)

IDL is a commercial language that requires the purchase of a licence. This limits the number of potential users and restricts the ability to share code. While examples of IDL code useful for forecast verification exist in various locations, there is no common repository for code.

### NCAR Command Language (NCL)

NCL is a programming language created and supported by the National Center for Atmospheric Research (NCAR) to address the unique challenges involved in managing and handling data from climate models. NCL is an open source language that runs on most popular versions of the LinuxSoftware system. Some features of climate model data include the following: large file sizes (a climate model experiment may generate terabytes of data); a global domain – which can create some challenges with projections and multiple layers vertically that extend both in the atmosphere and ocean. NCL addresses many of these issues by relying heavily on the structure of NetCDF files (http://www.unidata.ucar.edu/software/netcdf/). On the NCL website, there are many created functions and contributed functions to address data manipulation challenges such as forecast interpolation and extrapolation. Other functions create elegant graphics that permit many projections typical of those found in the IPCC (Intergovernmental Panel on Climate Change) reports.

NCL has the basic functions needed to calculate many of the verification statistics discussed in this book. For the scientist or researcher using NCL as a primary research tool, it would be logical to extend this to the verification stage of a project. However, NCL offers little in the way of explicit climate model verification functions. There is an implementation of the Taylor diagram contributed by Karl Taylor. This lack of verification functions may in part be due to the fact that the topic of climate model verification is in its infancy (Chapter 11). Certainly, as the methods and practices mature, NCL functions will be created and contributed to the NCL community.

A link to the main directory and contributed functions for NCL is: http://www.ncl.ucar.edu/.

### R statistical programming language

R is an open source programming language that over the last decade has become the dominant language in the area of statistical research. Its key features include that it is a free, open source language capable of running on all major operating systems. The base package contains functions equivalent to most statistical languages. These include functions for exploratory data analysis, linear models, graphics, matrix operations and more. While the language does permit some high level operations that support the creation of simple plots, statistical models and

summaries, the language also allows functions to be used at a very low level – where any detail on a graph may be modified and options set for the calculation of a model. Most of the figures presented in Chapters 8 and 10 were produced using R software.

The truly unique part of the R language is the contributed packages. In 2010 there were over 2400 contributed packages, which address nearly every type of scientific and statistical topic imaginable. Additionally, the Bioconductor project, an open source, open development effort to create tools to further genetic research and bioinformatics, uses R as a base platform. Nearly 400 additional packages are available from this project. This is relevant to research in verification because there is a large overlap between verification challenges found in the medical scientific community and the atmospheric science community. More information about the R Statistical Programming Language can be found at www.r-project.org. There are also several very active blogs that focus on many aspects of programming in R, as well as local user groups in many locations.

An excellent introduction to the applicability of R to atmospheric and verification research is the article 'Probabilistic weather forecasting in R' by Fraley et al. (2011). This article creates a probabilistic forecast from an ensemble by employing Bayesian model averaging. Of interest to this appendix are the verification methods included in the ensembleBMA package. These methods include functions to calculate mean absolute error, continuous ranked probability score, and Brier score in a spatial context. Options exist for forecasts created with exchangeable or unique forecasts. Techniques and results are illustrated using data from published papers.

Below we highlight a few packages in R relevant to verification.

**verification** (NCAR, 2010) provides basic verification functions including ROC plots (Chapter 3), attribute diagrams (Chapter 7), contingency table scores (Chapters 3 and 4) and more. In this package, the verify command is an overloaded function that calculates a variety of statistics, depending on the type of forecast and observation. For example, with a binary forecast and observation, contingency table statistics are calculated.

For probabilistic forecasts and binary outcomes, scores and statistics such as the Brier score are calculated. Continuous forecasts and continuous observations may be evaluated by statistics such as the mean squared error. Ensemble forecasts of full distributions and a single outcome may be verified by continuous ranked probability scores (crps). Spatial forecasts may be evaluated using a fractional skill score or spatial intensity-scale methods.

**ensembleBMA** (Fraley *et al.*, 2010) provides functions designed to specifically handle the verification of ensemble forecasts. This includes crps scores, rank histograms and probability integral transform values for forecasts expressed as a continuous distribution (Chapter 8).

**nnclust** (Lumley, 2010) provides tools for finding nearest neighbour and minimum spanning trees. This is useful for verifying multidimensional ensemble forecasts (Chapter 8).

**pROC** (Robin *et al.*, 2010) displays and analyses ROC curves (Chapter 3). The following is the description of the package: 'Tools for visualizing, smoothing and comparing receiver operating characteristic (ROC curves). (Partial) area under the curve (AUC) can be compared with statistical tests based on U-statistics or bootstrap. Confidence intervals can be computed for (p)AUC or ROC curves.'

**ROCR** (Sing *et al.*, 2009) creates and provides a variety of presentations for the information summarized in ROC plots (Chapter 3). The following is the description for ROCR: 'ROC graphs, sensitivity/specificity curves, lift charts, and precision/recall plots are popular examples of trade-off visualizations for specific pairs of performance measures. ROCR is a flexible tool for creating cutoff-parameterized 2D performance curves by freely combining two from over 25 performance measures (new performance measures can be added using a standard interface). Curves from different cross-validation or bootstrapping runs can be averaged by different methods, and standard deviations, standard errors or box plots can be used to visualize the variability across the runs. The parameterization can be visualized by printing cutoff values at the corresponding curve positions, or by coloring the curve according to cutoff. All components of a performance plot can be quickly adjusted using a flexible parameter dispatching mechanism. Despite its flexibility, ROCR is easy to use, with only three commands and reasonable default values for all optional parameters.'

## A.4  Institutional supported software

Challenges exist when verifying forecasts made across a large domain, at a high frequency or operationally where real-time verification is desired. Data handling and storage become much greater issues. In portions of this book, it is assumed that one has forecast and observations in pairs. In reality, it is often the case the forecasts and observations are different types or have different spatial and/or temporal resolution. In an operational setting, multiple users may need to view the verification statistics concurrently. In these instances, software engineering plays a greater role. Typically, this type of endeavour requires the effort and resources of an institution or team. Furthermore, an organization's mission statement and funding may limit their ability to share or support publicly available verification software.

Some examples of institutional verification systems and supported tools include the following.

### A.4.1  Model Evaluation Tool (MET)

MET was developed by and is supported at the National Center for Atmospheric Research (NCAR) by the Developmental Testbed Center (DTC) with funding from the US Air Force Weather Agency and NOAA. The tool has been designed to work directly with versions of the Weather Research and Forecasting (WRF) modelling system, but can be modified to be useful for other types of forecasts. Much of the analysis presented in Chapter 6 utilized MET.

MET provides a variety of verification techniques, including:

- Standard verification scores comparing gridded model data to point-based observations.
- Standard verification scores comparing gridded model data to gridded observations.
- Spatial verification methods comparing gridded model data to gridded observations using neighbourhood, object-based and intensity-scale decomposition approaches.
- Probabilistic verification methods comparing gridded model data to point-based or gridded observations.

Another powerful addition to MET is the MODE tool. MODE provides object-based methods to identify and match forecasts and observations – matching them by treating them as geometrical features. Figure 6.8 was created using MODE.

Strengths of MET include the strong institutional support provided to method development, instruction and research. This includes biannual MET tutorials, which accompany WRF tutorials taught at NCAR. Additionally, there is an annual MET workshop that focuses on new developments in MET. The project is open source and available to researchers from most countries (subject to US trade restrictions).

MET uses a number of libraries created by other organizations to read and process data. This requires that a number of libraries be installed locally on the user's machine. While many systems are supported, installation is an issue. In general, while MET may be an extremely useful tool for people familiar with operating and conducting research in numerical weather forecasts, the system has a steep learning curve for those interested in basic verification concepts.

### A.4.2 Ensemble Verification System (EVS)

EVS is an experimental prototype verification program developed by the Hydrological Ensemble Prediction (HEP) group of the US National Weather Service's Office of Hydrologic Development (OHD), (Brown *et al.*, 2010). Created in Java,

this program operates on most operating systems. Data need to be formatted in typical text files – one for observations, one for forecasts. Columns in these files specify the valid time and lead time for each forecast. From these files, a suite of verification plots is made including plots of root mean squared error and boxplots by pooling forecasts by observed values and errors by lead time. Without respect to lead time, ROC plots, Brier skill scores, reliability diagrams, continuous ranked probability skill scores and more are generated. EVS generates output in the form of plot files, text files and XML files. The software is well documented, but online support or a helpdesk is lacking. More information and download files can be found at http://amazon.nws.noaa.gov/ohd/evs/evs.html.

### A.4.3 EUMETCAL Forecast Verification Training Module

Since 2001, the European Virtual Organisation for Meteorological Training (www.eumetcal.org) has been developing and providing training modules for different topics in the meteorological sciences. Included in these topics is a module on forecast verification. This module interactively steps through the basic concepts of forecast verification. While not technically verification software, the training module has received continued institutional support and is worth inclusion in a discussion of institutional software.

## A.5  Displays of verification information

Examples of online displays of verification information exist. These sites and tools differ from verification software in that they are created to evaluate a fixed set of forecasts for a specific audience and so these tools are not suitable for research. However, they do provide good examples of how to communicate verification information to a variety of audiences. This list is limited in scope. Many organizations provide online and real-time verification information for internal use and this is frequently shared with research organizations and scientists. However, this information is not typically made available to the general public. Verification from commercial weather forecasting businesses is almost uniformly considered proprietary.

### A.5.1 National Weather Service Performance Management

To evaluate all warnings and forecasts produced by the US National Weather Service, the Performance Management group has created an online verification and evaluation tool. This tool analyses a wide variety of forecasts from an online archive. Forecasts may be stratified by forecast type (e.g. fire weather, aviation, public, etc.) and by forecast product. For public forecasts, a limited number of variables may be evaluated. These include minimum and maximum temperature and probability of precipitation. Warnings that can be evaluated include storm warnings, winter weather warnings, tornado warnings and many more. A suite of statistics and scores is produced, including mean squared error. Tools are also available for downloading data from the archived information. While this site serves as a good example that verification statistics may be useful for operational forecasters, one cannot add new

forecasts to the system in order to utilize the verification infrastructure. Additionally, the data contained in this system are limited to the USA. Registration is also required. (https://verification.nws.noaa.gov/).

### A.5.2 Forecast Evaluation Tool

The Forecast Evaluation Tool (FET) has been created to evaluate seasonal climate forecasts for temperature and precipitation produced by the National Weather Service Climate Prediction Center. These forecasts are made for lead times ranging from $1/2$ month to 13 months. For a chosen lead time, this tool makes summary maps of the USA showing areas of more and less skilful forecasts. Typically, forecasts are expressed by regions where warm/normal/cool or wet/normal/dry conditions exist. Information about the confidence levels within these regions is also expressed. FET is well documented and contains tutorials to help new users interpret the verification output (http://fet.hwr.arizona.edu/ForecastEvaluationTool/).

A limitation for researchers is that new forecasts cannot be loaded into the system to take advantage of the FET's methods. The domain evaluated by FET is the continental USA.

# Glossary

**David B. Stephenson**

Forecast verification has developed its own language with technical words and phrases that have specific meanings, although terminology is not always uniquely defined. The aim of this Glossary is to introduce some of this jargon, and provide some clarification. In particular, scores for categorical forecasts have often been confusingly renamed and so feature heavily here.

For the sake of uniqueness, mathematical definitions are given here in terms of *population* quantities such as expectation, which provide information on the 'true' performance of a forecasting system. The expectation of a quantity $B$ over all possible values of $A$ for cases where condition $C$ holds true is denoted by $E_A(B|C)$ (the conditional expectation of $B$ over all values of $A$ conditioned on $C$). The population variance of a quantity $B$ over all values of $A$ is denoted by $\text{var}_A(B)$.

In practice, measures of true performance are *estimated* in various different ways from paired *samples* of previous forecasts and observations, and it is these estimated values that are used by practitioners. The reader should refer to specific chapters of this book for detailed information about practical issues of estimation and interpretation.

## Acknowledgements

I wish to thank Ian Jolliffe, Simon Mason, Lenny Smith and Andreas Weigel for their useful comments, which helped me revise this Glossary from that presented in the first edition of this book. All errors and omissions are due to myself and these will be listed on the book website at www.emps.ex.ac.uk/fvb.

**accuracy** The average distance/error between deterministic forecasts and observations that depends on **bias**, **resolution** and **uncertainty** attributes. Often estimated using **mean squared error** but can be estimated more robustly using statistics such as **mean absolute error** that are less sensitive (more resistant) to large outlier errors.

**anomaly correlation coefficient (ACC)** The product moment correlation between forecast and observed anomalies (centred differences between 'raw' values and their long-term means), aggregated over all locations in the spatial domain (e.g. all grid points). It is commonly used to measure the skill of spatial field forecasts (see Chapter 6).

**artificial skill** An overestimate of the true skill of a forecasting system. This can arise for a variety of reasons, such as sampling variability in point estimates calculated for small samples of forecasts and observations; using the same data to evaluate the forecast skill as were used to develop/recalibrate the forecasting system; spurious correlation due the presence of long-term trends in the forecasts and observations, etc.

**association** The overall strength of the statistical relationship/dependency between deterministic forecasts and observations that is independent of the marginal distributions. Linear association is often measured using the **correlation coefficient**.

**attributes** Many different scalar attributes can affect **forecast quality** such as overall **bias**, **reliability/calibration** (Type 1 **conditional bias**), **uncertainty**, **sharpness/refinement**, **accuracy**, **association**, **resolution** and **discrimination**. These attributes provide useful complementary information about the performance of a forecasting system since *no single measure is sufficient for judging and comparing forecast quality*.

**base rate** The probability that an observed categorical event will occur. It is generally estimated by the relative frequency of such events in historical observations, i.e. the **sample climatology**. The base rate is a property of the observable world that is the same for all forecasting systems.

**bias** The difference between the expectation of the forecasts and the observations (also known as **overall bias**, **systematic bias** or **unconditional bias**). It is most commonly quantified using the **mean error**, $E(\hat{X}) - E(X)$, i.e. the difference between the means of the forecasts and the observations. For **categorical forecasts**, the **frequency bias** in marginal probabilities is estimated by the ratio of the total number of events forecast to the total number of events observed (i.e. $(a + b)/(a + c)$ for binary categorical forecasts – see **contingency table**).

**Brier probability score** The mean squared error of probability forecasts for a binary event $X = 0, 1$. It is defined as $B = E[(\hat{p} - X)^2]$ and is 0 for perfect (deterministic) forecasts and is 1 for forecasts that are always incorrect.

**calibration** See **reliability**, **sharpness**. Note that sometimes 'calibration' is used to signify correction of mean systematic errors in the forecast climatology (e.g. by tuning of the forecast model parameters) rather than recalibration of the forecasts by making a set of forecasts and then comparing them pairwise to past observations.

**calibration-refinement factorization** See **joint distribution**.

**categorical forecast** A forecast in which a discrete number of $K$ categories of events are forecast. Categories can be either **nominal** (no natural ordering – e.g. clear, cloudy, rain) or **ordinal** (the order matters – e.g. cold, normal, warm). Categorical forecasts can be either **deterministic** (a particular category, e.g. rain or no-rain tomorrow) or **probabilistic** (probabilities for each category, e.g. probability of 0.3 for rain and 0.7 for no-rain tomorrow).

**Clayton Skill Score (CSS)** A measure of categorical forecast performance equal to $a/(a + b) - c/(c + d)$, i.e. the ratio of hits to total number of events forecast minus the ratio of correct rejections to total number of non-events forecast (see **contingency table**). It is analogous to the **Peirce Skill Score** except it is stratified on the forecasts rather than the observations.

**conditional bias** The difference $E_A(A|B) - B$ between the conditional mean $E_A(A|B)$ (the average over all possible values of $A$ for a given value of $B$) of a random variable $A$ and the conditioning variable $B$. The conditional bias is zero when the linear regression of $A$ on $B$ has a slope equal to 1 and an intercept of 0. Type 1 conditional bias $E_X(X|\hat{X}) - \hat{X}$ is obtained by calculating the mean of the observations for particular values of the forecast, whereas type 2 conditional bias $E_{\hat{X}}(\hat{X}|X) - X$ is obtained by conditioning on the observed values. Measures of overall conditional bias can be obtained by averaging the mean squared bias over all possible values of the conditioning variable, e.g. $E_{\hat{X}}[(E_X(X|\hat{X}) - \hat{X})^2]$. Type 1 conditional bias is also known as **reliability** or **calibration** and is 0 for all values of $\hat{X}$ for a *perfectly reliable* (*well-calibrated*) forecasting system. A **reliability diagram** can be made by plotting $E_X(X|\hat{X})$ against $\hat{X}$.

**conditional distribution** The probability distribution $p(x|A)$ of a variable $X$ given that condition $A$ is true. It can be estimated by considering only values of $x$ at times when $A$ was true, i.e.

stratifying on $A$. The conditional distribution of the forecasts given the observations, $p(\hat{x}|x)$, determines the **discrimination** or **likelihood**. The conditional distribution of the observations given the forecasts, $p(x|\hat{x})$, determines the **calibration** or **reliability**. These two conditional distributions are related to each other via Bayes' theorem $p(x|\hat{x})p(\hat{x}) = p(\hat{x}|x)p(x)$.

**consistency** A deterministic performance measure is consistent with a directive if the measure is optimized by applying the directive. For example, for continuous variables the directive might be 'forecast the mean of the probability distribution'. The performance measure consistent with this directive is mean squared error, since this is minimized when forecasting the mean. This **metaverification** property is a modification of **propriety** to deterministic measures (see Chapter 3).

**contingency table** A two-way contingency table is a two-dimensional table that gives the discrete joint sample distribution of deterministic forecasts and categorical observations in terms of cell counts. For binary categorical events with only two possible outcomes, the $2 \times 2$ contingency table can be defined as shown:

**correct rejection** In categorical verification, a no-event forecast that occurs when no event is observed. See **contingency table**.

**correlation coefficient** A measure of the **association** between the forecasts and observations independent of the mean and variance of the marginal distributions. The Pearson product moment correlation coefficient is a measure of linear association and is invariant under any linear shifts or scalings of the forecast or observed variables. The Spearman rank correlation coefficient measures monotonicity in the relationship and is invariant under any monotonic transformations of either the forecast or observed variables. The Kendall tau rank correlation coefficient is a measure of the similarity of the orderings of ranks of the data.

**critical success index (CSI)** Also called **threat score** (TS) or **Gilbert score** (GS), the CSI is a verification measure of deterministic binary categorical forecast performance equal to $a/(a+b+c)$, i.e. the total number of correct event forecasts (hits) divided by the total number of event forecasts plus the number of misses (hits + false alarms + misses). The CSI is not affected by

|  |  | Event observed | | |
|---|---|---|---|---|
|  |  | Yes | No | Total forecast |
| Event forecast | Yes | $a$ (hits) | $b$ (false alarms) | $a+b$ |
|  | No | $c$ (misses) | $d$ (correct rejections) | $c+d$ |
|  | Total observed | $a+c$ | $b+d$ | $a+b+c+d = n$ |

Cell count $a$ is the number of event forecasts that correspond to event observations, or the number of **hits**; cell count $b$ is the number of event forecasts that do not correspond to observed events, or the number of **false alarms**; cell count $c$ is the number of no-event forecasts corresponding to observed events, or the number of **misses**; and cell count $d$ is the number of no-event forecasts corresponding to no events observed, or the number of **correct rejections**. Forecast quality for this ($2 \times 2$) binary situation can be assessed using a surprisingly large number of different measures (see Chapter 3).

the number of non-event forecasts that are not observed (correct rejections) and is therefore strongly dependent upon the **base rate**.

**deterministic forecasts** Non-probabilistic forecasts of either a specific category or particular value for either a discrete or continuous variable. Deterministic forecasts of continuous variables are **point forecasts** that fail to provide any estimates of possible uncertainty, and this leads to less optimal decision-making than can be obtained using **probabilistic forecasts**. Deterministic forecasts are often interpreted as probabilistic forecasts having only probabilities of 0 and 1 (i.e.

no uncertainty), yet it is more realistic to interpret them as probabilistic forecasts in which the uncertainty is not provided (i.e. unknown uncertainty). Sometimes (confusingly) referred to as categorical forecasts in the earlier literature.

**discrimination** The sensitivity of the likelihood $p(\hat{x}|x)$ to different observed values of $x$. It can be measured for a particular forecast value $\hat{x}$ by the **likelihood** ratio $p(\hat{x}|x_1)/p(\hat{x}|x_2)$. A single overall summary measure is provided by the variance $\text{var}_X[E_{\hat{X}}(\hat{X}|X)]$ of the means of the forecasts conditioned (stratified) on the observations.

**ensemble forecasts** A set of forecasts generated by either starting from slightly different initial conditions (initial condition ensemble), forecasts made using different model parameters (perturbed physics ensemble), or forecasts made using different models (multi-model ensemble). Such samples of forecasts are increasingly used in weather and climate science to help quantify the effect of uncertainty in initial conditions and model formulation.

**equitable/equitability** A **metaverification** property for screening of suitable scores for deterministic categorical forecasts (Gandin and Murphy, 1992). An equitable score takes the same, no-skill value for random forecasts and for unvarying forecasts of a constant category. This criterion is based on the principle that random forecasts or constant forecasts of a category should have the same expected no-skill score (Murphy and Daan, 1985).

**Equitable Threat Score** See **Gilbert Skill Score**.

**false alarm** A deterministic forecast of a categorical event that fails to be observed. See **contingency table**.

**false alarm rate (F)** A verification measure of categorical forecast performance equal to the number of false alarms divided by the total number of events observed. For the $(2 \times 2)$ verification problem in the definition of **contingency table**, $F = b/(b+d)$. Also known as **probability of false detection (POFD)** and not to be confused with **false alarm** *ratio*.

**false alarm ratio (FAR)** A verification measure of categorical forecast performance equal to the number of false alarms divided by the total number of events forecast. For the $(2 \times 2)$ verification problem in the definition of **contingency table**, FAR $= b/(a+b)$. Not to be confused with **false**

**alarm** *rate*, which is conditioned on observations rather than forecasts.

**forecast evaluation** In the meteorological context, forecast evaluation implies the study of **forecast value** involving user-specific losses rather than **forecast quality**, which is addressed in **forecast verification**.

**forecast quality** A statistical description of how well the forecasts match simultaneous observations that provides important feedback on the forecasting system. Unlike **forecast value**, it aims to provide an overall summary of the agreement between forecasts and observations that does not depend on any particular user's requirements. Forecast quality is multidimensional with many different **attributes** that provide useful information on the performance.

**forecast value** The economic utility of forecasts for a particular set of forecast users. This is often based on simplistic cost-loss models, and for deterministic forecasts is strongly dependent on the marginal distributions of the forecasts and observations e.g. the **frequency bias**.

**forecast verification** The process of summarizing and assessing the overall **forecast quality** of previous sets of forecasts. Philosophically, the word *verification* is a misnomer since all forecasts eventually fail and so can only be *falsified* not *verified*. Verification is often referred to as **forecast evaluation** in other areas such as economic forecasting. See **forecast evaluation**; **validation**.

**frequency bias** For deterministic **categorical forecasts**, the **frequency bias** in marginal probabilities is estimated by the ratio of the total number of events forecast to the total number of events observed (i.e. $(a+b)/(a+c)$ for binary categorical forecasts – see **contingency table**).

**Gerrity scores** A useful class of equitable scores for multi-category deterministic forecasts. See Chapter 4.

**Gilbert score (GS)** See **critical success index (CSI)**.

**Gilbert Skill Score** A score widely used for the verification of deterministic forecasts of rare events (e.g. precipitation amounts above a large threshold). It was developed by Gilbert (1884) as a modification of the **threat score** to allow for the number of hits that would have been obtained purely by chance – see Schaefer (1990) and Doswell *et al.* (1990). It is sometimes referred to as the **Equitable Threat Score (ETS)**, but this is misleading as the score is not **equitable** – see Chapter 3. In terms of raw cell counts it is defined as

$$\frac{a - a_r}{a - a_r + b + c}$$

where $a_r = (a + b)(a + c)/n$ is the number of hits expected for forecasts independent of observations (pure chance). Note that the appearance of $n$ in the expression for $a_r$ means that the Gilbert Skill Score (unlike the **threat score**) depends explicitly on the number of correct rejections, $d$.

**hedging** Issuing a forecast that differs from the forecaster's 'true judgment', in order to improve either the score awarded or its long-run expectation.

**Heidke Skill Score (HSS)** A skill score of categorical forecast performance based on the **proportion correct (PC)** that takes into account the number of hits due to chance. Hits due to chance is given as the event relative frequency multiplied by the number of event forecasts.

**hindcast** A forecast of an event that has occurred in the past (an *ex post* forecast in economics) as opposed to a real-time forecast of a future event yet to occur (an *ex ante* forecast in economics). Forecast verification is entirely based on hindcast performance, which may benefit, in principle, from future information that would not necessarily be available for genuine forecasts of the future (e.g. observed values that occur later than the forecast issue date).

**hit** A forecasted categorical event that is observed to occur. See **contingency table**.

**hit rate (H)** A categorical forecast score equal to the total number of correct event forecasts (hits) divided by the total number of events observed, i.e. $a/(a + c)$ in the $(2 \times 2)$ contingency table. Also known as the **probability of detection (POD)**.

**ignorance score** See **logarithmic probability score**.

**joint distribution** The probability distribution defined over two or more variables. For independent events (i.e. no serial or spatial dependency), the joint distribution of the forecasts and observations, $p(\hat{x}, x)$, contains all of the probabilistic information relevant to the verification problem.

The joint distribution can be factored into **conditional distributions** and **marginal distributions** in either of two ways:

- the **calibration-refinement** factorization $p(\hat{x}, x) = p(x|\hat{x})p(\hat{x})$;
- the **likelihood-base rate** factorization $p(\hat{x}, x) = p(\hat{x}|x)p(x)$.

See Murphy and Winkler (1987) for an exposition on this general and powerful framework.

**likelihood** The probability $p(\hat{x}|x)$ of a forecast value given a particular observed value. The sensitivity of the likelihood to the observed value determines the **discrimination** of the system. Note that the concept of likelihood is fundamental in much of statistical inference, but the usage of the term in that context is somewhat different.

**likelihood-base rate factorization** See **joint distribution**.

**Linear Error in Probability Space (LEPS)** A dimensionless score for deterministic forecasts of continuous variables defined as the mean absolute difference between the cumulative frequency of the forecast and the cumulative frequency of the observed value,

$$\frac{1}{n}\sum_{i=1}^{n}|F_X(\hat{x}_i) - F_X(x_i)|.$$

There is also a version of LEPS for ordered **categorical forecasts**. For more on LEPS and related scores see Chapters 4 and 5.

**logarithmic probability score** The logarithmic score is a proper score for probability forecasts of a binary event $X = 0, 1$. It is defined as $B = E[-X \log \hat{p} - (1 - X)\log(1 - \hat{p})]$. This score has been referred to more recently as the **ignorance score**. See **scoring rule**.

**marginal distribution** The probability distribution of either the observable variable or the forecast variable, i.e. $p(x)$ or $p(\hat{x})$. The **joint distribution** can be factored into the product of the conditional distribution and the marginal distribution, e.g. $p(x, \hat{x}) = p(\hat{x}|x)p(x)$. See also **joint distribution**, **uncertainty**, **sharpness**, **refinement**.

**mean absolute error (MAE)** The mean of the absolute differences between the forecasts and observations $E(|\hat{X} - X|)$. A more robust measure of forecast accuracy than **mean squared error** that is somewhat more resistant to the presence of large outlier errors. MAE can be made dimensionless and more stable by dividing by the mean absolute deviation of the observations $E(|X - E(X)|)$ to yield a **relative absolute error**.

**mean error (ME)** The mean of the differences of the forecasts and observations $E(\hat{X} - X) = E(\hat{X}) - E(X)$. It is an overall measure of the unconditional bias of the forecasts (see **reliability**).

**mean squared error (MSE)** The mean of the squares of the differences of the forecasts and observations $E[(\hat{X} - X)^2]$. It is a widely used measure of forecast **accuracy**, which depends on **bias**, **resolution** and **uncertainty**. Because it is a quadratic loss function, it can be overly sensitive to large outlier forecast errors and is therefore a non-resistant measure (see **mean absolute error**). The MSE can sometimes encourage forecasters to hedge towards forecasting smaller than observed variations in order to reduce the risk of making a large error.

**metaverification** Measures of goodness of scores such as **propriety**, **equitability**, etc. Such properties can help in avoiding scores that might be misleading.

**miss** See **contingency table**.

**non-probabilistic forecast** See **deterministic forecast**.

**Peirce Skill Score (PSS)** A measure of deterministic binary categorical forecast performance equal to $a/(a + c) - b/(b + d)$, i.e. the ratio of hits to total number of events observed minus the ratio of false alarms to total number of non-events observed (see **contingency table**). It is analogous to the **Clayton Skill Score** except it is stratified on the observations rather than the forecasts. Originally proposed by Peirce (1884), it has also been referred to as *Kuipers' performance index*, *Hansen and Kuipers' score*, the *True Skill Statistic* and the *Youden index*.

**point forecasts** See **deterministic forecasts**.

**predictand** The observable object $x$ that is to be forecast. In regression, the predictand is known as the **response variable**, which is predicted using explanatory variables known as **predictors**. Scalar predictands can be nominal categories (e.g. snow, foggy, sunny), ordinal categories (e.g.

cold, normal, hot), discrete variables (e.g. number of hurricanes), or continuous variables (e.g. temperature).

**predictor** A forecast of either the value $\hat{x}$ of a **predictand** (deterministic forecasts) or the probability distribution $\hat{p}(x)$ of a predictand (probabilistic forecasts). In regression, the word 'predictor' is used to denote an explanatory variable that is used to predict the predictand.

**probabilistic forecast** A forecast that specifies the future probability $\hat{p}(x)$ of one or more events $x$ occurring. The set of events can be discrete (categorical) or continuous. **Deterministic forecasts** can be considered to be the special case of probability forecasts in which the forecast probabilities are always either 0 or 1 – there is never any prediction uncertainty in the predictand. However, it is perhaps more realistic to consider deterministic forecasts to be forecasts in which the prediction uncertainty in the predictand is not supplied as part of the forecast rather than ones in which the prediction uncertainty is exactly equal to 0. Subjective probability forecasts can be constructed by eliciting expert advice.

**probability of detection (POD)** See **hit rate (H)**.

**probability of false detection (POFD)** See **false alarm rate (F)**.

**proper/propriety** A **metaverification** property for screening of suitable scores for probabilistic forecasts. A *strictly proper* score is one for which the best expected score is only obtained when the forecaster issues probability forecasts consistent with their beliefs (e.g. the forecasting model is correct). Proper scores discourage forecasters from **hedging** their forecasted probabilities towards probabilities that are likely to score more highly. See Chapter 7 for more precise mathematical definitions.

**proportion correct (PC)** The proportion of correct binary categorical forecasts (hits and correct rejections) equal to $(a + d)/(a + b + c + d)$ for the $(2 \times 2)$ problem (see **contingency table**). When expressed as a percentage it is known as *percentage correct*.

**Ranked Probability Score (RPS)** An extension of the **Brier Score** to probabilistic categorical forecasts having more than two *ordinal* categories. By using cumulative probabilities, it takes into account the ordering of the categories.

**refinement** A statistical property of the forecasts that has multiple definitions in the verification literature. It can mean the marginal probability distribution of the forecasts, $p(\hat{x})$, as used in the phrase *calibration-refinement factorization* (see **joint distribution**). However, often it is more specifically used to refer to the spread of the marginal probability distribution of the forecasts. In addition, refinement is also used synonymously to denote the **sharpness** of probability forecasts. However, in the Bayesian statistical literature refinement appears to be defined somewhat differently to sharpness (see DeGroot and Fienberg, 1983).

**relative absolute error (RAE)** A dimensionless measure obtained by dividing the **mean absolute error** by the mean absolute deviation of the observed values $E(|X - E(X)|)$. This rescaling can be useful when comparing forecasts of observations with widely differing variance (e.g. precipitation in different locations).

**relative (or receiver) operating characteristic (ROC)** A signal detection curve for binary forecasts obtained by plotting a graph of the **hit rate** (vertical axis) versus the **false alarm rate** (horizontal axis) over a range of different thresholds. For deterministic forecasts of a continuous variable, the threshold is a value of the continuous variable used to define the binary event. For probabilistic forecasts of a binary event, the threshold is a probability decision threshold that is used to convert the probabilistic binary forecasts into deterministic binary forecasts.

**reliability** The same as **calibration**. It is related to Type 1 **conditional bias** $E_X(X|\hat{X}) - \hat{X}$ of the observations given the forecasts. Systems with zero conditional bias for all $\hat{X}$ are *perfectly reliable* (*well-calibrated*) and so have no need to be recalibrated (bias corrected) before use. Forecasting systems can be made more reliable by posterior recalibration; e.g. the transformed forecast quantity $\hat{X}' = E_X(X|\hat{X})$ is perfectly reliable. Similar ideas also apply to probabilistic forecasts where predictors $\hat{X}$ are replaced by forecast probabilities $\hat{p}(x)$.

**reliability diagram** A diagram in which the conditional expectation of a predictand for given values of a continuous predictor is plotted against the value of the predictor. For deterministic forecasts of continuous variables, this is a plot of $f(\hat{x}) = E_X(X|\hat{X} = \hat{x})$ versus $\hat{x}$, i.e. the means of the observations stratified on cases with specific forecast values versus the forecast values. For probabilistic forecasts of binary events, this is a plot of $f(q) = E_X(X|\hat{p} = q)$ versus the forecast probability value $q$. Perfectly reliable forecasts have points that lie on the line $f(\hat{x}) = \hat{x}$ (deterministic forecasts of continuous variables) or $f(q) = q$ (probabilistic forecasts of binary variables). Given enough previous forecasts, recalibrated future forecasts can be obtained to good approximation by using the reliability curve to non-linearly transform the forecasts: $\hat{x}' = f(\hat{x})$ or $\hat{p}' = f(\hat{p})$. In practice, reliability diagrams are sample estimates of the above quantities and so are prone to sampling uncertainty, which can lead to perfect forecasts not being on the line $f(\hat{x}) = \hat{x}$.

**resistant measure** A verification measure not unduly influenced by the presence of very large or small outlier values in the sample, e.g. **mean absolute deviation**.

**resolution** The sensitivity of the conditional probability $p(x|\hat{x})$ to different forecast values of $\hat{x}$. If a forecasting system leads to identical probability distributions of observed values for different forecast values, i.e. $p(x|\hat{x}_1) = p(x|\hat{x}_2)$, then the system has no resolution. Resolution is essential for a forecasting system to be able to discriminate observable future events. A single overall summary measure is provided by the variance of the conditional expectation $\text{var}_{\hat{X}}[E_X(X|\hat{X})]$.

**robust measure** A verification measure that is not overly sensitive to assumptions about the shape/form of the probability distribution of the variables, e.g. normality.

**root mean squared error (RMSE)** The square root of the **mean squared error**.

**sample climatology** The empirical distribution function of the observable variable based on a sample of past observations, i.e. relative frequencies of past events. See **base rate**.

**scoring rule** A function for assigning 'rewards' to a single forecast and a corresponding observation. Verification scores can then be constructed by summing the rewards over all pairs of forecasts and corresponding observations. One of the simplest scoring rules is the quadratic scoring rule known as the **Brier Score** for probability forecasts of binary events. Note that not all verification measures may be written as summable scores (e.g. correlation).

**sharpness** An attribute of the marginal distribution of the forecasts that aims to quantify the ability of the forecasts to 'stick their necks out'. In other words, how much the forecasts deviate from the mean climatological value/category for deterministic forecasts, or from the climatological mean probabilities for probabilistic forecasts. Unvarying climatological forecasts take no great risks and so have zero sharpness; perfect forecasts are as sharp as the time-varying observations. For deterministic forecasts of discrete or continuous variables, sharpness is sometimes approximated by the variance $\text{var}(\hat{X})$ of the forecasts. For **perfectly calibrated** forecasts where $E_X(X|\hat{X}) = \hat{X}$, the sharpness $\text{var}(\hat{X})$ becomes identical to the **resolution** $\text{var}_{\hat{X}}[E_X(X|\hat{X})]$ of the forecasts. For probabilistic forecasts, although sharpness can also be defined by the variance $\text{var}(\hat{p})$, it is often frequently defined in terms of the *information content* (*negative entropy*) $I = E(\hat{p} \log \hat{p})$ of the forecasts. High-risk forecasts in which $\hat{p}$ is either 0 or 1 have maximum information content and are said to be *perfectly sharp*. By interpreting deterministic forecasts as probabilistic forecasts with zero prediction uncertainty in the predictand, deterministic forecasts may be considered to be perfectly sharp probabilistic forecasts. However, it is perhaps more realistic to consider deterministic forecasts to be ones in which the prediction uncertainty in the predictand is not supplied as part of the forecast rather than ones in which the prediction uncertainty is exactly equal to zero. Hence, a deterministic forecast can be considered to be a deterministic forecast with spread/sharpness $\text{var}(\hat{X})$, yet at the same time can also be considered to be a probability forecast with perfect sharpness. The word **refinement** is also sometimes used to denote sharpness.

**skill score** A relative measure of the quality of the forecasting system compared to some benchmark or reference forecast. Commonly used reference forecasts include mean climatology, persistence (random walk forecast) or output from an earlier version of the forecasting system. There are as many skill scores as there are possible scores and they are usually based on the expression

$$\text{SS} = \frac{S - S_0}{S_1 - S_0} \times 100\%$$

where $S$ is the forecast score, $S_0$ is the score for the benchmark forecast and $S_1$ is the best possible score. The skill scores generally lie in the range 0 to 1 but can in practice be negative when using good benchmark forecasts (e.g. previous versions of the forecasting system). Compared to raw scores, skill scores have the advantage that they help take account of non-stationarities in the system to be forecast. For example, improved forecast scores often occur during periods when the atmosphere is in a more persistent state.

**spread-skill relationship** A statistical **association** between the spread of a set of **ensemble forecasts** and the resulting skill of the forecasts. For example, a positive spread-skill association could be used to deduce that the ensemble forecasts will generally be more skilful when the ensemble forecast spread is small.

**success ratio (SR)** A categorical binary measure equal to the number of hits divided by the total number of events predicted $a/(a + b)$. It is conditioned on the forecasts, unlike the **hit rate**, which is conditioned on the observations.

**sufficiency** A concept introduced into forecast evaluation by DeGroot and Fienberg (1983), and developed by Ehrendorfer and Murphy (1988) and Krzysztofowicz and Long (1991) among others. When it can be demonstrated, sufficiency provides an unequivocal ordering on the quality of forecasts. When two forecasting systems, A and B, say, are being compared, A's forecasts are said to be sufficient for B's if forecasts with the same skill as B's can be obtained from A's by a stochastic transformation. Applying a stochastic transformation to A's forecasts is equivalent to randomizing the forecasts, or passing them through a noisy channel (DeGroot and Fienberg, 1983). Note that sufficiency is an important property in much of statistical inference, but the usage of the term is somewhat different in that context.

**threat score (TS)** See **critical success index (CSI)**.

**uncertainty** The natural variability of the observable variable. For example, in the **Brier Score** decomposition, uncertainty is the variance of the binary observable variable. It is an important aspect in the performance of a forecasting system, over which the forecaster has no control.

**validation** Forecast validation generally has the same meaning as forecast verification in meteorological forecasting. However, in other disciplines such as product testing (e.g. software engineering) there is often a clear distinction. Verification asks the question, 'Are we building the product right?' (i.e. quality assurance – do the products satisfy all the expected specifications and requirements?) whereas validation asks the question, 'Are we building the right product?' (i.e. quality control – does the actual product do what the user requires?). Forecast validation could refer to the detailed assessment of the forecast of one specific event, whereas **forecast verification** is an assurance of the overall quality of all the forecasts.

# References

Åberg, S., Lindgren, F., Malmberg, A. *et al.* (2005) An image warping approach to spatio-temporal modelling. *Environmetrics*, **16**, 833–848.

Agresti, A. (2007) *An Introduction to Categorical Data Analysis*, 2nd edn. New York: Wiley.

Agresti, A. and Coull, B.A. (1998) Approximate is better than 'exact' for interval estimation of binomial proportions. *Am. Stat.*, **52**, 119–126.

Ahijevych, D., Gilleland, E., Brown, E. and Ebert, E. (2009) Application of spatial forecast verification methods to gridded precipitation forecasts. *Weather Forecast.*, **24**, 1485–1497.

Allen, D.M. (1974) The relationship between variable selection and data augmentation. *Technometrics*, **16**, 125–127.

Altman, D.G. and Royston, P. (2000) What do we mean by validating a prognostic model? *Stat. Med.*, **19**, 453–473.

Anderson, D.L.T., Stockdale, T., Balmaseda, M. *et al.* (2007) Development of the ECMWF seasonal forecast System 3. *ECMWF Technical Memorandum*, **503**, 56 pp.

Anderson, J.L. (1996) A method for producing and evaluating probabilistic forecasts from ensemble model integrations. *J. Climate*, **9**, 1518–1530.

Anderson, T.W. and Darling, D.A. (1952) Asymptotic theory of certain "goodness of fit" criteria based on stochastic processes. *Ann. Math. Stat.*, **23**, 193–212.

Ångström, A. (1922) On the effectivity of weather warnings. *Nordisk Statistisk Tidskrift*, **1**, 394–408.

Armstrong, J.S. (2001) Evaluating forecasting methods. In: *Principles of Forecasting: A Handbook for Researchers and Practitioners* (ed. J.S. Armstrong). Norwell, MA: Kluwer, pp. 443–472.

Armstrong, J.S. and Collopy, F. (1992) Error measures for generalizing about forecasting methods: Empirical comparisons. *Int. J. Forecasting*, **8**, 69–80.

Armstrong, J.S. and Fildes, R. (1995) On the selection of error measures for comparisons among forecasting methods. *J. Forecasting*, **14**, 67–71.

Atger, F. (1999) The skill of ensemble prediction systems. *Mon. Weather Rev.*, **127**, 1941–1953.

Atger, F. (2001) Verification of intense precipitation forecasts from single models and ensemble prediction systems. *Nonlinear. Proc. Geoph.*, **8**, 401–417.

Atger, F. (2004) Estimation of the reliability of ensemble based probabilistic forecasts. *Q. J. Roy. Meteor. Soc.*, **130**, 627–646.

Baddeley, A.J. (1992) An error metric for binary images and a $L^p$ version of the Hausdorff metric. *Niuw Archief voor Wiskunde*, **10**, 157–183.

Baker, S.G. (2000) Identifying combinations of cancer markers for further study as triggers of early intervention. *Biometrics*, **56**, 1082–1087.

Bamber, D. (1975) The area above the ordinal dominance graph and the area below the receiver operating characteristic graph. *J. Math. Psychol.*, **12**, 387–415.

Barnes, L.R., Gruntfest, E.C., Hayden, M.H. *et al.* (2007) False alarms and close calls: a conceptual model of warning accuracy. *Weather Forecast.*, **22**, 1140–1147.

Barnes, L.R., Schultz, D.M., Gruntfest, E.C. *et al.* (2009) Corrigendum: False alarm rate or false alarm ratio? *Weather Forecast.*, **24**, 1452–1454.

Barnett, T.P. Preisendorfer, R.W., Goldstein, L.M. and Hasselmann, K. (1981) Origins and levels of monthly and seasonal forecast skill for United States surface temperature determined by canonical correlation analysis. *Mon. Weather Rev.*, **115**, 1825–1850.

Barnston, A.G. (1992) Correspondence among the correlation, RMSE, and Heidke forecast verification measures; refinement of the Heidke score. *Weather Forecast.*, **7**, 699–709.

Barnston, A.G. and Mason, S.J. (2011) Evaluation of IRI's seasonal climate forecasts for the extreme 15% tails. *Weather Forecast.*, **26**, 545–554.

Barnston, A.G., Li, S., Mason, S.J. *et al.* (2010) Verification of the first 11 years of IRI's seasonal climate forecasts. *J. Appl. Meteorol. Climatol.*, **49**, 493–520.

Benedetti, R. (2010) Scoring rules for forecast verification. *Mon. Weather Rev.*, **138**, 203–211.

Berliner, L.M., Wikle, C.K. and Cressie, N. (2000) Long-lead prediction of Pacific SSTs via Bayesian dynamic modeling. *J. Climate*, **13**, 3953–3968.

Bernardo, J.M. (1979) Expected information as expected utility. *Ann. Stat.*, **7**, 686–690.

Berrocal, V.J., Raftery, A.E., Gneiting, T. and Steed, R. (2010) Probabilistic forecasting for winter road maintenance. *J. Am. Stat. Assoc.*, **105**, 522–537.

Birdsall, T.G. (1966) The theory of signal detectability: ROC curves and their character. *Dissertation Abstracts International*, **28**, 1B.

Blackwell, D. and Girshick, M. (1979) *Theory of Games and Statistical Decisions*. New York: Dover Publications.

Bland, M. (1995) *An Introduction to Medical Statistics*. Oxford: Oxford University Press.

Blanford, H.H. (1884) On the connection of Himalayan snowfall and seasons of drought in India. *P. Roy. Soc. London*, **37**, 3–22.

Bao, L., Gneiting, T., Grimit, E.P. *et al.* (2010) Bias correction and Bayesian model averaging for ensemble forecasts of surface wind direction. *Mon. Weather Rev.*, **138**, 1811–1821.

Bowler, N.E. (2008) Accounting for the effect of observation errors on verification of MOGREPS. *Meteorol. Appl.*, **15**, 199–205.

Bradley, A.A., Hashino, T. and Schwartz, S.S. (2003) Distributions-oriented verification of probability forecasts for small data samples. *Weather Forecast.*, **18**, 903–917.

Breiman, L. (1973) *Probability*. Reading, MA: Addison-Wesley.

Brier, G.W. (1950) Verification of forecasts expressed in terms of probability. *Mon. Weather Rev.*, **78**, 1–3.

Brier, G.W. and Allen, R.A. (1951) Verification of weather forecasts. In: *Compendium of Meteorology* (ed. T.F. Malone). Boston: American Meteorological Society, pp. 841–848.

Briggs, W.M. and Levine, R.A. (1997) Wavelets and field forecast verification. *Mon. Weather Rev.*, **125**, 1329–1341.

Bröcker, J. (2008) On reliability analysis of multi-categorical forecasts. *Nonlinear Proc. Geoph.*, **15**, 661–673.

Bröcker, J. (2009) Reliability, sufficiency, and the decomposition of proper scores. *Q. J. Roy. Meteor. Soc.*, **135**, 1512–1519.

Bröcker, J. (2010) Regularized logistic models for probabilistic forecasting and diagnostics. *Mon. Weather Rev.*, **138**, 592–604.

Bröcker, J. and Smith, L.A. (2007a) Increasing the reliability of reliability diagrams. *Weather Forecast.*, **22**, 651–661.

Bröcker, J. and Smith, L.A. (2007b) Scoring probabilistic forecasts: The importance of being proper. *Weather Forecast.*, **22**, 382–388.

Bröcker, J. and Smith, L.A. (2008) From ensemble forecasts to predictive distribution functions. *Tellus A*, **60**, 663–678.

Bröcker, J., Siegert, S. and Kantz, H. (2011) Comment on "conditional exceedance probabilities". *Mon. Weather Rev.* (in press).

Brooks, H.E. (2004) Tornado-warning performance in the past and future. *B. Am. Meteorol. Soc.*, **85**, 837–843.

Brooks, H.E. and Doswell, C.A. (1996) A comparison of measures-oriented and distributions-oriented approaches to forecast verification. *Weather Forecast.*, **11**, 288–303.

Bross, I.D.J. (1953) *Design for Decision*. New York: Macmillan.

Brown, B.G. and Katz, R.W. (1991) Use of statistical methods in the search for teleconnections: past, present, and future. In: *Teleconnections Linking Worldwide Climate Anomalies: Scientific Basis and Societal Impact* (eds M.H. Glantz, R.W. Katz and N. Nicholls). Cambridge: Cambridge University Press, pp. 371–400.

Brown, J., Demargne J., Seo D. and Liu Y. (2010) The Ensemble Verification System (EVS): A software tool for verifying ensemble forecasts of hydrometeorological and hydrologic variables at discrete locations. *Environ. Modell. Softw.*, **25**, 854–872.

Brown, T.A. (1970) Probabilistic forecasts and reproducing scoring systems. Technical Report RM-6299-ARPA. Santa Monica, CA: RAND Corporation.

Buizza, R., Hollingsworth, A., LaLaurette, F. and Ghelli, A. (1999) Probabilistic predictions using the ECMWF ensemble prediction system. *Weather Forecast.*, **14**, 168–189.

Buizza, R., Barkmeijer, J., Palmer, T.N. and Richardson, D.S. (2000) Current status and future developments of the ECMWF Ensemble Prediction System. *Meteorol. Appl.*, **7**, 163–175.

Caminade, C. and Terray, L. (2010) Twentieth century Sahel rainfall variability as simulated by the ARPEGE AGCM, and future changes. *Clim. Dynam.*, **35**, 75–94.

Campbell, S.D. and Diebold, F.X. (2005) Weather forecasting for weather derivatives. *J. Am. Stat. Assoc.*, **100**, 6–16.

Candille, G. and Talagrand, O. (2005) Evaluation of probabilistic prediction systems for a scalar variable. *Q. J. Roy. Meteor. Soc.*, **131**, 2131–2150.

Casati, B. (2010) New developments of the intensity-scale technique within the Spatial Verification Methods Inter-Comparison Project. *Weather Forecast.*, **25**, 113–143.

Casati, B. and Wilson, L. (2007) A new spatial-scale decomposition of the Brier Score: Application to the verification of lightning probability forecasts. *Mon. Weather Rev.*, **135**, 3052–3069.

Casati, B., Ross, G. and Stephenson, D.B. (2004) A new intensity-scale approach for the verification of spatial precipitation forecasts. *Met. Apps.* **11**, 141–154.

Casati, B., Wilson, L.J., Stephenson, D.B. *et al.* (2008) Forecast verification: Current status and future directions. *Meteorol. Appl.*, **15**, 3–18.

Centor, R.M. (1991) Signal detectability: the use of ROC curves and their analyses. *Med. Decis. Making*, **11**, 102–106.

Cervera, J.L. and Muñoz, J. (1996) Proper scoring rules for fractiles. In: *Bayesian Statistics 5*, (eds J.M. Bernardo, J.O. Berger, A.P. Dawid and A.F.M. Smith). Oxford: Oxford University Press, pp. 513–519.

Chatfield, C. (2001) *Time Series Forecasting*. Boca Raton, FL: Chapman & Hall/CRC Press.

Choulakian, V., Lockhart, R.A., and Stephens, M.A. (1994) Cramér-von Mises statistics for discrete distributions. *Can. J. Stat.*, **22**, 125–137.

Christoffersen, P.F. (1998) Evaluating interval forecasts. *Int. Econ. Rev.*, **39**, 841–862.

Clayton, H.H. (1934) Rating weather forecasts. *B. Am. Meteorol. Soc.*, **15**, 279–283.

Cohen, J. (1960) A coefficient of agreement for nominal scales. *Educ. Psychol. Meas.*, **20**, 37–46.

Considine, T.J., Jablonowski, C., Posner, B. and Bishop, C.H. (2004) The value of hurricane forecasts to oil and gas producers in the Gulf of Mexico. *J. Appl. Meteor.*, **43**, 1270–1281.

Cressie, N. (1993) *Statistics for Spatial Data*. New York: Wiley.

Crosby, D.S., Ferraro, R.R. and Wu, H. (1995) Estimating the probability of rain in an SSM/I FOV using logistic regression. *J. Appl. Meteorol.*, **34**, 2476–2480.

Daan, H. (1984) *Scoring Rules in Forecast Verification*. WMO Short- and Medium-Range Weather Prediction Research Publication Series No. 4.

Daniels, H.E. (1944) The relation between measures of correlation in the universe of sample permutations. *Biometrika*, **33**, 129–135.

Davis, C.A., Brown, B.G. and Bullock, R.G. (2006a) Object-based verification of precipitation forecasts, Part I: Methodology and application to mesoscale rain areas. *Mon. Weather Rev.* **134**, 1772–1784.

Davis, C.A., Brown, B.G. and Bullock, R.G. (2006b) Object-based verification of precipitation forecasts, Part II: Application to convective rain systems. *Mon. Weather Rev.* **134**, 1785–1795.

Davis, C.A., Brown, B.G., Bullock, R.G. and Halley-Gotway, J. (2009) The Method for Object-based Diagnostic Evaluation (MODE) applied to WRF forecasts from the 2005 NSSL/SPC Spring Program. *Weather Forecast.*, **24**, 1252–1267.

Davis, R. (1976) Predictability of sea surface temperature and sea level pressure anomalies over the North Pacific Ocean. *J. Phys. Oceanogr.*, **6**, 249–266.

Davison, A.C. and Hinkley, D.V. (1997) *Bootstrap Methods and their Application*. New York: Cambridge University Press.

Dawid, A.P. (1982) The well-calibrated Bayesian. *J. Am. Stat. Assoc.*, **77**, 605–610.

Dawid, A.P. (1986) Probability forecasting. In: *Encyclopedia of Statistical Sciences, Vol. 7* (eds. S. Kotz, N.L. Johnson and C.B. Read). New York: Wiley, pp. 210–218.

de Finetti, B. (1970) Logical foundations and measurement of subjective probability. *Acta Psychol.*, **34**, 129–145.

DeGroot, M.H. (1986) *Probability and Statistics*, 2nd edn. Reading, MA: Addison-Wesley.

DeGroot, M.H. and Fienberg, S.E. (1982) Assessing probability assessors: calibration and refinement. In: *Statistical Decision Theory and Related Topics III, Vol. 1* (eds S.S. Gupta and J.O. Berger). New York: Academic Press, pp. 291–314.

DeGroot, M.H. and Fienberg, S.E. (1983) The comparison and evaluation of forecasters. *Statistician*, **32**, 12–22.

DelSole, T. and Shukla, J. (2010) Model fidelity versus skill in seasonal forecasting. *J. Climate*, **23**, 4794–4806.

Déqué, M. (1991) Removing the model systematic error in extended range forecasting. *Ann. Geophys.*, **9**, 242–251.

Déqué, M. (1997) Ensemble size for numerical seasonal forecasts. *Tellus*, **49A**, 74–86.

Déqué, M. and Royer, J.F. (1992) The skill of extended-range extratropical winter dynamical forecasts. *J. Climate*, **5**, 1346–1356.

Déqué, M., Royer, J.F. and Stroe, R. (1994) Formulation of Gaussian probability forecasts based on model extended-range integrations. *Tellus*, **46A**, 52–65.

Diebold, F.X. and Lopez, J. (1996) Forecast evaluation and combination. In: *Handbook of Statistics* (eds G.S. Maddala and C.R. Rao). Amsterdam: North-Holland, pp. 241–268.

Diebold, F.X. and Mariano, R.S. (1995) Comparing predictive accuracy. *J. Bus. Econ. Stat.*, **13**, 253–263.

Diebold, F.X., Hahn, J. and Tay, A. (1999) Multivariate density forecast evaluation and calibration in financial risk management: High-frequency returns on foreign exchange. *Rev. Econ. Stat.*, **81**, 661–673.

Doblas-Reyes, F.J., Hagedorn, R., Palmer, T.N. and Morcrette, J-J. (2006) Impact of increasing greenhouse gas concentrations in seasonal ensemble forecasts. *Geophys. Res. Lett.*, **33**, article number L07708.

Doblas-Reyes, F.J., Coelho, C.A.S. and Stephenson, D.B. (2008) How much does simplification of probability forecasts reduce forecast quality? *Meteorol. Appl.*, **15**, 155–162.

Doblas-Reyes, F.J., Weisheimer, A., Déqué, M. *et al.* (2009) Addressing model uncertainty in seasonal and annual dynamical ensemble forecasts. *Q. J. Roy. Meteor. Soc.*, **135**, 1538–1559.

Donaldson, R.J., Dyer, R.M. and Kraus, M.J. (1975) An objective evaluator of techniques for predicting severe weather events. Preprints, Ninth Conference on Severe Local Storms*, Norman, Oklahoma. American Meteorological Society*, pp. 321–326.

Dong, B-W., Sutton, R.T., Jewson, S.P. *et al.* (2000) Predictable winter climate in the North Atlantic sector during the 1997-1999 ENSO cycle. *Geophys. Res. Lett.*, **27**, 985–988.

Doolittle, M.H. (1885) The verification of predictions. *Bull. Philos. Soc. Washington*, **7**, 122–127.

Doolittle, M.H. (1888) Association ratios. *Bull. Philos. Soc. Washington*, **10**, 83–87, 94–96.

Doswell, C.A., Davies-Jones, R. and Keller, D.L. (1990) On summary measures of skill in rare event forecasting based on contingency tables. *Weather Forecast.*, **5**, 576–585.

Dowd, K. (1998) *Beyond Value at Risk: The New Science of Risk Management*. Chichester: Wiley.

Dowd, K. (2005) *Measuring Market Risk*, 2nd edn. Chichester: Wiley.

Dowd, K. (2007) Validating multi-period density-forecasting models *J. Forecasting*, **26**, 251–270.

Draper, D. (1995) Assessment and propagation of model uncertainty. *J. Roy. Stat. Soc. B*, **57**, 45–97.

Draper, N.R. and Smith, H. (1998) *Applied Regression Analysis*, 3rd edn. New York: Wiley.

Dunstone, N.J. and Smith, D.M. (2010) Impact of atmosphere and sub-surface ocean data on decadal climate prediction. *Geophys. Res. Lett.*, **37**, article number L02709.

Ebert, E.E. (2008) Fuzzy verification of high resolution gridded forecasts: A review and proposed framework. *Meteorol. Appl.*, **15**, 51–64.

Ebert, E.E. (2009) Neighbourhood verification: a strategy for rewarding close forecasts. *Weather Forecast.*, **24**, 1498–1510.

Ebert, E.E. and McBride, J.L. (2000) Verification of precipitation in weather systems: Determination of systematic errors. *J. Hydrol.*, **239**, 179–202.

Efron, B. and Tibshirani, R.J. (1993) *An Introduction to the Bootstrap*. London: Chapman & Hall.

Ehrendorfer M. (2006) The Liouville equation and atmospheric predictability. In: *Predictability of Weather and Climate* (eds T. Palmer and R. Hagedorn). Cambridge: Cambridge University Press, pp. 59–98.

Ehrendorfer, M. and Murphy, A.H. (1988) Comparative evaluation of weather forecasting systems: sufficiency, quality and accuracy. *Mon. Weather Rev.*, **116**, 1757–1770.

Elmore, K.L. (2005) Alternatives to the chi-square test for evaluating rank histograms from ensemble forecasts. *Weather Forecast.*, **20**, 789–795.

Elsner, J.B. and Schmertmann, C.P. (1994) Assessing forecast skill through cross validation. *Weather Forecast.*, **9**, 619–624.

Epstein, E.S. (1966) Quality control for probability forecasts. *Mon. Weather Rev.*, **94**, 487–494.

Epstein, E.S. (1969) A scoring system for probability forecasts of ranked categories. *J. Appl. Meteorol.*, **8**, 985–987.

Everitt, B.S. (1992) *The Analysis of Contingency Tables*, 2nd edn. London: Chapman & Hall.

Fawcett, R. (2008) Verification techniques and simple theoretical forecast models. *Weather Forecast.*, **23**, 1049–1068.

Feller, W. (1966) *An Introduction to Probability Theory and Its Applications, Vol. 1*. New York: Wiley.

Ferro, C.A.T. (2007a) Comparing probabilistic forecasting systems with the Brier Score. *Weather Forecast.*, **22**, 1076–1088.

Ferro, C.A.T. (2007b) A probability model for verifying deterministic forecasts of extreme events. *Weather Forecast.*, **22**, 1089–1100.

Ferro, C.A.T. and Stephenson, D.B. (2011) Extremal Dependence Indices: improved verification measures for deterministic forecasts of rare binary events. *Weather Forecast.* (in press).

Ferro, C.A.T., Richardson, D.S. and Weigel, A.P. (2008) On the effect of ensemble size on the discrete and continuous ranked probability scores. *Meteorol. Appl.*, **15**, 19–24.

Fildes, R. and Kourentzes, N. (2011) Validation and forecasting accuracy in models of climate change. *Int. J. Forecasting*, **27**, 968–1005.

Finley, J.P. (1884) Tornado predictions. *Am. Meteorol. J.*, **1**, 85–88.

Flueck, J.A. (1987) A study of some measures of forecast verification. *Preprints, 10th Conference on Probability and Statistics in Atmospheric Science,* Edmonton, Canada. American Meteorological Society, pp. 69–73.

Fraley, C., Raftery, A., Sloughter, J.M. and Gneiting, T. (2010) ensembleBMA: Probabilistic Forecasting using Ensembles and Bayesian Model Averaging. R package version 4.5 (http://CRAN.R-project.org/package=ensembleBMA).

Fraley, C., Raftery A., Gneiting, T. *et al.* (2011) Probabilistic weather forecasting in R. *R Journal* 3, 55–63.

Gabriel, K.R. (2000) Parallels between statistical issues in medical and meteorological experimentation. *J. Appl. Meteorol.*, **39**, 1822–1836.

Gallus, W.A. (2009) Application of object-based verification techniques to ensemble precipitation forecasts. *Weather Forecast.*, **25**, 144–158.

Gandin, L.S. and Murphy, A.H. (1992) Equitable scores for categorical forecasts. *Mon. Weather Rev.*, **120**, 361–370.

Garthwaite, P.H., Jolliffe, I.T. and Jones, B. (2002) *Statistical Inference,* 2nd edn. Oxford: Oxford University Press.

Gates, W.L. (1992) AMIP: the Atmospheric Model Intercomparison Project. *B. Am. Meteorol. Soc.*, **73**, 1962–1970.

Gent, P.R., Yeager, S.G., Neale, R.B. *et al.* (2010) Improvements in a half degree atmosphere/land version of the CCSM. *Clim. Dynam.*, **34**, 819–833.

Gerrity, J.P. (1992) A note on Gandin and Murphy's equitable skill score. *Mon. Weather Rev.*, **120**, 2709–2712.

Gesù, V.D. and Starovoitov, V. (1999) Distance-based functions for image comparison, *Pattern Recogn. Lett.*, **20**, 207–214.

Giannini, A. (2010) Mechanisms of climate change in the semiarid African Sahel: the local view. *J. Climate*, **23**, 743–756.

Gibson, J.K., Kallberg, P., Uppala, S. *et al.* (1997) *ERA Description.* ECMWF Reanalysis project report series. Shinfield Park, Reading, UK: ECMWF.

Gilbert, G.K. (1884) Finley's tornado predictions. *Am. Meteorol. J.*, **1**, 166–172.

Gilleland, E. (2011) Spatial forecast verification: Baddeley's Delta metric applied to the ICP test cases. *Weather Forecast.*, **26**, 409–415.

Gilleland, E., Lee, T.C.M., Halley-Gotway, J. *et al.* (2008) Computationally efficient spatial forecast verification using Baddeley's Δ image metric. *Mon. Weather Rev.* **136**, 1747–1757.

Gilleland, E., Ahijevych, D., Brown, B.G. *et al.* (2009) Inter-comparison of spatial verification methods. *Weather Forecast.*, **24**, 1416–1430.

Gilleland, E., Ahijevych, D., Brown, B.G. and Ebert, E.E. (2010a) Verifying forecasts spatially. *B. Am. Meteorol. Soc.*, **91**, 1365–1373.

Gilleland, E., Lindström, J. and Lindgren, F. (2010b) Analyzing the image warp forecast verification method on precipitation fields from the ICP. *Weather Forecast.*, **25**, 1249–1262.

Glahn, B. (2005) Tornado-warning performance in the past and future – another perspective. *B. Am. Meteorol. Soc.*, **86**, 1135–1141.

Glahn, H.R. (2004) Discussion of verification concepts in *Forecast Verification: A Practitioner's Guide in Atmospheric Science. Weather Forecast.*, **19**, 769–775.

Glahn, H.R. and Jorgensen, D.L. (1970) Climatological aspects of the Brier P-score. *Mon. Weather Rev.*, **98**, 136–141.

Gneiting, T. (2011) Evaluating point forecasts. *J. Am. Stat. Assoc.*, **106**, 746–762.

Gneiting, T. and Raftery, A.E. (2007) Strictly proper scoring rules, prediction and estimation. *J. Am. Stat. Assoc.*, **102**, 359–378.

Gneiting, T., Raftery, A., Westveld, A.H. and Goldmann, T. (2005) Calibrated probabilistic forecasting using ensemble model output statistics and minimum CRPS estimation. *Mon. Weather Rev.*, **133**, 1098–1118.

Gneiting, T., Stanberry, L.I., Grimit, E.P. *et al.* (2008) Assessing probabilistic forecasts of multivariate quantities, with an application to ensemble predictions of surface winds. *Test*, **17**, 211–235.

Gombos, D., Hansen, J.A., Du, J. and McQueen, J. (2007) Theory and applications of the minimum spanning tree rank histogram *Mon. Weather Rev.*, **135**, 1490–1505.

Good, I.J. (1952) Rational decisions. *J. Roy. Stat. Soc.*, **14**, 107–114.

Goodman, L.A. and Kruskal, W.H. (1959) Measures of association for cross classifications. II: Further discussion and references. *J. Am. Stat. Assoc.*, **54**, 123–163.

Goodman, L.A. and Kruskal, W.H. (1979) *Measures of Association for Cross Classifications.* New York: Springer.

Granger, C.W.J. and Jeon, Y. (2003) A time-distance criterion for evaluating forecasting models. *Int. J. Forecasting*, **19**, 199–215.

Granger, C.W.J. and Pesaran, M.H. (1996) A decision theoretic approach to forecast evaluation. Discussion paper 96-23. Department of Economics, University of California, San Diego.

Grant, A. and Johnstone, D. (2010) Finding profitable forecast combinations using probability scoring rules. *Int. J. Forecasting*, **26**, 498–510.

Gringorten, I.I. (1951) The verification and scoring of weather forecasts. *J. Am. Stat. Assoc.*, **46**, 279–296.

Gu, W. and Pepe, M.S. (2009) Estimating the capacity for improvement in risk prediction with a marker. *Biostatistics*, **10**, 172–186.

Hagedorn, R. and Smith, L.A. (2009) Communicating the value of probabilistic forecasts with weather roulette. *Meteorol. Appl.*, **16**, 143–155.

Halmos, P. (1974) *Measure Theory*. New York: Springer.

Hamill, T.M. (1999) Hypothesis tests for evaluating numerical precipitation forecasts. *Weather and Forecast.*, **14**, 155–167.

Hamill, T.M. (2001) Interpretation of rank histograms for verifying ensemble forecasts. *Mon. Weather Rev.*, **129**, 550–560.

Hamill, T.M. and Colucci, S.J. (1997) Verification of Eta-RSM short-range forecasts. *Mon. Weather Rev.*, **125**, 1312–1327.

Hamill, T.M. and Colucci, S.J. (1998) Evaluation of Eta-RSM ensemble probabilistic precipitation forecasts. *Mon. Weather Rev.*, **126**, 711–724.

Hamill, T.M. and Juras, J. (2006) Measuring forecast skill: is it real skill or is it the varying climatology? *Q. J. Roy. Meteor. Soc.*, **132**, 2905–2923.

Hamill, T.M. and Whitaker, J.S. (2006) Probabilistic quantitative precipitation forecasts based on reforecast analogs: Theory and application. *Mon. Weather Rev.*, **134**, 3209–3229

Hamill, T.M., Whitaker, J.S. and Wei, X. (2004) Ensemble reforecasting: improving medium-range forecast skill using retrospective forecasts. *Mon. Weather Rev.*, **132**, 1434–1447.

Hansen, J.W., Mason, S.J., Sun, L. and Tall, A. (2011) Review of seasonal climate forecasting for agriculture in sub-Saharan Africa. *Exp. Agr.*, **47**, 205–240.

Hanssen, A.W. and Kuipers, W.J.A. (1965) On the relationship between the frequency of rain and various meteorological parameters. *Mededelingen en Verhandelingen*, **81**, 2–15.

Harris, D., Foufoula-Georgiou, E., Droegemeier, K.K. and Levit, J.J. (2001) Multiscale statistical properties of a high-resolution precipitation forecast. *J. Hydrometeorol.*, **2**, 406–418.

Harvey, L.O. Jr, Hammond, K.R., Lusk, C.M. and Mross, E.F. (1992) The application of signal detection theory to weather forecasting behaviour. *Mon. Weather Rev.*, **120**, 863–883.

Hastie, T., Tibshirani, R. and Friedman, J. (2001) *The Elements of Statistical Learning*. New York: Springer.

Hawkins, E. and Sutton, R. (2009) The potential to narrow uncertainty in regional climate predictions. *B. Am. Meteorol. Soc.*, **90**, 1095–1107.

Heffernan, J.E. (2000) A directory of coefficients of tail dependence. *Extremes*, **3**, 279–290.

Heidke, P. (1926) Berechnung der erfolges und der gute der windstarkevorhersagen im sturmwarnungdienst. *Geografiska Annaler*, **8**, 301–349.

Hersbach, H. (2000) Decomposition of the continuous ranked probability score for ensemble prediction systems. *Weather Forecast.*, **15**, 559–570.

Hoffrage, U., Lindsey, S., Hertwig, R. and Gigerenzer, G. (2000) Communicating statistical information. *Science*, **290**, 2261–2262.

Hoffschildt, M., Bidlot, J-R., Hansen, B. and Janssen, P.A.E.M. (1999) Potential benefit of ensemble forecasts for ship routing. ECMWF Technical Memorandum 287, Reading, UK: ECMWF.

Hogan, R.J., O'Connor, E.J., Illingworth, A.J. (2009) Verification of cloud fraction forecasts. *Q. J. Roy. Meteor. Soc.*, **135**, 1494–1511.

Hogan, R.J., Ferro, C.A.T., Jolliffe, I.T. and Stephenson, D.B. (2010) Equitability revisited: Why the 'equitable threat score' is not equitable. *Weather Forecast.*, **25**, 710–726.

Hogg, R.V. and Tanis, E.A. (1997) *Probability and Statistical Inference*, 5th edn. Upper Saddle River, NJ: Prentice Hall.

Hohenegger, C. and Schär, C. (2007) Atmospheric predictability at synoptic versus cloud-resolving scales. *B. Am. Meteorol. Soc.*, **88**, 1783–1793.

Holloway, J.L. and Woodbury, M.A. (1955) *Application of Information Theory and Discriminant Function to Weather Forecasting and Forecast Verification*. Philadelphia: University of Pennsylvania, Institute for Cooperative Research.

Huberty, C.J. (1994) *Applied Discriminant Analysis*. New York: Wiley.

Hyndman, R.J. and Koehler, A.B. (2006) Another look at measures of forecast accuracy. *Int. J. Forecasting*, **22**, 679–688.

Jakob, C. (2010) Accelerating progress in global model development through improved parameterizations: challenges, opportunities, and strategies. *B. Am. Meteorol. Soc.*, **91**, 869–875.

Jeffrey, R. (2004) *Subjective Probability*. Cambridge: Cambridge University Press.

Jewson, S. (2003) Moment based methods for ensemble assessment and calibration. arXiv:physics; available online at: http://arxiv.org/PS_cache/physics/pdf/0309/0309042v1.pdf/.

Joanes, D.N. and Gill, C.A. (1998) Comparing measures of sample skewness and kurtosis. *Statistician*, **47**, 183–189.

Johansson, A. and Saha, S. (1989) Simulation of systematic error effects and their reduction in a simple model of the atmosphere. *Mon. Weather Rev.*, **117**, 1658–1675.

Johnson, C. and Bowler, N. (2009) On the reliability and calibration of ensemble forecasts. *Mon. Weather Rev.*, **137**, 1717–1720.

Jolliff, J.K., Kindle, J.C., Shulman, I. *et al.* (2009) Summary diagrams for coupled hydrodynamic-ecosystem model skill assessment. *J. Marine Syst.*, **76**, 64–82.

Jolliffe, I.T. (1999) An example of instability in the sample median. *Teaching Statistics*, **21**, 29.

Jolliffe, I.T. (2004) P stands for.... *Weather*, **59**, 77–79.

Jolliffe, I.T. (2007) Uncertainty and inference for verification measures. *Weather Forecast.*, **22**, 637–650.

Jolliffe, I.T. (2008) The impenetrable hedge: a note on propriety, equitability and consistency. *Meteorol. Appl.*, **15**, 25–29.

Jolliffe, I.T. and Foord, J.F. (1975) Assessment of long-range forecasts. *Weather*, **30**, 172–181.

Jolliffe, I.T. and Jolliffe, N. (1997) Assessment of descriptive weather forecasts. *Weather*, **52**, 391–396.

Jolliffe, I.T. and Primo, C. (2008) Evaluating rank histograms using decompositions of the chi-square test statistic. *Mon. Weather Rev.*, **136**, 2133–2139.

Jolliffe I.T. and Stephenson D.B. (2005) Comments on "Discussion of verification concepts in Forecast Verification A Practitioner's Guide in Atmospheric Science". *Weather Forecast.*, **20**, 796–800.

Jolliffe, I.T. and Stephenson, D.B. (2008) Proper scores for probability forecasts can never be equitable. *Mon. Weather Rev.*, **136**, 1505–1510.

Jonathan, P., Krzanowski, W.J. and McCarthy, W.V. (2000) On the use of cross-validation to assess performance in multivariate prediction. *Stat. Comput.*, **10**, 209–229.

Jones, R.H. (1985) Time series analysis – time domain. In: *Probability, Statistics and Decision Making in the Atmospheric Sciences* (eds A.H. Murphy and R.W. Katz). Boulder, CO: Westview Press, pp. 223–259.

Jose, V.R.R. and Winkler, R.L. (2009) Evaluating quantile assessments. *Oper. Res.*, **57**, 1287–1297.

Jose, V.R.R., Nau, R.F. and Winkler, R.L. (2008) Scoring rules, generalized entropy, and utility maximization. *Oper. Res.*, **56**, 1146–1157.

Jose, V.R.R., Nau, R.F. and Winkler, R.L. (2009) Sensitivity to distance and baseline distributions in forecast evaluation. *Manage. Sci.*, **55**, 582–590.

Joslyn, S. and Savelli, S. (2010) Communicating forecast uncertainty: public perception of weather forecast uncertainty. *Meteorol. Appl.*, **16**, 180–195.

Judd, K., Smith, L.A. and Weisheimer, A. (2007) How good is an ensemble at capturing truth? Using bounding boxes for forecast evaluation. *Q. J. Roy. Meteor. Soc.*, **133**, 1309–1325.

Jung, T. and Leutbecher, M. (2008) Scale-dependent verification of ensemble forecasts. *Q. J. Roy. Meteor. Soc.*, **134**, 973–984.

Kaas, E., Guldberg, A., May, W. and Déqué, M. (1999) Using tendency errors to tune the parameterisation of unresolved dynamical scale interactions in atmospheric general circulation models. *Tellus*, **51A**, 612–629.

Kalnay, E., Hunt, B., Ott, E. and Szunyogh, I. (2006) Ensemble forecasting and data assimilation: two problems with the same solution? In: *Predictability of Weather and Climate* (eds T. Palmer and R. Hagedorn). Cambridge: Cambridge University Press, pp. 157–180.

Kang, I.S. and Yoo, J.H. (2006) Examination of multimodel ensemble seasonal prediction methods using a simple climate system. *Clim. Dynam.*, **26**, 285–294.

Katz, R.W. (1988) Use of cross correlations in the search for teleconnections. *J. Climatol.*, **8**, 241–253.

Katz, R.W. and Murphy, A.H. (eds.) (1997a) *Economic Value of Weather and Climate Forecasts*. Cambridge: Cambridge University Press.

Katz, R.W. and Murphy, A.H. (1997b) Forecast value: prototype decision-making models. In: *Economic Value of Weather and Climate Forecasts* (eds R.W. Katz and A.H. Murphy). Cambridge: Cambridge University Press, pp. 183–217.

Katz, R.W., Murphy, A.H. and Winkler, R.L. (1982) Assessing the value of frost forecasts to orchardists: A dynamic decision-making approach. *J. Appl. Meteorol.*, **21**, 518–531.

Keenlyside, N.S., Latif, M., Jungclaus, J. *et al.* (2008) Advancing decadal-scale climate prediction in the North Atlantic sector. *Nature*, **453**, 84–88.

Keil, C. and Craig, G.C. (2007) A displacement-based error measure applied in a regional ensemble forecasting system. *Mon. Weather Rev.*, **135**, 3248–3259.

Keil, C. and Craig, G.C. (2009) A displacement and amplitude score employing an optical flow technique. *Weather Forecast.*, **24**, 1297–1308.

Keith, R., (2003) Optimization of value of aerodrome forecasts. *Weather Forecast.*, **18**, 808–824.

Kendall, M. and Stuart, A. (1979) *The Advanced Theory of Statistics*, 4th edn. London: Griffin.

Kharin, V.V. and Zwiers, F.W. (2003) On the ROC score of probability forecasts. *J. Climate*, **16**, 4145–4150.

Knutti, R., Furrer, R., Tebaldi, C. *et al.* (2010) Challenges in combining projections from multiple climate models. *J. Climate*, **23**, 2739–2758.

Köppen, W. (1884) Eine rationelle method zur prüfung der wetterprognosen. *Meteorol. Z.*, **1**, 39–41.

Krzanowski W.J. and Hand, D.J. (2009) *ROC Curves for Continuous Data*. Boca Raton, FL: CRC Press.

Krzysztofowicz, R. (1983) Why should a forecaster and a decision maker use Bayes theorem? *Water Resour. Res.* **19**, 327–336.

Krzysztofowicz, R. (2001) The case for probabilistic forecasting in hydrology. *J. Hydrol.*, **249**, 1–4.

Krzysztofowicz R. and Long D. (1991) Beta likelihood models of probabilistic forecasts. *Int. J. Forecasting*, **7**, 47–55.

Lack, S., Limpert, G.L. and Fox, N.I. (2010) An object-oriented multiscale verification scheme. *Weather Forecast.*, **25**, 79–92.

Lalaurette, F. (2003) Early detection of abnormal weather conditions using a probabilistic extreme forecast index. *Q. J. Roy. Meteor. Soc.*, **129**, 3037–3057.

Lambert, N. and Shoham, Y. (2009) Eliciting truthful answers to multiple choice questions. In: *Proceedings of the Tenth ACM Conference on Electronic Commerce, Stanford, California*, pp. 109–118.

Lambert, N., Pennock, D.M. and Shoham, Y. (2008) Eliciting properties of probability distributions. In: *Proceedings of the Ninth ACM Conference on Electronic Commerce*, Chicago, Illinois, pp. 129–138.

Landman, W.A. and Goddard, L. (2002) Statistical recalibration of GCM forecasts over southern Africa using model output statistics. *J. Climate*, **21**, 2038–2055.

Landman, W.A., Mason, S.J., Tyson, P.D. and Tennant, W.J. (2001) Retro-active skill of multi-tiered forecasts of summer rainfall over southern Africa. *Int. J. Climatol.*, **21**, 1–19.

Ledford, A.W. and Tawn, J.A. (1996) Statistics for near independence in multivariate extreme values. *Biometrika*, **83**, 169–187.

Ledford, A.W. and Tawn, J.A. (1997) Modelling dependence within joint tail regions. *J. Roy. Stat. Soc. B*, **59**, 475–499.

Lee, J.Y., Wang, B., Kang, I.S. *et al.* (2010) How are seasonal prediction skills related to models' performance on mean state and annual cycle? *Clim. Dynam.*, **35**, 267–283.

Lee, P.M. (1997) *Bayesian Statistics: An Introduction*, 2nd edn. London: Arnold.

Leigh, R.J. (1995) Economic benefits of Terminal Aerodrome Forecasts (TAFs) for Sydney Airport, Australia. *Meteorol. Appl.*, **2**, 239–247.

Leith, C.E. (1974) Theoretical skill of Monte-Carlo forecasts. *Mon. Weather Rev.*, **102**, 409–418.

Levi, K. (1985) A signal detection framework for the evaluation of probabilistic forecasts. *Organ. Behav. Hum. Dec.*, **36**, 143–166.

Liljas, E. and Murphy, A.H. (1994) Anders Angstrom and his early papers on probability forecasting and the use/value of weather forecasts. *B. Am. Meteorol. Soc.*, **75**, 1227–1236.

Livezey, R.E. (1999) Field intercomparison. In: *Analysis of Climate Variability: Applications of Statistical Techniques*, 2nd edn (eds H. von Storch and A. Navarra). Berlin: Springer, pp. 161–178.

Livezey, R.E. and Chen, W.Y. (1983) Statistical field significance and its determination by Monte Carlo techniques. *Mon. Weather Rev.*, **111**, 46–59.

Livezey, R.E. and Timofeyeva, M.M. (2008) The first decade of long-lead US seasonal forecasts – insights from a skill analysis. *B. Am. Meteorol. Soc.*, **89**, 843–854.

Loeve, M. (1963) *Probability Theory*, 3rd edn. New York: Van Nostrand.

Lorenz, E.N. (1993) *The Essence of Chaos*. Seattle: University of Washington Press.

Luce, R.D. (1963) Detection and recognition. In: *Handbook of Mathematical Psychology* (eds R.D. Luce, R.R. Bush and E. Galanter). New York: Wiley, pp. 103–189.

Lumley, T. (2010) nnclust: Nearest-neighbour tools for clustering. R package version 2.2 (http://CRAN.R-project.org/package=nnclust).

Lyon, B. and Mason, S.J. (2009) The 1997–98 summer rainfall season in southern Africa. Part II: atmospheric model simulations and coupled model predictions. *J. Climate*, **22**, 3812–3818.

Macadam, I., Pitman, A.J., Whetton, P.H. and Abramowitz G. (2010) Ranking climate models by performance using actual values and anomalies: implications for climate change impact assessments. *Geophys. Res. Lett.*, **37**, article number L16704.

Mailier, P.J., Jolliffe, I.T. and Stephenson, D.B. (2008) Assessing and reporting the quality of commercial weather forecasts. *Meteorol. Appl.*, **15**, 423–429.

Manzato, A. (2005) An odds ratio parameterization for ROC diagram and skill score indices. *Weather Forecast.*, **20**, 918–930.

Marsigli, C., Montani, A. and Pacagnella. T. (2008) A spatial verification method applied to the evaluation of high-resolution ensemble forecasts. *Meteorol. Appl.*, **15**, 125–143.

Marzban, C. (1998) Scalar measures of performance in rare-event situations. *Weather Forecast.*, **13**, 753–763.

Marzban, C. and Sandgathe, S. (2009) Verification with variograms. *Weather Forecast.*, **24**, 1102–1120.

Marzban, C., Wang, R., Kong, F. and Leyton, S. (2011) On the effect of correlations on rank histograms: reliability of temperature and wind-speed forecasts from fine-scale ensemble reforecasts. *Mon. Weather Rev.*, **139**, 295–310.

Mason, I.B. (1979) On reducing probability forecasts to yes/no forecasts. *Mon. Weather Rev.*, **107**, 207–211.

Mason, I.B. (1980) Decision-theoretic evaluation of probabilistic predictions. *WMO Symposium on Probabilistic and Statistical Methods in Weather Forecasting*, Nice, France, pp. 219–228.

Mason, I.B. (1982a) A model for assessment of weather forecasts. *Aust. Meteorol. Mag.*, **30**, 291–303.

Mason, I.B. (1982b) On scores for yes/no forecasts. *Preprints Ninth AMS Conference on Weather Forecasting and Analysis*, Seattle, Washington, pp. 169–174.

Mason, I. (1989) Dependence of the Critical Success Index on sample climate and threshold probability. *Aust. Meteorol. Mag.*, **37**, 75–81.

Mason, S.J. (2004) On using "climatology" as a reference strategy in the Brier and ranked probability skill scores. *Mon. Weather Rev.*, **132**, 1891–1895.

Mason, S.J. (2008) Understanding forecast verification statistics. *Meteorol. Appl.*, **15**, 31–40.

Mason, S.J. (2011) *Guidance on Verification of Operational Seasonal Climate Forecasts*. World Meteorological Organization, Commission for Climatology XIV Technical Report.

Mason, S.J. and Baddour, O. (2008) Statistical modeling. In: *Seasonal Climate Variability: Forecasting and Managing Risk* (eds A. Troccoli, M.S.J. Harrison, D.L.T. Anderson and S.J. Mason). Dordrecht: Springer, pp. 163–201.

Mason, S.J. and Graham, N.E. (1999) Conditional probabilities, relative operating characteristics, and relative operating levels. *Weather Forecast.*, **14**, 713–725.

Mason, S.J. and Graham, N.E. (2002) Areas beneath the relative operating characteristic (ROC) and relative operating levels (ROL) curves: statistical significance and interpretation. *Q. J. Roy. Meteorol. Soc.*, **128**, 2145–2166.

Mason, S.J. and Mimmack, G.M. (2002) Comparison of some statistical methods of probabilistic forecasting of ENSO. *J. Climate*, **15**, 8–29.

Mason, S.J. and Stephenson, D.B. (2008) How do we know whether seasonal climate forecasts are any good? In: *Seasonal Climate: Forecasting and Managing Risk* (eds A. Troccoli, M. Harrison, D.L.T. Anderson and S.J. Mason). Dordrecht: Springer, pp. 259–289.

Mason, S.J. and Weigel, A.P. (2009) A generic forecast verification framework for administrative purposes. *Mon. Weather Rev.*, **137**, 331–349.

Mason, S.J., Galpin, J.S., Goddard, L. *et al.* (2007) Conditional exceedance probabilities. *Mon. Weather Rev.*, **135**, 363–372.

Mason, S.J., Tippett, M.K., Weigel, A.P. *et al.* (2011) Reply to "Comment on 'Conditional Exceedance Probabilities'". *Mon. Weather Rev.*, doi: 10.1175/2011MWR3659.1.

Mass, C.F., Ovens, D., Westrick, K. and Colle, B.A. (2002) Does increasing horizontal resolution produce more skilful forecasts? The results of two years of real-time numerical weather prediction over the Pacific Northwest. *B. Am. Meteorol. Soc.*, **83**, 407–430.

Matheson, J.E. and Winkler, R.L. (1976) Scoring rules for continuous probability distributions. *Manage. Sci.*, **22**, 1087–1096.

Matthews, R.A.J. (1996a) Base-rate errors and rain forecasts. *Nature*, **382**, 766.

Matthews, R.A.J. (1996b) Why are weather forecasts still under a cloud? *Math. Today*, **32**, 168–170.

Matthews, R.A.J. (1997) Decision-theoretic limits on earthquake prediction. *Geophys. J. Int.*, **131**, 526–529.

McCollor, D. and Stull, R. (2008) Hydrometeorological short-range ensemble forecasts in complex terrain. Part II: economic evaluation. *Weather Forecast.*, **23**, 557–574.

McCoy, M.C. (1986) Severe-storm-forecast results from the PROFS 1983 forecast experiment. *B. Am. Meteorol. Soc.*, **67**, 155–164.

McCullagh, P. and Nelder, J. (1989) *Generalized Linear Models*. London: Chapman & Hall.

McGill, R., Tukey, J.W. and Larsen, W.A. (1978) Variations of box plots. *Am. Stat.*, **32**, 12–16.

McLachlan, G.J. (1992) *Discriminant Analysis and Statistical Pattern Recognition*. New York: Wiley.

Meehl, G.A., Goddard, L., Murphy, J. *et al.* (2009) Decadal prediction: can it be skilful? *B. Am. Meteorol. Soc.*, **90**, 1467–1485.

Meehl, G.A., Hu, A.X. and Tebaldi, C. (2010) Decadal prediction in the Pacific region. *J. Climate*, **23**, 2959–2973.

Mesinger, F. and Black, T.L. (1992) On the impact on forecast accuracy of the step-mountain (eta) vs. sigma coordinate. *Meteorol. Atmos. Phys.*, **50**, 47–60.

Meyer, S. (1975) *Data Analysis for Scientists and Engineers*. New York: Wiley.

Michaelsen, J. (1987) Cross-validation in statistical climate forecast models. *J. Clim. Appl. Meteorol.*, **26**, 1589–1600.

Micheas, A., Fox, N.I., Lack, S.A. and Wikle, C.K. (2007) Cell identification and verification of QPF ensembles using shape analysis techniques. *J. Hydrol.*, **344**, 105–116.

Miller, J. and Roads, J.O. (1990) A simplified model of extended-range predictability. *J. Climate*, **3**, 523–542.

Millner, A. (2009) What is the true value of forecasts? *Weather, Climate, Society*, **1**, 22–37.

Mittermaier, M. and Roberts, N. (2010) Inter-comparison of spatial forecast verification methods: Identifying skillful spatial scales using the Fractions Skill Score. *Weather Forecast.*, **25**, 343–354.

Miyakoda, K., Hembree, G.D., Strickler, R.F. and Shulman, I. (1972) Cumulative results of extended forecast experiments. I: Model performance for winter cases. *Mon. Weather Rev.*, **100**, 836–855.

Mochizuki, T., Ishii, M., Kimoto, M. *et al.* (2010) Pacific decadal oscillation hindcasts relevant to near-term climate prediction. *P. Natl. Acad. Sci. USA*, **107**, 1833–1837.

Moise, A.F. and Delage, F. (2011) New climate metric based on object-orientated pattern matching of rainfall. *J. Geophys. Res.*, **116**, D12108.

Molteni, F., Buizza, R., Palmer, T.N. and Petroliagis, T. (1996) The ECMWF ensemble prediction system: methodology and validation. *Q. J. Roy. Meteorol. Soc.*, **122**, 73–119.

Mood, A.M., Graybill, F.A. and Boes, D.C. (1974) *Introduction to the Theory of Statistics*. New York: McGraw-Hill.

Moron, V., Robertson, A.W. and Ward, M.N. (2006) Seasonal predictability and spatial coherence of rainfall characteristics in the tropical setting of Senegal. *Mon. Weather Rev.*, **134**, 3246–3260.

Moron, V., Lucero, A., Hilario, F. *et al.* (2009) Spatiotemporal variability and predictability of summer monsoon onset over the Philippines. *Clim. Dynam.*, **33**, 1159–1177.

Morss, R.E., Lazo, J.K. and Demuth, J.L. (2010) Examining the user of weather forecasts in decision scenarios: results from a US survey with implications for uncertainty communication. *Meteorol. Appl.*, **17**, 149–162.

Muller, R.H. (1944) Verification of short-range weather forecasts (a survey of the literature). *B. Am. Meteorol. Soc.*, **25**, 18–27, 47–53, 88–95.

Müller, W.A., Appenzeller, C., Doblas-Reyes, F.J. and Liniger, M.A. (2005) A debiased ranked probability skill score to evaluate probabilistic ensemble forecasts with small ensemble sizes. *J. Climate*, **18**, 1513–1523.

Murphy, A.H. (1966) A note on the utility of probabilistic predictions and the probability score in the cost-loss ratio decision situation. *J. Appl. Meteorol.*, **5**, 534–537.

Murphy, A.H. (1971) A note on the ranked probability score. *J. Appl. Meteorol.*, **10**, 155.

Murphy, A.H. (1972) Scalar and vector partitions of the ranked probability score. *Mon. Weather Rev.*, **100**, 701–708.

Murphy, A.H. (1973) A new vector partition of the probability score. *J. Appl. Meteorol.*, **12**, 595–600.

Murphy, A.H. (1977) The value of climatological, categorical and probabilistic forecasts in the cost-loss ratio situation. *Mon. Weather Rev.*, **105**, 803–816.

Murphy, A.H. (1978) Hedging and the mode of expression of weather forecasts. *B. Am. Meteorol. Soc.*, **59**, 371–373.

Murphy, A.H. (1988) Skill scores based on the mean square error and their relationships to the correlation coefficient. *Mon. Weather Rev.*, **16**, 2417–2424.

Murphy, A.H. (1991a) Forecast verification: its complexity and dimensionality. *Mon. Weather Rev.*, **119**, 1590–1601.

Murphy, A.H. (1991b) Probabilities, odds, and forecasts of rare events. *Weather Forecast.*, **6**, 302–307.

Murphy, A.H. (1993) What is a good forecast? An essay on the nature of goodness in weather forecasting. *Weather Forecast.*, **8**, 281–293.

Murphy, A.H. (1996a) The Finley affair: a signal event in the history of forecast verification. *Weather Forecast.*, **11**, 3–20.

Murphy, A.H. (1996b) General decompositions of MSE-based skill scores: Measures of some basic aspects of forecast quality. *Mon. Weather Rev.*, **124**, 2353–2369.

Murphy, A.H. (1997) Forecast verification. In: *The Economic Value of Weather and Climate Forecasts* (eds R.W. Katz and A.H. Murphy). Cambridge: Cambridge University Press, pp. 19–74.

Murphy, A.H. (1998). The early history of probability forecasts: some extensions and clarifications. *Weather Forecast.*, **13**, 5–15.

Murphy, A.H. and Daan, H. (1985) Forecast evaluation. In: *Probability, Statistics and Decision Making in the Atmospheric Sciences* (eds A.H. Murphy and R.W. Katz). Boulder, CO: Westview Press, pp. 379–437.

Murphy, A.H. and Ehrendorfer, M. (1987) On the relationship between the accuracy and value of forecasts in the cost-loss ratio situation. *Weather Forecast.*, **2**, 243–251.

Murphy, A.H. and Epstein, E.S. (1967a) Verification of probability predictions: a brief review. *J. Appl. Meteorol.*, **6**, 748–755.

Murphy, A.H. and Epstein, E.S. (1967b) A note on probability forecasts and "hedging". *J. Appl. Meteorol.*, **6**, 1002–1004.

Murphy, A.H. and Epstein, E.S. (1989) Skill scores and correlation coefficients in model verification. *Mon. Weather Rev.*, **117**, 572–581.

Murphy, A.H. and Winkler, R.L. (1974) Credible interval temperature forecasting: Some experimental results. *Mon. Weather Rev.*, **102**, 784–794.

Murphy, A.H. and Winkler, R.L. (1977). Reliability of subjective probability forecasts of precipitation and temperature. *Appl. Stat.*, **26**, 41–47.

Murphy, A.H. and Winkler, R.L. (1984) Probability forecasting in meteorology. *J. Am. Stat. Assoc.*, **79**, 489–500.

Murphy, A.H. and Winkler, R. L. (1987) A general framework for forecast verification. *Mon. Weather Rev.*, **115**, 1330–1338.

Murphy, A.H. and Winkler, R.L. (1992) Diagnostic verification of probability forecasts. *Int. J. Forecasting*, **7**, 435–455.

Murphy, A.H., Brown, B.G. and Chen, Y-S. (1989) Diagnostic verification of temperature forecasts. *Weather Forecast.*, **4**, 485–501.

Nachamkin, J.E. (2004) Mesoscale verification using meteorological composites. *Mon. Weather Rev.*, **132**, 941–955.

Nachamkin, J.E., Chen, S. and Schmidt, J. (2005) Evaluation of heavy precipitation forecasts using composite-based methods: a distributions-oriented approach. *Mon. Weather Rev.*, **133**, 2165–2177.

NCAR (2010) verification: Forecast verification utilities. R package version 1.32 (http://CRAN.R-project.org/package=verification).

Nicholls, N. (1992) Recent performance of a method for forecasting Australian seasonal tropical cyclone activity. *Aust. Meteorol. Mag.*, **40**, 105–110.

Nicholls, N. (2001) The insignificance of significance testing. *B. Am. Meteorol. Soc.*, **82**, 981–986.

North, G.R., Kim, K.Y., Chen, S.P. and Hardin, J.W. (1995) Detection of forced climate signals: Part I: filter theory. *J. Climate*, **8**, 401–408.

Ogallo, L.J., Bessemoulin, P., Ceron, J-P. *et al.* (2008) Adapting to climate variability and change: the Climate Outlook Forum process. *J. World Meteorol. Org.*, **57**, 93–102.

Owens, B.F. and Landsea, C.W. (2003) Assessing the skill of operational Atlantic seasonal tropical cyclone forecasts. *Weather Forecast.*, **18**, 45–54.

Palmer, T.N., Molteni, F., Mureau, R. and Buizza, R. (1993). Ensemble prediction. *ECMWF Seminar Proceedings 'Validation of Models over Europe: Vol. 1'*. Reading, UK: ECMWF

Palmer, T.N., Brankovic, C. and Richardson, D.S. (2000) A probability and decision-model analysis of PROVOST seasonal multi-model integrations. *Q. J. R. Meteorol. Soc.*, **126**, 2013–2033.

Palmer, T.N., Alessandri, A., Anderson, U. *et al.* (2004) Development of a European multimodel ensemble system for seasonal-to-interannual prediction (DEMETER). *B. Am. Meteorol. Soc.*, **85**, 853–872.

Palmer, T.N., Buizza, R., Leutbecher, M. *et al.* (2007) The ECMWF Ensemble Prediction System: recent and ongoing developments. ECMWF Technical Memorandum 540.

Palmer, T.N., Doblas-Reyes, F.J., Weisheimer, A. and Rodwell, M.J. (2008) Towards seamless prediction: calibration of climate change projections using seasonal forecasts. *B. Am. Meteorol. Soc.*, **89**, 459–470.

Palmer, T.N., Doblas-Reyes, F.J., Weisheimer, A. and Rodwell, M.J. (2009) Towards seamless prediction: calibration of climate change projections using seasonal forecasts. Reply. *B. Am. Meteorol. Soc.*, **90**, 1551–1554.

Palmer, W.C. and Allen, R.A. (1949) Note on the accuracy of forecasts concerning the rain problem. Washington, D.C.: U.S. Weather Bureau manuscript.

Park, Y.-Y., Buizza, R. and Leutbecher, M. (2008) TIGGE: Preliminary results on comparing and combining ensembles. *Q. J. Roy. Meteorol. Soc.*, **134**, 2029–2050.

Peel, S. and Wilson, L.J. (2008) Modeling the distribution of precipitation forecasts from the Canadian ensemble prediction system using kernel density estimation. *Weather Forecast.*, **23**, 575–595.

Peirce, C.S. (1884) The numerical measure of the success of predictions. *Science*, **4**, 453–454.

Pepe, M.S. (2003) *The Statistical Evaluation of Medical Tests for Classification and Prediction*. Oxford: Oxford University Press.

Persson, A. and Grazzini, F. (2005) User guide to ECMWF forecast products. Technical report. Reading, UK: European Centre for Medium Range Weather Forecasts.

Pickup, M.N. (1982) A consideration of the effect of 500mb cyclonicity on the success of some thunderstorm forecasting techniques. *Meteorol. Mag.*, **111**, 87–97.

Potts, J.M., Folland, C.K., Jolliffe, I.T. and Sexton, D. (1996) Revised "LEPS" scores for assessing climate model simulations and long-range forecasts. *J. Climate*, **9**, 34–53.

Press, W.H., Teukolsky, S.A., Vetterling, W.T. and Flannery, B.P. (2007) *Numerical Recipes 3rd Edition: The Art of Scientific Computing*. Cambridge: Cambridge University Press.

Primo, C. and Ghelli, A. (2009) The effect of the base rate on the extreme dependency score. *Meteorol. Appl.*, **16**, 533–535.

Primo, C., Ferro, C.A.T., Jolliffe, I.T. and Stephenson, D.B. (2009) Calibration of probabilistic forecasts of binary events. *Mon. Weather Rev.*, **137**, 1142–1149.

Raftery, A.E., Gneiting, T., Balabdaoui, F. and Polakowski, M. (2005) Using Bayesian model averaging to calibrate forecast ensembles. *Mon. Weather Rev.*, **133**, 1155–1174.

Ramos, A. and Ledford, A. (2009) A new class of models for bivariate joint tails. *J. Roy. Stat. Soc. B*, **71**, 219–241.

R Development Core Team (2010) *R: A Language and Environment for Statistical Computing*. Vienna, Austria: R Foundation for Statistical Computing (http://www.R-project.org).

Rencher, A. and Pun, F.C. (1980) Inflation of $R^2$ in best subset regression. *Technometrics*, **22**, 49–53.

Richardson, D.S. (2000) Skill and relative economic value of the ECMWF Ensemble Prediction System. *Q. J. Roy. Meteor. Soc.*, **126**, 649–668.

Richardson, D.S. (2001) Measures of skill and value of Ensemble Prediction Systems, their interrelationship and the effect of ensemble size. *Q. J. Roy. Meteor. Soc.*, **127**, 2473–2489.

Roberts, N.M. and Lean, H.W. (2008) Scale-selective verification of rainfall accumulations from high-resolution forecasts of convective events. *Mon. Weather Rev.*, **136**, 78–97.

Robertson, A.W., Moron, V. and Swarinoto, Y. (2009) Seasonal predictability of daily rainfall statistics over Indramayu district, Indonesia. *Int. J. Climatol.*, **29**, 1449–1462.

Robin, X., Turck, N., Sanchez, J. and Müller M. (2010) pROC: Tools for visualizing, smoothing and comparing receiver operating characteristic (ROC curves). R package version 1.3.2 (http://CRAN.R-project.org/package=pROC).

Rockafellar, R.T. (1970) *Convex Analysis*. Princeton, NJ: Princeton University Press.

Rodenberg, C. and Zhou, X-H. (2000) ROC curve estimation when covariates affect the verification process. *Biometrics*, **56**, 1256–1262.

Rodwell, M.J., Richardson, D.S., Hewson, T.D. and Haiden, T. (2010) A new equitable score suitable for verifying precipitation in numerical weather prediction. *Q. J. Roy. Meteor. Soc.*, **136**, 1344–1363.

Roebber, P.J. (2009) Visualizing multiple measures of forecast quality. *Weather Forecast.*, **24**, 601–608.

Roebber, P.J. and Bosart, L.F. (1996) The complex relationship between forecast skill and forecast value: a real-world analysis. *Weather Forecast.*, **11**, 544–559.

Roeckner, E., Arpe, K., Bengtsson, L. *et al.* (1996) The atmospheric general circulation model ECHAM4: Model description and simulation of present-day climate. Max Planck Institut für Meteorologie Report No. 218. Hamburg: MPI für Meteorologie.

Roquelaure, S. and Bergot, T. (2008) A local ensemble prediction system for fog and low clouds: construction, Bayesian model averaging calibration, and validation. *J. Appl. Meteorol. Clim.*, **47**, 3072–3088.

Rossa, A., Nurmi, P. and Ebert, E. (2008) Overview of methods for the verification of quantitative precipitation forecasts. In: *Precipitation: Advances in Measurement, Estimation and Prediction* (ed. S. Michaelides). Berlin: Springer, pp. 419–452.

Roulston, M.S. and Smith, L.A. (2002) Evaluating probabilistic forecasts using information theory. *Mon. Weather Rev.*, **130**, 1653–1660.

Roulston, M.S. and Smith, L.A. (2003) Combining dynamical and statistical ensembles. *Tellus*, **55A**, 16–30.

Roulston, M.S., Bolton, G.E., Kleit, A.N., Sears-Collins, A.L. (2006) A laboratory study of the benefits of including uncertainty information in weather forecasts. *Weather Forecast.*, **21**, 116–122.

Royer, J-F., Roeckner, E., Cubasch, U. *et al.* (2009) Production of seasonal to decadal hindcasts and climate change scenarios. In: *ENSEMBLES: Climate Change and its Impacts: Summary of Research and Results from the ENSEMBLES Project* (eds P. van der Linden and J.F.B. Mitchell). Exeter, UK: Met Office Hadley Centre, pp. 35–46.

Saha, P. and Heagerty, P.J. (2010) Time-dependent predictive accuracy in the presence of competing risks. *Biometrics*, **66**, 999–1011.

Sanders, F. (1958) The evaluation of subjective probability forecasts. Scientific Report No. 5. Massachusetts Institute of Technology, Department of Meteorology.

Sanders, F. (1963) On subjective probability forecasting. *J. Appl. Meteor.*, **2**, 191–201.

Saporta, G. (1990) *Probabilités, analyse de données et statistiques*. Paris: Technip.

Savage, L.J. (1971) Elicitation of personal probabilities and expectations. *J. Am. Stat. Assoc.*, **66**, 783–810.

Savage, L.J. (1972) *Foundations of Statistics*. New York: Dover Publications.

Scaife, A., Buontempo, C., Ringer, M. *et al.* (2009) Comment on "Toward seamless prediction: calibration of climate change projections using seasonal forecasts." *B. Am. Meteorol. Soc.*, **90**, 1549–1551.

Schaefer, J.T. (1990) The critical success index as an indicator of warning skill. *Weather Forecast.*, **5**, 570–575.

Schervish, M.J. (1989) A general method for comparing probability assessors. *Ann. Stat.*, **17**, 1856–1879.

Schowengerdt, R.A. (1997) *Remote Sensing, Models, and Methods for Image Processing*. New York: Academic Press.

Scott, D.W. (1979) On optimal and data-based histograms. *Biometrika*, **66**, 605–610.

Seaman, R., Mason, I. and Woodcock, F. (1996) Confidence intervals for some performance measures of yes/no forecasts. *Aust. Meteorol. Mag*, **45**, 49–53.

Seillier-Moiseiwitsch, F. and Dawid, A.P. (1993) On testing the validity of sequential probability forecasts. *J. Am. Stat. Assoc.*, **88**, 355–359.

Sheskin, D.J. (2007) *Handbook of Parametric and Nonparametric Statistical Procedures*. Boca Raton, FL: Chapman & Hall/CRC Press.

Shongwe, M.E., Landman, W.A. and Mason, S.J. (2006) Performance of recalibration systems of GCM forecasts for Southern Africa. *Int. J. Climatol.*, **26**, 1567–1585.

Silverman, B.W. (1986) *Density Estimation for Statistics and Data Analysis*. London: Chapman & Hall.

Sing, T., Sander, O., Beerenwinkel, N. and Lengauer, T. (2009) ROCR: Visualizing the performance of scoring classifiers. R package version 1.0-3 (http://CRAN.R-project.org/package=ROCR).

Slonim, M.J. (1958) The trentile deviation method of weather forecast evaluation. *J. Am. Stat. Assoc.*, **53**, 398–407.

Sloughter, J.M., Raftery, A.E., Gneiting, T., and Fraley, C. (2007) Probabilistic quantitative precipitation forecasting using Bayesian model averaging. *Mon. Weather Rev.*, **135**, 3209–3220.

Smith, D.M., Cusack, S., Colman, A.W. *et al.* (2008) Improved surface temperature prediction for the coming decade from a global climate model. *Science*, **317**, 796–799.

Smith, L.A. (2001) Disentangling uncertainty and error: On the predictability of nonlinear systems. In: *Nonlinear Dynamics and Statistics* (ed. A.I. Mees). Boston: Birkhäuser Press, pp. 31–64.

Smith, L.A. and Hansen, J.A. (2004) Extending the limits of ensemble forecast verification with the minimum spanning tree. *Mon. Weather Rev.*, **132**, 1522–1528.

Stainforth, D.A., Allen, M.R., Tredger, E.R. and Smith, L.A. (2007) Confidence, uncertainty and decision-support relevance in climate predictions. *Philos. T. Roy. Soc. A*, **365**, 2145–2161.

Stanski, H.R., Wilson, L.J. and Burrows, W.R. (1989) Survey of common verification methods in meteorology. World Weather Watch Technical Report No. 8. Geneva: World Meteorological Organisation.

Stephenson, D.B. (1997) Correlation of spatial climate/weather maps and the advantages of using the Mahalanobis metric in predictions. *TELLUS A*, **49**, 513–527.

Stephenson, D.B. (2000) Use of the 'odds ratio' for diagnosing forecast skill. *Weather Forecast.*, **15**, 221–232.

Stephenson, D.B. and Doblas-Reyes, F.J. (2000) Statistical methods for interpreting Monte Carlo forecasts. *Tellus*, **52A**, 300–322.

Stephenson, D.B., Coelho, C.A.S., Doblas-Reyes, F.J. and Balmaseda, M. (2005) Forecast assimilation: A unified framework for the combination of multi-model weather and climate predictions. *Tellus*, **57A**, 253–264.

Stephenson, D.B., Casati, B., Ferro, C.A.T. and Wilson, C.A. (2008a) The Extreme Dependency Score: a non-vanishing measure for forecasts of rare events. *Meteorol. Appl.*, **15**, 41–50.

Stephenson, D.B., Coelho, C.A.S. and Jolliffe, I.T. (2008b) Two extra components in the Brier score decomposition. *Weather Forecast.*, **23**, 752–757.

Stephenson, D.B, Jolliffe, I.T., Ferro, C.A.T. *et al.* (2010) White Paper review on the verification of warnings. Met Office Forecasting Research Technical Report No. 546. Available from: http://www.metoffice.gov.uk/media/pdf/e/n/FRTR546.pdf.

Stockdale, T.N., Anderson, D.L.T., Alves, J.O.S. and Balmaseda, M.A. (1998) Global seasonal rainfall forecasts using a coupled ocean-atmosphere model. *Nature*, **392**, 370–373.

Stone, J.V. (2004) *Independent Component Analysis: A Tutorial Introduction*. Cambridge, MA: MIT Press.

Stone, M. (1974) Cross-validatory choice and assessment of statistical predictions. *J. Roy. Stat. Soc. B*, **36**, 111–147.

Sturges, H.A. (1926) The choice of a class interval. *J. Am. Stat. Assoc.*, **21**, 65–66.

Swets, J.A. (1973) The relative operating characteristic in psychology. *Science*, **182**, 990–1000.

Swets, J.A. (1986a) Indices of discrimination or diagnostic accuracy: their ROCs and implied models. *Psychol. Bull.*, **99**, 100–117.

Swets, J.A. (1986b) Form of empirical ROCs in discrimination and diagnostic tasks: implications for theory and measurement of performance. *Psychol. Bull.*, **99**, 181–198.

Swets, J.A. (1988) Measuring the accuracy of diagnostic systems. *Science*, **240**, 1285–1293.

Swets, J.A. (1996) *Signal Detection Theory and ROC Analysis in Psychology and Diagnostics: Collected Papers*. Mahwah, NJ: Lawrence Erlbaum.

Swets, J.A. and Pickett, R.M. (1982) *Evaluation of Diagnostic Systems: Methods from Signal Detection Theory*. New York: Academic Press.

Talagrand, O., Vautard, R. and Strauss, B. (1998) Evaluation of probabilistic prediction systems. In: *Proceedings of ECMWF Workshop on Predictability, 20–22 October 1997*, pp. 1–25.

Tanner, W.P. Jr and Birdsall, T.G. (1958) Definitions of $d'$ and $\eta$ as psychophysical measures. *J. Acoustical Soc. Amer.*, **30**, 922–928.

Tartaglione, N., Mariani, S., Accadia, C. *et al.* (2008) Objective verification of spatial precipitation forecasts. In: *Precipitation: Advances in Measurement, Estimation and Prediction* (ed. S. Michaelides). Berlin: Springer, pp. 453–472.

Tay, A.S. and Wallis, K.F. (2000) Density forecasting: a survey. *J. Forecasting*, **19**, 235–254.

Taylor, J.W. (1999) Evaluating volatility and interval forecasts. *J. Forecasting*, **18**, 111–128.

Taylor, J.W. and Buizza, R. (2003) Using weather ensemble predictions in electricity demand forecasting. *Int. J. Forecasting*, **19**, 57–70.

Taylor, K.E. (2001). Summarizing multiple aspects of model performance in a single diagram. *J. Geophys. Res.*, **106**, 7183–7192.

Teweles, S. and Wobus, H.B. (1954) Verification of prognostic charts. *B. Am. Meteorol. Soc.*, **35**, 455–463.

Thompson, J.C. (1952) On the operational deficiencies in categorical weather forecasts. *B. Am. Meteorol. Soc.*, **33**, 223–226.

Thompson, J.C. (1966) A note on meteorological decision making. *J. Appl. Meteorol.*, **5**, 532–533.

Thompson, J.C. and Carter, G.M. (1972) On some characteristics of the S1 score. *J. Appl. Meteorol.*, **11**, 1384–1385.

Thornes, J.E. and Stephenson, D.B. (2001) How to judge the quality and value of weather forecast products. *Meteorol. Appl.*, **8**, 307–314.

Tippett, M.K. (2008) Comments on "The discrete Brier and ranked probability skill scores." *Mon. Weather Rev.*, **136**, 3629–3633.

Tippett, M.K. and Barnston, A.G. (2008) Skill of multimodel ENSO probability forecasts. *Mon. Weather Rev.*, **136**, 3933–3946.

Tippett, M.K., Barnston, A.G., DeWitt, D.G. and Zhang, R-H. (2005) Statistical correction of tropical Pacific sea surface temperature forecasts. *J. Climate*, **18**, 5141–5162.

Tippett, M.K., Barnston, A.G. and Robertson, A.W. (2007) Estimation of seasonal precipitation tercile-based categorical probabilities from ensembles. *J. Climate*, **20**, 2210–2228.

Troccoli, A. and Palmer, T.N. (2007) Ensemble decadal predictions from analysed initial conditions. *Philos. T. Roy. Soc. A*, **365**, 2179–2191.

Tukey, J.W. (1977) *Exploratory Data Analysis*. Reading, MA: Addison Wesley.

Uppala, S.M., Kallberg, P.W., Simmons, A.J. *et al.* (2005) The ERA-40 Re-Analysis. *Q. J. Roy. Meteor. Soc.*, **131**, 2961–3012.

van den Dool, H., Huang, J. and Fan, Y. (2003) Performance and analysis of the constructed analogue method applied to US soil moisture over 1981–2001. *J. Geophys. Res-Atmos.*, **108**, article number 8617.

van Oldenborgh, G.J., Balmaseda, M.A., Ferranti, L. *et al.* (2005) Evaluation of atmospheric fields from the ECMWF seasonal forecasts over a 15-year period. *J. Climate*, **18**, 3250–3269.

Venkatraman, E.S. (2000) A permutation test to compare Receiver Operating Characteristic curves. *Biometrics*, **56**, 1134–1138.

Venugopal, V., Basu, S. and Foufoula-Georgiou, E. (2005) A new metric for comparing precipitation patterns with an application to ensemble forecasts. *J. Geophys. Res-Atmos.*, **110**, article number D08111.

Vitart F. (2006) Seasonal forecasting of tropical storm frequency using a multi-model ensemble. *Q. J. Roy. Meteor. Soc.*, **132**, 647–666.

Vogel, C. and O'Brien K. (2006) Who can eat information? Examining the effectiveness of seasonal climate forecasts and regional climate-risk management strategies. *Climate Res.*, **33**, 111–122.

von Storch, H. and Zwiers, F.W. (1999) *Statistical Analysis in Climate Research*. Cambridge: Cambridge University Press.

Wallis, K.F. (1995) Large-scale macroeconomic modelling. In: *Handbook of Applied Econometrics* (eds M.H. Pesaran and M.R. Wickens). Oxford: Blackwell, pp. 312–355.

Wandishin, M.S. and Brooks, H.E. (2002) On the relationship between Clayton's skill score and expected value for forecasts of binary events. *Meteorol. Appl.*, **9**, 455–459.

Wang, H., Schemm, J.K.E., Kumar, A. *et al.* (2009) A statistical forecast model for Atlantic seasonal hurricane activity based on the NCEP dynamical seasonal forecast. *J. Climate*, **22**, 4481–4500.

Wang, X. and Bishop, C.H. (2005) Improvement of ensemble reliability with a new dressing kernel. *Q. J. Roy. Meteor. Soc.*, **131**, 965–986.

Ward, M.N. and Folland, C.K. (1991) Prediction of seasonal rainfall in the north Nordeste of Brazil using eigenvectors of sea-surface temperature. *Int. J. Climatol.*, **11**, 711–743.

Watson, G.S. (1961) Goodness-of-fit tests on a circle. *Biometrika*, **48**, 109–114.

Watterson, I.G. (1996) Non-dimensional measures of climate model performance. *Int. J. Climatol.*, **16**, 379–391.

Weigel, A.P. and Bowler, N.E. (2009) Comment on "Can multi-model combination really enhance the prediction skill of probabilistic ensemble forecasts?" *Q. J. Roy. Meteor. Soc.*, **135**, 535–539.

Weigel, A.P. and Mason, S.J. (2011) The Generalized Discrimination Score for ensemble forecasts. *Mon. Weather Rev.*, doi: 10.1175/MWR-D-10-05069.1.

Weigel, A.P., Liniger, M.A. and Appenzeller, C. (2007a) The discrete Brier and ranked probability skill scores. *Mon. Weather Rev.*, **135**, 118–124.

Weigel, A.P., Liniger, M.A. and Appenzeller, C. (2007b) Generalization of the discrete Brier and ranked probability skill scores for weighted multimodel ensemble forecasts. *Mon. Weather Rev.*, **135**, 2778–2785.

Weigel, A.P., Liniger, M.A. and Appenzeller, C. (2008a) Can multi-model combination really enhance the prediction skill of probabilistic ensemble forecasts? *Q. J. Roy. Meteor. Soc.*, **134**, 241–260.

Weigel, A.P., Baggenstos, D., Liniger, M.A. *et al.* (2008b) Probabilistic verification of monthly temperature forecasts. *Mon. Weather Rev.*, **136**, 5162–5182.

Weigel, A.P., Liniger, M.A. and Appenzeller, C. (2009) Seasonal ensemble forecasts: are recalibrated single models better than multimodels? *Mon. Weather Rev.*, **137**, 1460–1479.

Weigel, A.P., Knutti, R., Liniger, M.A. and Appenzeller, C. (2010) Risks of model weighting in multimodel climate projections. *J. Climate*, **23**, 4175–4191.

Weijs, S.V., van Nooijen, R. and van de Giesen, N. (2010) Kullback–Leibler divergence as a forecast skill score with classic reliability-resolution-uncertainty decomposition. *Mon. Weather Rev.*, **138**, 3387–3399.

Weisheimer, A., Smith, L.A. and Judd, K. (2005) A new view of seasonal forecast skill: bounding boxes from the DEMETER ensemble forecasts. *Tellus*, **57A**, 265–279.

Weisheimer, A., Doblas-Reyes, F.J., Palmer, T.N. *et al.* (2009) ENSEMBLES: A new multi-model ensemble for seasonal to annual predictions – Skill and progress beyond DEMETER in forecasting tropical Pacific SSTs. *Geophys. Res. Lett.*, **36**, article number L21711.

Wernli, H., Paulat, M., Hagen, M. and Frei, C. (2008) SAL – A novel quality measure for the verification of quantitative precipitation forecasts. *Mon. Weather Rev.*, **136**, 4470–4487.

Wernli, H., Hofmann, C. and Zimmer, M. (2009) Spatial forecast verification methods inter-comparison project – application of the SAL technique. *Weather Forecast.*, **24**, 1472–1484.

Whetton, P., Macadam, I., Bathols, J. and O'Grady, J. (2007) Assessment of the use of current climate patterns to evaluate regional enhanced greenhouse response patterns of climate models. *Geophys. Res. Lett.*, **34**, article number L14701.

Whitaker, J.S. and Loughe, A.F. (1998) The relationship between ensemble spread and ensemble mean skill. *Mon. Weather Rev.*, **126**, 3292–3302.

Wilkinson, L. and Dallal, G.E. (1981) Tests of significance in forward selection regression with an *F*-to-enter stopping rule. *Technometrics*, **23**, 377–380.

Wilks, D.S. (2000) Diagnostic verification of the Climate Prediction Center long-lead outlooks, 1995–98. *J. Climate*, **13**, 2389–2403.

Wilks, D.S. (2002) Smoothing forecast ensembles with fitted probability distributions. *Q. J. Roy. Meteor. Soc.*, **128**, 2821–2836.

Wilks, D.S. (2004) The minimum spanning tree histogram as a verification tool for multidimensional ensemble forecasts. *Mon. Weather Rev.*, **132**, 1329–1340.

Wilks, D.S. (2006a) On "field significance" and the false discovery rate. *J. Appl. Meteorol. Climatol.*, **45**, 1181–1189.

Wilks, D.S. (2006b) *Statistical Methods in the Atmospheric Sciences: An Introduction*, 2nd edn. Amsterdam: Academic Press.

Wilks, D.S. and Godfrey, C.M. (2002) Diagnostic verification of the IRI Net Assessment forecasts, 1997–2000. *J. Climate*, **15**, 1369–1377.

Wilks, D.S. and Hamill, T.M. (1995) Potential economic value of ensemble forecasts. *Mon. Weather Rev.*, **123**, 3565–3575.

Wilks, D.S. and Hamill, T.M. (2007). Comparison of ensemble-MOS methods using GFS reforecasts. *Mon. Weather Rev.*, **135**, 2379–2390.

Wilson, E.B. (1927) Probable inference, the law of succession, and statistical inference. *J. Am. Stat. Assoc.*, **22**, 209–212.

Wilson, L. (2000) Comments on 'Probabilistic predictions of precipitation using the ECMWF ensemble prediction system'. *Weather Forecast.*, **15**, 361–364.

Wilson, L.J., Burrows, W.R. and Lanzinger, A. (1999) A strategy for verification of weather element forecasts from an ensemble prediction system. *Mon. Weather Rev.*, **127**, 956–970.

Winkler, R.L. (1969) Scoring rules and the evaluation of probability assessors. *J. Am. Stat. Assoc.*, **64**, 1073–1078.

Winkler, R.L. (1996) Scoring rules and the evaluation of probabilities. *Test*, **5**, 1–60 (including discussion).

Winkler, R.L. and Murphy, A.H. (1968) "Good" probability assessors. *J. Appl. Meteorol.*, **7**, 751–758.

Winkler, R.L. and Murphy, A.H. (1985) Decision analysis. In: *Probability, Statistics and Decision Making in the Atmospheric Sciences* (eds. A.H. Murphy and R.W. Katz). Boulder, CO: Westview Press, pp. 493–524.

Woodcock, F. (1976) The evaluation of yes/no forecasts for scientific and administrative purposes. *Mon. Weather Rev.*, **104**, 1209–1214.

Woodworth, R.S. (1938) *Experimental Psychology*. New York: Holt, Rinehart and Winston.

World Meteorological Organization (1992) *Manual on the Global Data-Processing and Forecasting System (GDPFS)*. WMO-No. 485, Attachment II.8.

World Meteorological Organization (2007) Standard Verification System (SVS) for Long-Range Forecasts (LRF). *Manual on the Global Data Processing and Forecasting Systems*. WMO-No. 485, II.8.1–II.8.17, available at: http://www.bom.gov.au/wmo/lrfvs/Attachment_II-8.doc.

Xie, P. and Arkin, P.A. (1996) Analyses of global monthly precipitation using gauge observations, satellite estimates, and numerical model predictions. *J. Climate*, **9**, 840–858.

Xu, Q-S. and Liang, Y-Z. (2001) Monte Carlo cross validation. *Chemometr. Intell. Lab.*, **56**, 1–11.

Yates, E.S., Anquetin, S., Ducrocq, V. *et al.* (2006) Point and areal validation of forecast precipitation fields. *Meteorol. Appl.*, **13**, 1–20.

Youden, W.J. (1950) Index for rating diagnostic tests. *Cancer*, **3**, 32–35.

Yule, G.U. (1900) On the association of attributes in statistics. *Philos. T. Roy. Soc.*, **194A**, 257–319.

Yule, G.U. (1912) On the methods of measuring association between two attributes. *J. Roy. Stat. Soc.*, **75**, 579–652.

Zacharov, P. and Rezacova, D. (2009) Using the fractions skill score to assess the relationship between an ensemble QPF spread and skill. *Atmos. Res.*, **94**, 684–693.

Zepeda-Arce, J., Foufoula-Georgiou, E. and Droegemeier, K.K. (2000) Space-time rainfall organization and its role in validating quantitative precipitation forecasts. *J. Geophys. Res.*, **105**, 10129–10146.

Zhu, Y., Toth, Z., Wobus, R. *et al.* (2002) The economic value of ensemble-based weather forecasts. *B. Am. Meteor. Soc.*, **83**, 73–83.

Zingerle, C. and Nurmi, P. (2008) Monitoring and verifying cloud forecasts originating from numerical models. *Meteorol. Appl.*, **15**, 325–330.

# Index